M. Altmaier, V. Montoya, L. Duro, A. Valls (Hrsg.)

Proceedings of the First Annual Workshop of the HORIZON 2020 CEBAMA Project

Karlsruhe Institute of Technology
KIT SCIENTIFIC REPORTS 7734

Proceedings of the First Annual Workshop of the HORIZON 2020 CEBAMA Project

Herausgegeben von
M. Altmaier, V. Montoya, L. Duro, A. Valls

KIT Scientific Publishing

Report-Nr. KIT-SR 7734

Impressum

KIT Scientific Publishing

Karlsruher Institut für Technologie (KIT)
KIT Scientific Publishing
Straße am Forum 2
D-76131 Karlsruhe

KIT Scientific Publishing is a registered trademark
of Karlsruhe Institute of Technology.
Reprint using the book cover is not allowed.

www.ksp.kit.edu

Print on Demand 2017 – Gedruckt auf FSC-zertifiziertem Papier

ISSN 1869-9669
ISBN 978-3-7315-0660-7
DOI 10.5445/KSP/1000068889

FOREWORD

The present document contains the proceedings of the First Annual Workshop (AWS) of the EURATOM H2020 Collaborative Project CEBAMA (Cement-based materials, properties, evolution, barrier functions). The electronic version of these proceedings is available on the webpage of the project (http://www.cebama.eu/) and at KIT Scientific Publishing (www.ksp.kit.edu). The project started in June 2015 and has a duration of four years. The project is implemented by a consortium with 27 Beneficiaries, from 9 EURATOM Signatory States, Japan and Switzerland. National Waste Management Organizations contribute to the running project by participation in the End-User Group, by co-funding Beneficiaries, and provide for knowledge and information transfer.

These proceedings serve several purposes. The key purpose is to document and make available the progress of the CEBAMA project to a broad scientific community. For this purpose, a considerable part of the project activity is reported by the proceedings, together with the scientific-technical contributions containing details of the work and the "Topical Sessions" dealing with different subjects of interest for the development of the project. Being the first annual workshop proceedings, the progress shown corresponds to the first year of life of the activities, and we are very proud to see that all groups have started their research and that good results have already been obtained. The project facilitates training mobility measures for students, and the workshop also served to put in place and communicate the procedures for application of these grants. Additional purposes of the proceedings are to ensure on-going documentation of the project outcome, promote systematic scientific-technical development throughout the project and to allow thorough review of the project progress.

All Scientific and Technical papers (S&T) included in the proceedings have been reviewed by the Work Package leaders and the EUG (End-User-Group) of CEBAMA. The EUG is specifically set up within the project representing the interest of the national waste management and/or national regulatory organizations who may use the results of the project for their Safety Cases. For this reason, this EUG had a strong involvement on guiding the priorities of the activities of the project and it is a very useful tool to guarantee that the research is done in a way that allows for future knowledge transfer.

The proceedings give only very brief information about the project structure and the different activities around the project. This type of information is available in detail under http://www.cebama.eu/.

The editors of the proceedings thank all contributors to the project, especially those submitting Scientific and Technical contributions, and the workpackage leaders who provided the summary of the different workpackages for publication in these proceedings. Special thanks is given to the reviewers, the members of the EUG, whose effort and hard work reflected their commitment and dedication to the project and contribute to a high quality of the research performed within CEBAMA.

TABEL OF CONTENTS

THE PROJECT

Cebama is a Collaborative Project funded by the European Commission under the Horizon 2020 Reasearch and Training Programme of the European Atomic Energy Community (EURATOM) (H2020-NFRP-2014/2015), section B - Contribute to the Development of Solutions for the Management of Ultimate Radioactive Waste, Topic NFRP 6 – 2014: Supporting the implementation of the first-of-the-kind geological repositories).

The Collaborative Project Cebama addresses key issues of relevance for long term safety and key scientific questions related to the use of cement-based materials in nuclear waste disposal applications. These materials are key components in the barrier system of repositories, independent on the actual host rocks. They are used as waste forms, liners and structural components as well as sealing materials in a broad variety of applications. Waste forms and their behaviour as well as the technical feasibility and long-term performance of repository components are key topics detailed in the Strategic Research Agenda (SRA) of the Implementing Geological Disposal Technology Platform (IGD-TP) and and cover studies related with: (a) the release of radionuclides, (b) long-term behaviour of seals and plugs, (c) evolution of cement-based seals, (d) interaction of cement with clays, and (e) optimisation aspects.

The overall objective of Cebama is to support the implementation of geological disposal by improving significantly the knowledge base for the Safety Case for European repository concepts.

The research planned in Cebama is largely independent of specific disposal concepts and addresses different types of host rocks in addition to bentonite. Cebama is not focusing on one specific cement material, but aims to study a variety of important cement-based materials in order to provide insight on general processes and phenomena.

The ambition of Cebama is the development of a comprehensive model for predicting the transport characteristics such as porosity, permeability and diffusion parameters of cement-based materials in contact with the engineered and natural barriers of repositories in crystalline and argillaceous host rocks. Dedicated studies on radionuclide retention processes and on the impact of chemical alterations on these processes are part of this advanced approach.

The project started on 1st June 2015 and will last 4 years. The project is implemented by a consortium with 27 Beneficiaries, from 9 EURATOM Signatory States, Japan and Switzerland. National Waste Management Organizations contribute to the running project by participation in the End-User Group, by co-funding Beneficiaries, and provide for knowledge and information transfer.

Indispensable for CEBAMA is the documentation of the scientific and technical state of knowledge and the dissemination of the generated knowledge not only to the WMO but also to the general scientific community and other interested parties. This dissemination strategy will be implemented by presentations at international conferences and publications in peer-reviewed journals. Interaction with social stakeholders is of relevance for facilitating global understanding for the need of research in nuclear waste management, and CEBAMA aims at helping to bridge this communication process. The aspect of training and education of the next generation of waste management professionals is specifically addressed in CEBAMA.

1st ANUAL WORKSHOP

The 1st Annual Workshop of the Collaborative Project CEBAMA was held in Barcelona (Spain), 11-13 May 2016. The workshop was hosted by Amphos 21 at the Barceló Atenea Mar Hotel. There were 58 attendees at the workshop, representing beneficiaries, the End-User-Group and project external organizations. The workshop was organized in three days (see general agenda in Figure 1).

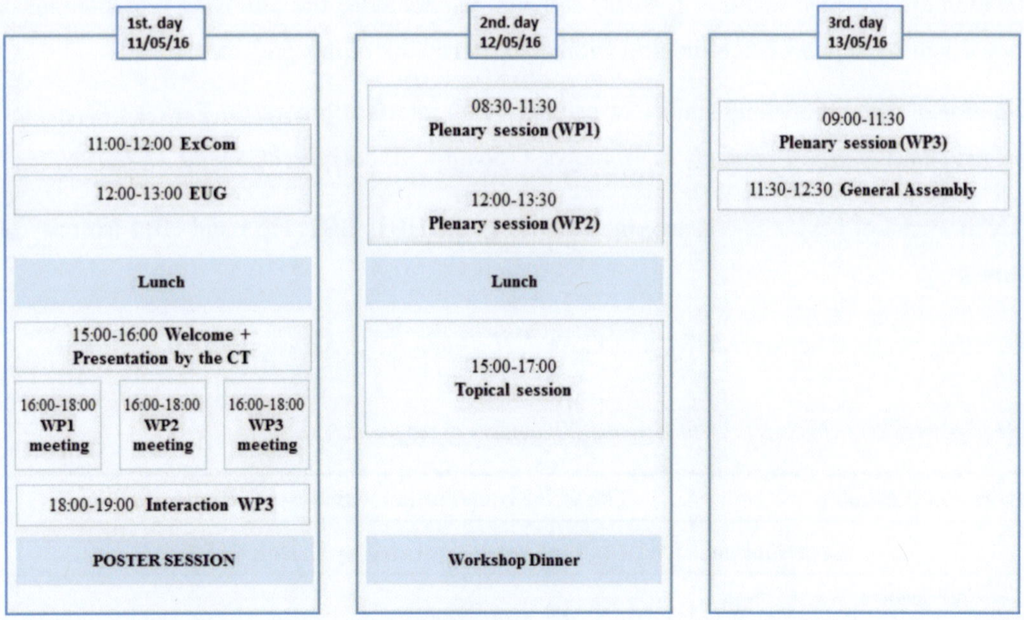

Figure 1: *General agenda of the 1st Annual Workshop of Cebama.*

The objectives of the 1st Annual Workshop of this project are presented below.

- Ensuring that all project partners are aware of the project objectives, work program and reporting obligations
- Promoting joint research activities within Cebama
- Providing for detailed agreements on the work program, activities and training measures
- Communicating the status of the different project activities, work and state of progress between all project partners including non-technical stakeholders
- Discussing the status of the work programme and decide on next steps, including deviations from original work planning, if required
- Preparing for periodic reporting, including preparation of public Workshop Proceedings for effective documentation and communication of the project achievements
- Preparing for finalization of the project, including final reporting

This workshop was a cluster of several meetings and sessions organized to achieve the above cited objectives. The *ExCom and EUG meeting* took place during the first day of the wotkshop. The rest of the day was focused on the internal WP management. Firstly, *individual WP sessions* were conducted by the corresponding WP leaders aimed on internally discussions on their work progress. Then, a common session for all WP allowed enhancing the interaction between technical WP.

During the *plenary sessions* (2nd and 3rd day), the WP leaders presented a summary of the work progress performed within each workpackage taking into account the contribution of each partner. The contents of each WP presentation are provided in Table 1. These sessions also included the following presentations carried out by the young scientists as part of their training within the framework of this project.

- Experimental and modelling studies of cement / clay interface processes. Task 1 : Characterization of low pH cements. *N. Ait Mouheb, V. Montoya, T. Schäfer, E. Soballa, D. Schild, C. Borkel.*
- THCm model of bentonite-concrete interactions at the HB1, HB2, HB3 and HB4 lab tests. *A. Mon, J. Samper.*
- Modeling transport across reactive interfaces. *L.H. Damiani, G. Kosakowski.*

Table 1: *List of parters contributions presented by the WP leaders at the plenary sessions.*

Institution (Partner nº)	Tite of the contribution within the WP presentation
Contributions of WP1 (all the work presented by E. Holt (VTT)	
WP leaders (BRGM, VTT, UNIBERN)	WP1 Overview – 1st Year Progress
BRGM (3)	BRGM contribution
BGS (4)	Experiments on Interface Processes at the Cement/Callovo-Oxfordian Claystone Interface and the Impact on Physical Properties
CIEMAT (5)	Interaction concrete/FEBEX bentonite: outlining experimental conditions and characterization approaches at laboratory and *in situ* scale
TUDELFT (6)	Characterisation of porosity and permeability in relation to the interface processes between concrete and Boom Clay in the Dutch context: preliminary results
RWMC (10)	RWMC: Observation of growth of the altered zone regarding cement-bentonite interaction by using Ca-XAFS analysis
SCK·CEN (11)	SCK•CEN: Changes in μ-structure, mineralogy and transport properties of Concrete – Boom Clay interface
UJV (13) and CTU (16)	Interaction between cement and Czech bentonite under temperature load and in in-situ conditions – 1st year progress
USFD (17)	Interactions of high-pH backfill cement and low-pH cement with groundwater
VTT (18)	VTT contribution
HZDR (19)	Spatiotemporal visualization of transport processes

Institution (Partner n°)	Tite of the contribution within the WP presentation
LML (20)	Experimental investigation and modeling of hydromechanical behavior of concrete fracture and interface – Preliminary results
UAM (21)	Definition of sampling and characterization of in-situ FEBEX-OPC concrete contact and performance of new experiments studying surface interface reactivity
CSIC (22)	Characterisation of concrete aging due to interaction with groundwaters in contact with different EBS
ANDRA (23)	Low pH concretes – In situ experiment in the french URL (Bure – Meuse/Haute-Marne, France)
UNIBERN (25)	Core infiltration experiments with aged claystone-mortar interfaces coupled with X-ray CT and PET.
IRSN (26)	Geochemical and microstructural evolution of cementitious materials in contact with a clayey rock at 70°C
Contributions of WP2 (all the work presented by B. Grambow (ARMINES)	
WP leader (ARMINES)	WP2 Overview – 1st Year Progress
KIT (1)	Contribution of KIT-INE to WP2 (1st year): Solubility, hydrolysis and sorption of beryllium in cementitious systems
AMPHOS21 (2)	Molybdenum behaviour in cementitious materials
BRGM (3)	AFm sorption and dissolution kinetics
JUELICH (7)	Understanding the behaviour of safety relevant radionuclides in cementitious systems: First results on the uptake of radium and technetium by CSH, AFm and AFt
RATEN (8)	Carbon-14 transport parameters through CEM II mortar
ARMINES (12)	Investigation of the diffusion properties of inorganic ^{14}C species through unsaturated hardened cement paste: Influence of water saturation conditions
SURRAY (15)	Procedures For Assessing Radionuclide Retention in Cementitious Systems and Single Mineral Phases
EMPA/PSI (24)	Uptake of Sulfur, Selenium and Iodine in AFm-phases
Contributions of WP3 (all the work presented by A. Idiart (AMPHOS21)	
WP leader (AMPHOS21)	WP3 Overview – 1st Year Progress
KIT (1)	Experimental and modelling studies on low pH cement / clay interface processes
AMPHO21 (2)	Amphos 21 – WP3 scientific contribution
BRGM (3)	Surface electrical properties of calcite and low pH cements
JULICH (7)	Pore-scale reactive transport modelling based on the Lattice-Boltzmann-Method
NRG (9)	NRG Contribution
RWMC (10)	Effect of selection of secondary minerals on H-M-C coupling calculation
SCK·CEN (11)	SCK-CEN Contribution to WP3
CTU (16)	Methods of evaluation of through-diffusion experiments on sandwich bentonite-cement layers in a simple experimental set-up (WP3)

Institution (Partner n°)	Tite of the contribution within the WP presentation
VTT (18)	VTT Contribution
LML (20)	Micromechanics-based models for cement-based materials submitted to chemical degradations and hydro-mechanical modeling pf interface
ANDRA (23)	Interpretation and modelling

In addition to these activities, the workshop held a poster session and a topical session. The poster session was held at the end of the first day of the workshop at the same venue. Beneficiaries had the opportunity of presenting their results. Following, the list of the presented poster is provided. All of them are included in those proceedings in the Section *POSTERS*.

- Experiments on Interface Processes at the Cement/Callovo-Oxfordian Claystone Interface and the Impact on Physical Properties; Progress. *R. Cuss, A. Wiseall, J. Harrington, J. Talandier, X. Bourbon.*
- Observation of growth of the altered zone regarding cement-bentonite interaction by using Ca-XAFS analysis. *H. Owada, N. Fujii, D. Hayashi.*
- Interaction between cement and Czech bentonite under temperature load and in in-situ conditions: an overview of experimental program. *P. Večerník, L. Hausmannová, R. Červinka, R. Vašíček, M. Roll, J. Hloušek, V. Havlová.*
- Characterisation of UK and French geological disposal cement materials prior to groundwater leaching. *R.GW.. Vasconcelos, N.C. Hyatt, J.L. Provis, C.L. Corkhill.*
- Kinetic parameters for assessing the pH evolution of low-pH concretes. *T. Vehmas, M. Leivo, E. Holt.*
- Baseline Reference Mix Design and Castings. *T. Vehmas, A. Schindler, M. Löija1, M. Leivo, E. Holt.*
- PET/CT during degradation processes at the cement-clay interface and derivation of process parameters. *J. Kulenkampff, U. Mäder, M. Gründig, S. Eichelbaum, J. Lippmann-Pipke.*
- Development of Surface Reactivity Interface Experiments (SERIE) for studying FEBEX bentonite-concrete interaction. *D. González, R. Fernández, A.I. Ruiz, J. Cuevas.*
- Characterisation of Concrete Aging due to interaction with groundwaters in contact with different EBS. *M.C. Alonso, J.L. García Calvo, V. Flor-Laguna.*
- Solubility, hydrolysis and sorption of beryllium in cementitious systems. *X. Gaona, M. Böttle, T. Rabung, M. Altmaier.*
- Dissolution kinetics of AFm-Cl as a function of pH and at room temperature. *N.C.M. Marty, S. Grangeon.*
- Characterization of Hydrated Cement Paste (CEM II) by Selected Instrumental Methods and a Study of ^{85}Sr Uptake. *J. Kittnerová, B. Drtinová, D. Vopálka.*

- Effect of selection of secondary minerals on H-M-C coupling calculation. *H. Owada, D. Hayashi, A. Iizuka.*
- Modeling Transport Across Reactive Interfaces. *L.H. Damiani, G. Kosakowski.*

The *topical session* was focused on "Lessons learned from previous and current projects on cementitious barriers". The main results and lessons learned from three different projects were presented. The details of these presentations are presented below.

- **CI** Experiment: Cement-Clay-Interaction: 5-year evolution of cement-clay interfaces. U. Mäder (UNIBERN), V. Cloet (Nagra).
- DOPAS PROJECT – Experiences and lessons learned 2012- 2016. J. Hansen (Posiva), M. Vuorio (Posiva).
- Results, lessons learned and residual uncertaintics from the ECOCLAY II project. (Effect of Cement on Clay Barriers Performance – Phase II). N. Michau (ANDRA).

At the end of the workshop, Seif Ben Hadj Hassine (EUG chairman) gave some statements on behalf of the EUG about the progress of the project. Following, the general assembly closed the workshop.

Structure of the proceedings

The present proceedings are divided into three different sections:

1) **WP Overview:** An overview of the work conducted in the frame of the technical Work Packages.
2) **S+T contributions:** Scientific and technical manuscripts. All these contributions submitted by the beneficiaries were reviewed by the EUG members.
3) **Posters:** Posters presented during the poster session of the workshop.

WP OVERVIEW

Overview of Work Package 1

Experiments on interface processes and the impact on physical properties

WP leaders:
Erika Holt (VTT)
Francis Claret (BRGM)
Urs Mäder *(UNIBERN)*

The 1st workshop provided the third extensive discussion group for workpackage 1, after our earlier focused WP1 meeting which took place in London (November 2015) and the project kick-off meeting (Brussels, July 2015). This 1st workshop in June 2016 had participants from 16 of the 19 partner organizations of WP1. Some partners had multiple persons attending, including participation of new junior staff or students getting integrated to the project and waste management topics. During the WP1 plenary session, there were scientific progress updates provided by 17 partners during the 3 hour session. The proceedings include 18 papers and 7 posters from WP1. The previous day at the workshop during WP1 detailed discussions, progress was reviewed regarding aspects such as cooperative benchmark studies, exchange of samples and synergy between experimental methods. There were also inputs from the End Users group regarding the applicability of methods and selected materials. The topical session on lessons learned from other projects on cementitious barriers also provided complimentary information to be linked with CEBAMA, such as:

- DOPAS project (Marja Vuorio, Posiva) where concrete and bentonite have been used in tunnel end plugs with initial state material design, construction, modelling and performance monitoring.
- ECOCLAY-II (Nicolas Michau, ANDRA) with compatibilities between clay (granite) and cementitious material and their evolution.

A major outcome from earlier discussions at the WP1 meeting in London had been the need for a common baseline concrete material for comparative studies with national recipes. The 1st workshop then shared about the agreed-upon requirements for this mixture, its laboratory production and distribution to partners. A ternary low-pH mixture was produced, using both silica fume and blast furnace slag with a calcium-silica ratio of the binder being 0.61. The produced mixture had workability (slump) of 180 mm, air content of 0.9% and density of 2 450 kg/m^3, with the appearance as shown in Figure 1. An additional paste reference mixture was also made. Approximately 40 samples were distributed in spring and early summer to 6 different partners, and more sample were available upon request. Other partners have used the benchmark methodologies for preparing local samples, for instance in the Czech Republic (Figure 2). The expectation is that the benchmark low-pH concrete and paste samples will allow for comparable test result interpretation and better cooperative development of methodologies for interpreting material performance that also comes from existing aged samples (i.e., Figure 3). More comprehensive results and first results from the benchmark studies will be available by the second workshop and ready for inputs to WP3 modelling.

Figure 1. *Slump of CEBAMA WP1 benchmark reference concrete, produced at VTT.*

Figure 2. *Preparation of bentonite suuspension, at UJV.*

Figure 3. *Example of investigation of aged concrete-bentonite interface within a concrete plug in a granite gallery context (study by UAM).*

At the workshop, WP1 partners gave updates about their resources, schedule and dissemination activities. During the first year, some of the partners have been in the process of hiring students and staff to fulfil the work plans and ensure competence development of the next generation of nuclear waste management experts. Partners have been interacting with end users groups to determine the highest priority for cementitious material recipes and exposure conditions in the experimental studies, which will best serve the quantitative modelling and long-term safety evaluations. Regarding dissemination plans, many WP1 partners participated and presented at the "Mechanisms and Modeling of Waste/Cement Interactions" workshop in Murten, Switzerland on

May 22 - 25, 2016 (see www.empa.ch/cement2016). Some of the partners also shared at the FP7 DOPAS project ending seminar (Turku, Finland on May 25-27, 2016) on “Full-Scale Demonstration of Plugs and Seals”, where concrete-bentonite interaction may also be an issue for long-term safety and performance (see http://www.posiva.fi/en/dopas/dopas_2016_seminar). Preparations are underway for participation to the Clay Barriers conference (Switzerland, September 2017) for further dissemination of WP1 achievements (see http://www.clayconferencedavos2017.com/).

The activities of Work-package 1 during the first year of the project have addressed the synergy between various research methods and experimental studies of cementitious materials used in the repositories. At the start of the project an internal survey of WP1 participants was done to gather more details that served as the basis for the deliverable reports of the first year. This has included discussions and information on experimental methods and cementitious materials being studied, to ensure cooperation and implementation by the end users group. The first Deliverable report D1.01 provided detailed scientific work summary plans from each partner, serving as a basis for future reference. The next Deliverables D1.02 on systems to be studied, D1.03 as a state-of-the-art report on WP1 topics to be used as external reference by the general public, D1.04 as Experimental method boundary conditions to be used in WP1 studies, and D1.05 on Experimental materials to be used in the WP1 program were all available by or near the workshop time.

Overview of Work Package 2

Radionuclide retention in high pH concrete

WP leader:
Bernd Grambow (ARMINES*)*

The objective of WP2 is the study radionuclide retention processes in high pH concrete environments. The work conducted is largely independent of specific disposal concepts and addresses different types of host rocks. The aim is to provide insight on general processes and phenomena and their couplings in overall interaction mechanism, which can then be transferred to different disposal situations and water access scenarios in a high pH repository environment with cementitious materials. It also assesses the impact of chemical alterations (e.g., high pH concrete ageing, carbonation, transition from oxidizing to reducing conditions) on radionuclide retention.

A large number of studies are conducted, including leaching and solubility tests, diffusion tests, sorption experiments and coprecipitation studies. Variables are water composition, redox state and solid/water ratios. Also interaction with gaseous radionuclides (C14) is studied. Most experiments consist in exposing a selected radionuclide to fresh or preequilibrated solid/water systems for extended periods of times. Solids are a variety of important high pH cement-based materials: fresh and water aged cement pastes CEM I and CEM V, but as well individual phases such as CSH phases of various Ca/Si ratios, LDH and OH, SO_4, Cl of CO_3 type AFm and AFt phases. Radionuclides which have high priority from the scientific and applied perspective are selected. Some radionuclides are replaced by non-radionuclide isotopes. The corresponding chemical elements of interest are in particular Be, C, Cl, Ca, Se, Mo, I, and Ra. The number of systems to study is large. To allow for compatibility between individual approaches the meeting participants agreed to use as far as possible and feasible similar experimental protocols, boundary conditions or similar solids. Experimental programs are coupled to various modelling approaches, allowing to pave a way how for example to go from individual cement phases to the overall cement pastes.

All partners have started their experimental program and/or the respective development of protocols for preparation of individual cement phases. The following presentations of the various partners at the first CEBAMA workshop in Barcelona include a description of a detailed refinement of the experimental procedures, characterization of solid phases and of protocols for synthesis of cement phases, as well as first results of solubility and retention studies of Be, Ra and Tc on concrete and key cementitious phases.

Overview of Work Package 3

Interpretation & modelling

WP leader:
Andrés Idiart (AMPHOS21)

In recent years, significant advances have been achieved regarding the understanding of coupled physical and chemical processes affecting the performance of cement-based barriers. However, important gaps still exist on the: (1) coupling between chemical changes of cement paste microstructure and its physical properties, (2) relation of classical macroscopic reactive transport models to the microstructural features of cement and concrete, and (3) confidence in extrapolating numerical models of cementitious systems for long-term conditions.

The main goal of Work Package 3 (WP3) is to contribute in filling these gaps by modelling and interpretation of experimental results generated within the project. The focus is mainly on physical and chemical processes that can lead to changes in transport properties both in the cementitious systems as well as their interface with clays or compacted bentonite. Specific objectives of WP3 are (1) to improve the validity of existing numerical models to predict changes in transport properties of cementitious systems, (2) to give support to the advanced data interpretation and process modelling of WP1 and WP2 experiments by mechanistic modelling of chemically-induced changes in transport properties, and (3) to contribute to our ability to extrapolate models of system-level to modelling for Safety Case application (length and time upscaling).

An essential basis for WP3 is the outcome of the experiments and their characterization performed within the project. Every WP3 partner has already established collaborations with Work Packages 1 and 2 to model specific sets of WP1 and WP2 experiments. During the 1st Annual Workshop, new collaborations have been defined. It is expected that more interaction will result from the next Annual Workshops, as results start becoming available.

WP3 activities have started right after the kick-off of the project. However, experimental results from the project will in general not be available until much later. Therefore, each WP3 partner is using existing experimental data, relevant to their objectives within the project, in order to test and verify their modelling approaches and main developments during the first stages of the project.

As presented in the scientific and technical contributions from each partner, the models developed in this project mainly focus on:

1) cement/concrete matrices and cement/clay interactions,
2) reactive transport models, also coupled to hydro-mechanical models,
3) mostly at the laboratory length scale, but also at pore- and macro-scale,
4) time scales mostly covering the duration of lab tests, but also long-term predictions.

During the 1st Annual Workshop in Barcelona, all WP3 partners from 13 organizations presented their progress during the first year of the project. From these, 8 scientific and technical contributions are presented in this Workshop Proceedings.

KIT S&T presents the conceptual model to be used in continuum-scale reactive transport simulations of dedicated experiments of the interaction between low pH cement paste and bentonite water. The model will be implemented in the iCP modelling platform, coupling Comsol Multiphysics and Phreeqc. As presented during the workshop, VTT will also model degradation reactions in low-pH cement systems, while NRG will model in 1D the reactive transport processes at the interface between cement paste and Boom clay, using the open-source framework ORCHESTRA. CTU S&T presents an experimental setup for measuring diffusion coefficients in a sandwiched bentonite cement arrangement and its interpretation through 1D modelling using the GoldSim framework. During the workshop, the working plan to implement a new code for modelling reactive transport processes in cement/clay interfaces was presented by PSI, including electrochemical couplings (Nernst-Planck equations).

BRGM S&T presents the working plan to perform streaming potential and spectral induced polarization experiments to describe the surface electrical, petrophysical, and mineralogical properties of low pH cement pastes. The method to interpret experimental measurements with surface complexation and transport models at the pore and sample scales is described with the final goal of improving the predictions of reactive transport properties.

JUELICH S&T presents their working plan and preliminary developments of a new implementation of a pore-scale reactive transport model. The goal is then to use this development for studying degradation processes in the cementitious matrix. The numerical framework is presented that makes use of the existing software Palabos and Phreeqc, and a verification example is shown. During the workshop, SCK·CEN presented their plans about pore-scale reactive transport modelling using their numerical tool YANTRA, which is in a more advance development stage. Different experiments will be interpreted by each partner, giving new insights in reactive transport processes at a more fundamental scale than continuum models.

RWMC S&T introduces an application case of the modelling of the effect of secondary minerals on cement-clay and cement-bentonite interactions under repository conditions using. A coupled hydro-chemo-mechanical framework is used and its main features are presented, coupling the reactive transport code Phreeqc-TRANS to the hydro-mechanical code DACSAR-BA. A different approach for modelling coupled chemo-mechanical processes in concrete is presented by LML in their S&T, based on homogenization theory and advanced nonlinear constitutive models. The models are applied to simulate several degradation processes under laboratory conditions, and a comparison to experimental data is done. In the same line, Amphos 21 S&T describes the chemo-mechanical model implemented in iCP, coupling reactive transport with mechanical and chemical damage models. They present preliminary results of the simulation of existing experiments. In turn, UDC S&T focuses on reactive transport processes of concrete bentonite interaction experiments at non-isothermal conditions. The model takes into account the thermo-hydro-chemical couplings and their impact on mechanical behaviour using the code INVERSE-FADES-CORE developed at UDC. Preliminary results of the model applied to the interaction of OPC concrete and FEBEX bentonite are presented.

Finally, ANDRA presented a working plan during the workshop to model reactive transport processes in low-pH concrete components at the macro-scale and their interface with clayey materials. They will integrate data acquired within CEBAMA to conduct Performance Assessment models at a larger spatial scale and focusing on long-term predictions.

S + T CONTRIBUTIONS

Experimental studies on low pH cement / clay interface processes: Characterization of low pH cements

Naila Ait Mouheb[1*], Vanessa Montoya[1], Christoph Borkel[1], Thorsten Schäfer[1]

[1] KIT, Institut für Nukleare Entsorgung (DE)

* Corresponding author: naila.ait-mouheb@kit.edu

Abstract

Preparation and characterization of three low pH cement pastes is described in this work. SEM-EDS analyses show after 90 days of hydration that the main hydrated phases are C-S-H gels with a Ca:Si ratio between 0.8 - 1.0. DTA and XRD indicate the presence of calcite in the mixtures where limestone filler has been used, but due to the low crystallinity of the samples, additional techniques were needed to identify the hydrated solid phases (i.e., FTIR). Equilibrium water analysis included measurements by IC, ICP-OES and ICP-MS. Porosity and pore size distribution was characterized by MIP.

Introduction

The deep geological repository concept is based on the confinement of the radioactive waste over a long period of time by multiple barriers. Many of the concepts developed in the internationally use concrete and clay as confinement barriers (ANDRA, 2005; ENRESA, 1995). The different barriers are not in geochemical equilibrium and during the prolonged period post disposal, cementitious material will undergo alterations possibly changing the chemical and physical properties of this barrier. One important process that could reduce the durability of a concrete barrier is the leaching/degradation of the solid especially at the cement/clay interface due to the contact of clay pore water with the cementitious material. In order to minimize the interaction between the "classical" cement materials and the bentonite porewater (pH ~ 7.0 - 7.5) low pH cements were developed within the nuclear waste disposal context in the late 90's.

The main characteristic of these low pH cements is the absence of portlandite as hydrated solid phase, which reduces the pH to ~11 compared to the "classical" cements (pH $\geq$ 12). Thus, to obtain 'low pH' cement, the clinker content in the binder is substituted for supplementary materials like silica fume or fly ashes (Cau-dit-Coumes et al., 2006). From a chemical point of view the Ca/Si ratio composition in the hydrated low pH cement is less than 1.0 and C-S-H phases are the main solids present.

In the framework of the CEBAMA project, KIT-INE will study the interaction between low pH cement materials and bentonite pore water assessing the impact of mineralogical and microstructural modifications of the cement on transport properties and its effect on the mobility of radionuclides. To evaluate time dependent changes on transport properties and mineralogical composition in the geochemical perturbed system, it is crucial to have a detailed physicochemical characterization of the cementitious starting materials under equilibrium conditions. The aim of this contribution is to describe the preparation of the low pH cements and their characterization by different methods, namely X-ray diffraction (XRD), thermogravimetric - differential thermal analysis (TG-DTA), Fourier-transform infrared spectroscopy (FTIR), scanning electron Microscopy - energy dispersive X-ray spectroscopy (SEM-EDX) and mercury intrusion porosimetry (MIP).

Materials and methods

Samples preparation

Three different low pH cement pastes (MIX 3C, 3D and 3E) consisting either of binary or ternary binder composition with sulphate resistant portland cement (CEM I 52.5N SR, from Lafarge), silica fume (microsilica ELKEM 940U) and limestone filler ($CaCO_3$) have been prepared (see Table 1). An inorganic superplasticizer (SioxX®) was used to achieve good workability in the MIX 3C. The elemental composition of the different initial materials calculated on the oxide basis is summarized in Table 2.

Table 1: *Composition (w.w%) of the mix solid proportions used in this work.*

	MIX 3C (%)	MIX 3D (%)	MIX 3E (%)
OPC (CEM I 52.5)	39	40	50
Silica fume (ELKEM 940U)	39	40	50
Limestone filler	19	20	-
Superplasticizer (SioxX)	3	-	-

Table 2: *Chemical composition (wt. %) of the OPC, superplasticizer and the silica fume.*

	SiO_2 (%)	Al_2O_3 (%)	Fe_2O_3 (%)	MgO (%)	CaO (%)	Na_2O (%)	K_2O (%)	SO_3 (%)	P_2O_5 (%)	Cr_2O_3 (%)	LOI (%)	Sum (%)
OPC (CEM I 52.5)[a]	20.62	3.22	4.94	1.12	64.47	0.08	0.36	2.56	0.06	0.024	2.18	99.89
Superplasticizer[b]	60 - 80	15 - 30										
Silica fume[b]	100											

[a] obtained by X-ray fluorescence (XRF)
[b] provided by the manufacturer

The mixtures have been hydrated with de-ionised water using a water/binder ratio (w/b) = 0.6 and the samples have been cured in a chamber at 98% relative humidity (RH) and 21 ± 2°C over a duration of 90 days.

Different techniques were then used to characterize the three mixtures. A protocol was followed in order to obtain representative samples for analysis from the whole block (see Figure 1). Initially, 5 mm from the surface of the cement block were removed to obtain undisturbed samples concerning potential air CO_2 influence.

For the analysis of pore water pH and chemical composition the *ex situ* leaching method described by Alonso et al. (2011) was used. The evolution of the hydration process was monitored via pH and chemical composition analysis as a function of hydration time (30 - 90 days).

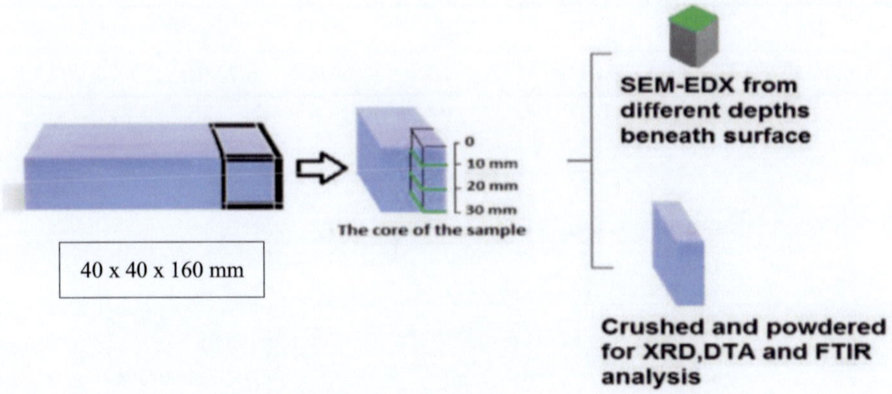

Figure 1: *Scheme of the sample preparation for the characterization.*

For analysis by SEM-EDX, three polished cubes (6 cm^2) from different depth (10, 20 and 30 mm beneath the surface), were analysed, using an ESEM (model FEI Quanta 650 FEG), equipped with an energy-dispersive X-ray spectrometer (model Thermo Scientific NORAN System 7) for elemental microanalysis. The electron accelerating voltage was set to 20 kV and the working distance for analysis was 10 mm. Observations were performed on polished sections either coated by chromium or uncoated using the variable pressure mode (70 Pa water atmosphere) of the ESEM.

For the analysis by X-Ray Diffraction (XRD), Thermogravimetric analysis (TG-DTA) and Fourier-transform infrared (FTIR) spectroscopy samples of 2.3 x 3 x 3 cm are crushed and powdered in a grinder for 2 min. XRD was performed with a Bruker D8 ADVANCE diffractometer using Cu Kα radiation under controlled N_2 atmosphere (O_2 and CO_2 concentrations < 5 ppm) with a step width of 0.02° per 2 s time step. TG-DTA was conducted using a heating rate of 10 °C/min from 25 to 1 200°C with a STA409 (Netzsch Gerätebau GmbH) on 20 - 30 mg powder samples. FTIR spectroscopy was performed with a few milligrams of solid using an Attenuated Total Reflection (ATR) Spectrometer (IFS 55 EQUINOX Spectrometer, Bruker) equipped with a DTGS detector.

Results and discussion

Evolution of pH and contact water chemistry

The evolution of the pH as a function of hydration time shows the asymptotical decrease to values around pH 11 after 90 days of hydration for all mixtures investigated (see Figure 2).

The contact solution composition during the hydration process has also been analyzed. In Table 3, the aqueous composition after 90 days of hydration of MIX 3E is shown.

Table 3: *Chemical composition of the contact water solution (mol/L) (MIX 3E).*

Si	Na	Ca	K	Mg	Al	Fe	Cl	S
$2.79\cdot10^{-3}$	$4.64\cdot10^{-3}$	$1.78\cdot10^{-2}$	$3.81\cdot10^{-3}$	$3.18\cdot10^{-6}$	$9.99\cdot10^{-8}$	$6.73\cdot10^{-8}$	$1.94\cdot10^{-3}$	$1.24\cdot10^{-2}$

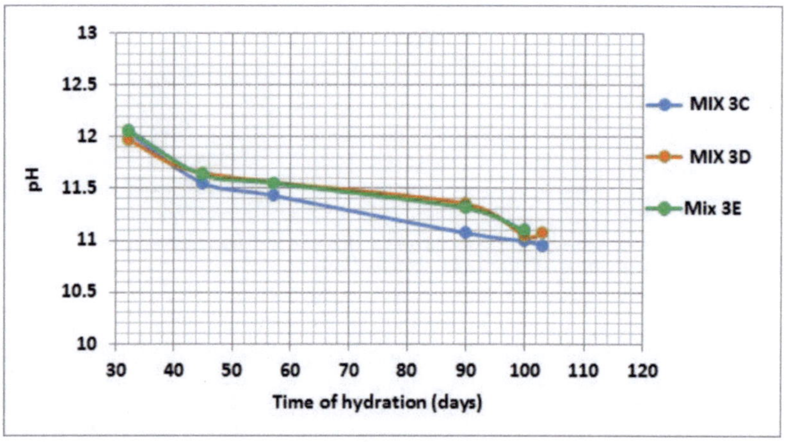

Figure 2: *pH evolution of contact water as a function of hydration time for the three different crushed mixtures (3C, 3D and 3E).*

X-Ray diffraction (after 90 days of hydration)

Figure 3 presents the X-Ray diffractograms of the three mixtures. This technique in general provides information of crystalline phases and the identification of poorly crystalline hydrated phases present in cement pastes is more challenging.

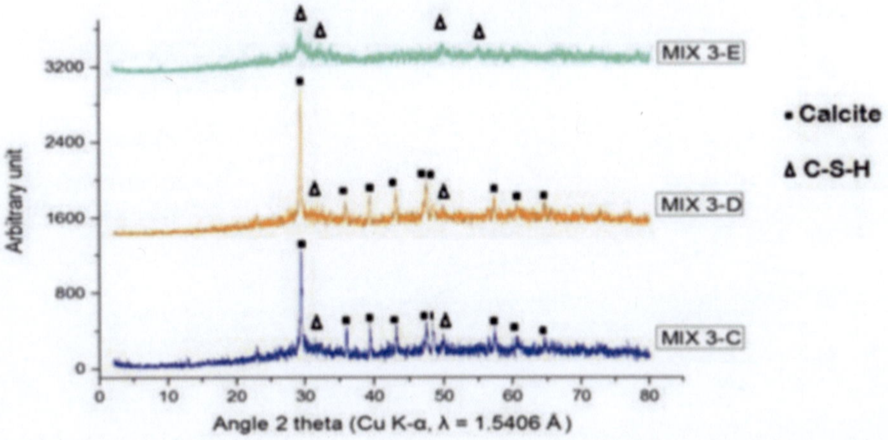

Figure 3 : *Powder XRD pattern of the samples (MIX 3C, blue line; MIX 3D, orange line; MIX 3E, green line).*

The diffractograms of sample 3C and 3D are very similar and calcite has been identified (diffraction peaks marked with black square in Figure 3) as the only crystalline phase present in the solid. Calcite is initially added in the blend of these two cement pastes as limestone filler. On the other hand, no calcite could be identified in sample 3E.

C-S-H phases could not be detected in any of the samples probably due to its amorphous nature. However, looking in detail to the diffractogram of mixture 3E, without limestone filler, small peaks indicate the presence of crystalline C-S-H phases (Chen et al., 2004).

Neither portlandite nor clinker is detected by XRD in any of the samples suggesting that the samples are almost fully hydrated and portlandite has been consumed in the hydration process by the well-known pozzolanic reaction in agreement with the measured pH. Furthermore, ettringite ($Ca_6Al_2(SO_4)_3(OH)_{12} \cdot 26H_2O$) with a characteristic peak at pic ($2\theta = 9.10$ Cu K-α, $\lambda = 9.7$ Å) has not been identified.

TG - DTA after 90 days of hydration

DTA can give quantitative information of the different phases present in the samples by measuring the temperature dependent mass loss (% wt) and therefore giving the advantage of characterizing the amorphous or low-crystalline hydrated phases not detectable by XRD.

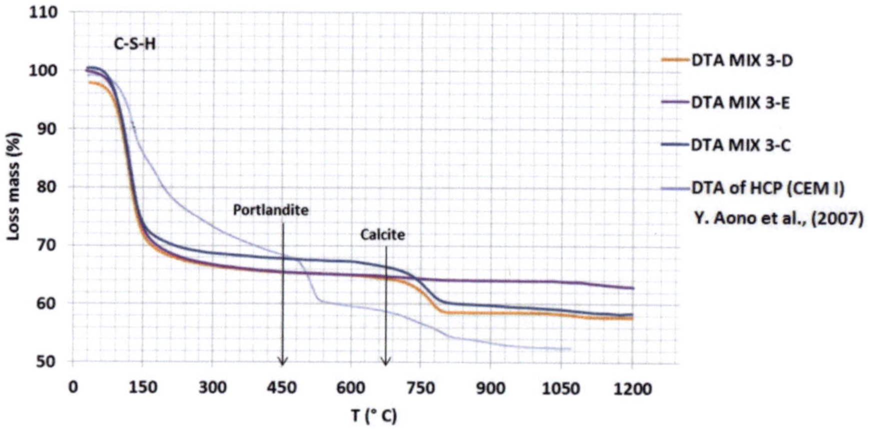

Figure 4: *DTA curves of MIX 3C, 3D and 3E after 90 days of hydration together with the analysis of cement paste (CEM I) from Aono et al. (2007).*

In Figure 4, the DTA data of the three mixtures in comparison to the measurement of CEM I with w/c = 0.35 (Aono et al., 2007) is presented. In all samples, the mass loss around 100 - 300°C is attributed to the presence of C-S-H phases and ettringite. However, in this temperature range the mass loss can also be attributed to mineral surface bond water making an exact quantification of the of C-S-H phases difficult.

The weight loss at 450°C indicative for portlandite is not observed in any of the investigated samples confirming the absence of this solid phase. Finally, for mix 3C and 3D a mass loss between 700 - 850°C is observed which is attributed to the presence of calcite in the initial mixture.

FT-IR spectroscopy analysis after 90 days of hydration

Figure 5 shows the FT-IR spectra of the three different mixtures. In the range between 4 000 – 2 800 cm^{-1} appears the band corresponding to the stretching vibrations of O–H groups in H_2O or hydroxyls. The band associated to the presence of portlandite (3 640 cm^{-1}) is not observed in any of the samples (Chollet et al., 2011; Hartmann et al., 2014; Yu et al., 1999) in agreement with the observations by XRD and DTA.

The main bands corresponding to C–S–H phases are detected at 958 cm^{-1} and 439 cm^{-1}. There is evidence of the presence of C–S–H gels with tobermorite structure (low Ca/Si ratio). The band at 439 cm^{-1} corresponds to the internal deformation of tetrahedral SiO_4. It is true that quartz from silica fume has a band at similar wavenumbers (470 - 460 cm^{-1}), but it is sharper than the band around 450 - 460 cm^{-1} detected in the low-pH pastes spectra (Garcia Calvo et al., 2013).

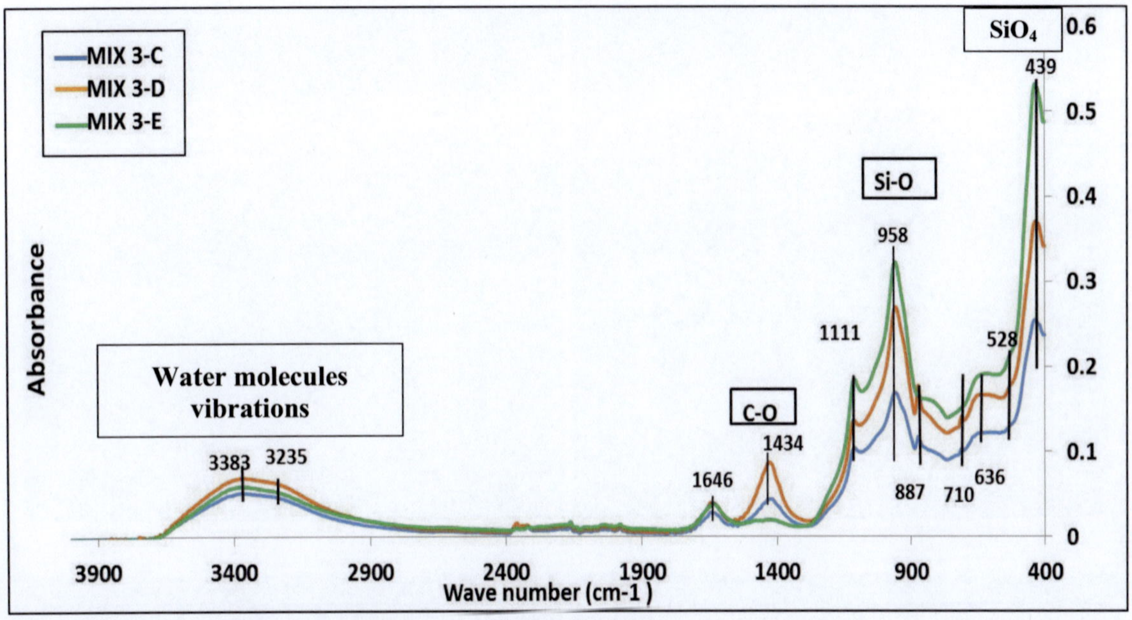

Figure 5: *FT-IR spectra of the MIX 3C ,3D and 3E after 90 days of hydration.*

Calcite has been clearly detected in mix 3C and 3D with a band at 1 434 cm^{-1} assigned to the C-O stretching vibration and two bands at wavenumbers of 710 and 887 cm^{-1} corresponding to the C-O bending vibrations. Bands in mix 3E for calcite were not detected.

Some characteristic SO_4^{2-} bands from ettringite (calcium sulfo-aluminate hydrate) are located at 1 111 cm^{-1} and have been identified in all samples.

SEM-EDX analyses after 90 days of hydration

The three pastes were also analyzed by SEM-EDX to observe the morphology and chemical composition of the different hydrated phases. This technique opens the opportunity to determine the Ca/Si ratio of the C-S-H phases present which is not possible to be done with the other techniques used in this study so far.

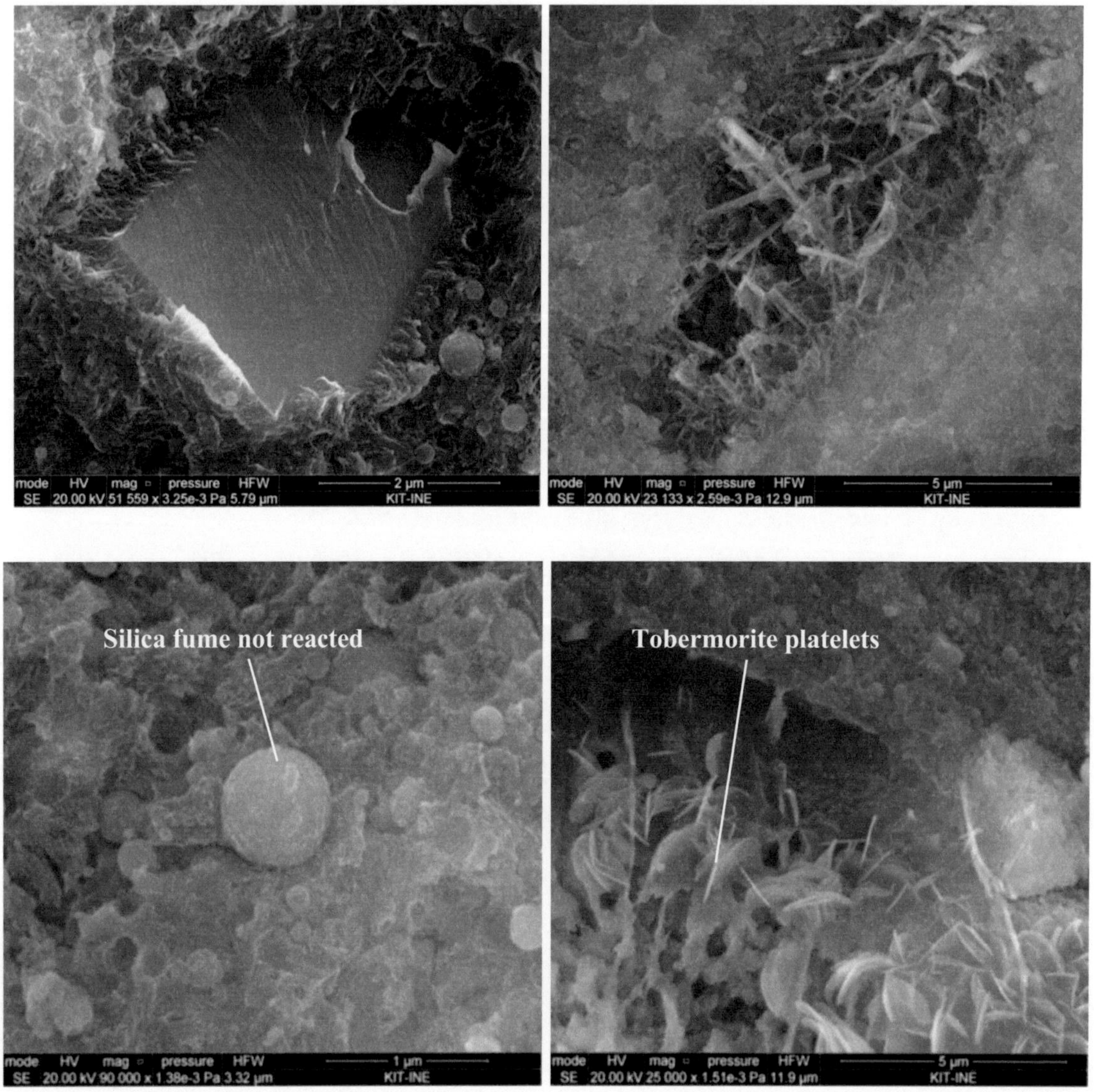

Figure 6: *SEM pictures of mix 3C (**top-left**: presence of calcite, **top-right**: C-S-H phases, **bottom-left**: silica fume not reacted, **bottom- right**: tobermorite platelets).*

Figure 6 and Figure 7 show images of the mixtures at different positions. The main phases identified are C-S-H in all samples. Calcite can also be clearly identified in samples 3C and 3D (see Figure 6 top-left and Figure 7 top-right). Additionally, non-reacted silica fume is present in all the mixtures (Figure 6, bottom-left and Figure 7 bottom-left) with its characteristic spherical structure (Nochaiya et al., 2010). The presence of platy, disc-shaped crystals in sample 3-C (Figure 6, bottom-right and Figure 7 top-left) has been identified as crystalline C-S-H phases of tobermorite-type structure (Franus et al., 2015).

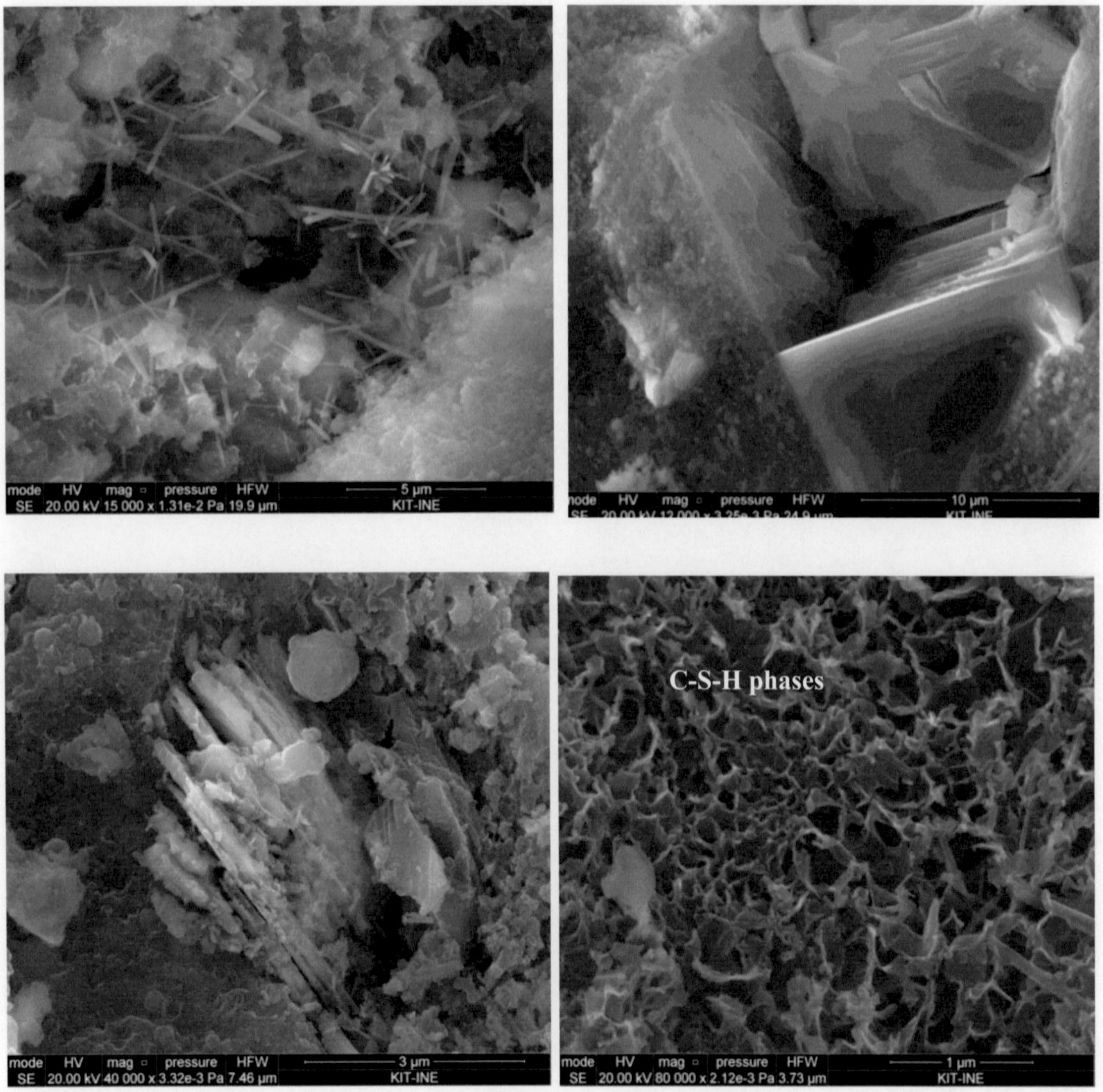

Figure 7: *SEM pictures of mix 3D (upper row) and mix 3E (lower row) (**top-left**: crystalline C-S-H phases of tobermorite-type structure, **top-right**: presence of calcite, **bottom-left**: silica fume not reacted and calcite, **bottom- right**: C-S-H phases).*

In order to determine the different phases present and the Ca/Si ratio of the C-S-H phases a multi-elementary mapping of the surface was conducted by EDX. Twenty different spots varying the beam penetration depth by changing the acceleration voltage have been investigated to determine the atomic ratio of Ca, Al, Si, C, Fe, S, Na, Mg and K. In Figure 8, the data analysis for the sample 3E is presented. The same analysis will be performed in the future for mix 3C and 3D.

Figure 8 shows the plot of Al/Ca versus Si/Ca atomic ratios analyzed at several points of the mix 3E surface. According to Codina et al. (2008) and Winter (2012) using this type of data presentation the presence of C-S-H phases are found in the chemical range (Al/Ca: 0.04 - 0.08, Si/:Ca: 1.5 - 4) and, portlandite near to the origin with very low Al/Ca and Si/Ca ratio. The presence of AFm (monosulfoaluminate)

$(CaO)_3(Al_2O_3)(CaSO_4)\cdot 12H_2O$ is located at (Al/Ca: 0.5, Si/Ca: 0), AFt (ettringite) at (Al/Ca: 0.35, Si/Ca: 0). According to this interpretation and looking to Figure 8, the analysis shows that mix 3E is composed essentially by C-S-H phases. The presence of high Si/Ca ratio can be related to the presence of non-reacted silica fume as observed by SEM. In order to determine the Ca/Si ratio of solid phases a frequency histogram is shown in Figure 9. As can be observed, the range of the Ca/Si ratios spans values from 0.2 to 1.2 with a clear maximum around 0.9. The presence of C-S-H phases with a Ca/Si ratio = 0.8 - 0.9 is in agreement with an equilibrium solution of pH = 11 as measured in this work (Sugiyama et al., 2006; Atkinson et al., 1987; Fujii et al., 1981).

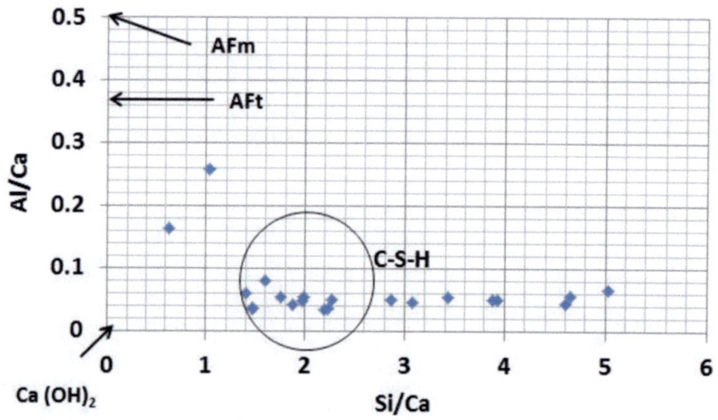

Figure 8: *Al/Ca versus Si/Ca atomic ratios for different microanalyses spots of the mixture 3E after 90 days of hydration.*

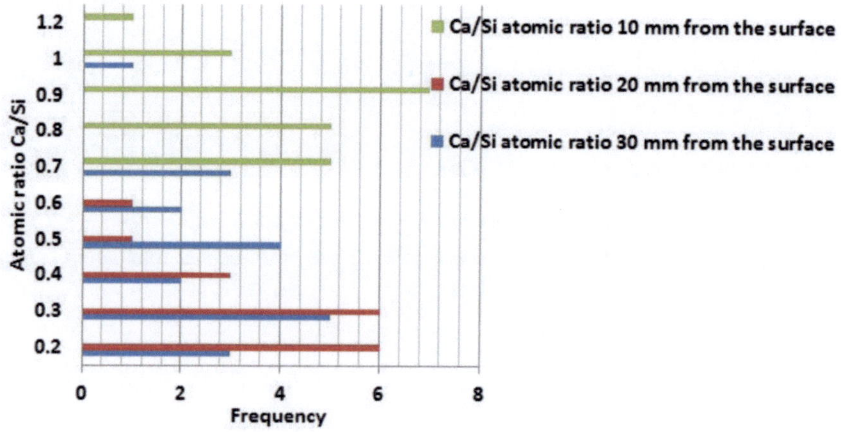

Figure 9: *Ca/Si frequency histogram for mixture 3E measured at different points determined by EDX.*

Finally, to complete the interpretation of the data and to show the homogeneity of the sample, Figure 10 shows the spatial resolved element distribution of silica (in red), oxygen (in blue), calcium (in yellow) and iron (in dark blue). The atom percent increases with the intensity of the colour (maximum silica content 58%, calcium 24%, iron 8%), whereas black areas represent locations of element absence. Superposition of the calcium, silica and iron content as RGB image allows to clearly distinguish three different zones (see Figure 10, right image). The orange color in the RGB image represents areas of C-S-H phase predominance, whereas purple – blue colors shows non- reacted OPC and the red color means that there is unreacted silica fume.

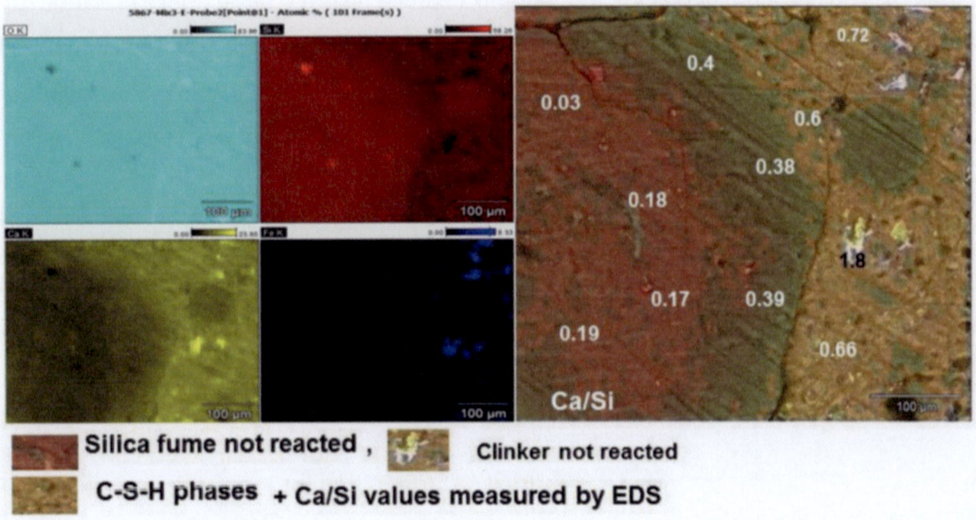

Figure 10: *Multi- element mapping carried out by EDS for mix 3E in element distribution maps of Ca, Si, Fe and O **(left)** and as combined RGB image **(right)**.*

Mercury intrusion porosimetry (MIP) after 90 days of hydration

Various methods can be employed for pore size analyses. Mercury intrusion porosimetry has initially been selected for studying the porosity and pore connectivity of the three different cement pastes. After 90 days of hydration the measured porosity of mix 3C is 24%, 43% for mix 3D and 35% for mix 3E, respectively. According to this results, it seems that the addition of superplasticizer decrease the porosity of the cement paste. This is further cooperated by the significant decrease of nano- porosity. However, although MIP is frequently used for porosity characterisation this technique has limitations when applied to materials with irregular pore geometry, especially for materials that contain C-S-H phases with microstructural pores. As MIP can be applied to measure connected pores in the size range of 0.005 - 10 µm complementary techniques including image analysis and NMR will be applied in the future to characterize the full pore size distribution.

Conclusions and Future work

The chemical and physical properties of the low pH cement pastes prepared mainly with Ordinary Portland cement (OPC) and silica fume (40 - 50%) are clearly different from the ones typically found with only OPC. The main hydrated phases formed after 90 days are the calcium silicate hydrates (C-S-H) with a low Ca/Si ratio which varies between 0.8 and 1. Unreacted silica fume has also been detected in all the samples and calcite is present in two of the samples where limestone filler was added. Additionally, as seen by MIP, the presence of superplasticizer clearly decreases the porosity and the smallest connected pore size detectable of the cement pastes.

In the future, beside the completion of the sample characterization further interpretation of the results including geochemical modelling will be performed. Additional analytical techniques will be employed including e.g., solid-state nuclear magnetic resonance spectroscopy to thoroughly characterize the samples.

Acknowledgement

The research leading to these results has received funding from the European Union's European Atomic Energy Community's (Euratom) Horizon 2020 Programme (NFRP-2014/2015) under grant agreement, 662147 – Cebama.

References

ANDRA (2005). Andra Research on the Geological Disposal of Highlevel Long-lived Radioactive Waste– Results and Perspectives.

Alonso, M.C., García Calvo, J.L., Walker, C., Naito, M., Pettersson, S., Puigdomenech, I., Cuñado, M.A., Vuorio, M., Weber, H., Ueda, H. and others (2011). Development of an accurate pH measurement methodology for the pore fluids of low pH cementitious materials. Proceedings of the 13th International Congress on the Chemistry of Cement, Madrid, Spain.

Aono, Y., Matsushita, F., Shibata, S., Hama, Y. (2007). Nano-structural Changes of C-S-H in Hardened Cement Paste during Drying at 50ºC. Journal of Advanced Concrete Technology, 5(3), 313-323.

Atkinson, A., Hearne, J.A., Knights, C.F. (1989). Aqueous chemistry and thermodynamic modelling of $CaO-SiO_2-H_2O$ gels. Journal of the Chemical Society, Dalton Transactions, 12, 2371-2379.

Cau-dit-Coumes, C., Courtois S., Nectoux, D., Leclercq, S., Bourbon, X. (2006). Formulating a low-alkalinity, high-resistance and low-heat concrete for radioactive waste repositories. Cement and Concrete Research, 36(12), 2152-2163.

Chen, J.J., Thomas, J.J., Taylor, H.F.W., Jennings, H.M. (2004). Solubility and structure of calcium silicate hydrate, Cement and Concrete Research, 34(9), 1499-1519.

Codina, M., Cau-dit-Coumes, C., Le Bescop, P., Verdier, J., Ollivier, J.P. (2008). Design and characterization of low-heat and low-alkalinity cements. Cement and Concrete Research, 38 (4), 437-448.

ENRESA (1995). Almacenamiento geológico profundo de residuos radiactivos de alta actividad (AGP). Diseños conceptuales genéricos. Publicación Técnica ENRESA, 11/95.

Chollet, M. and Horgnies, M. (2011). Analyses of the surfaces of concrete by Raman and FT-IR spectroscopies: comparative study of hardened samples after demoulding and after organic post-treatment. Surface and Interface Analysis, 43(3), 714-725.

Franus, W., Panek, R., Wdowin, M. (2015). SEM investigation of microstructures in hydration products of portland cement. 2nd International Multidisciplinary Microscopy and Microanalysis Congress, 105-112.

Fujii, K. and Kondo, W. (1981). Heterogeneous equilibrium of calcium silicate hydrate in water at 30°C. Journal of the Chemical Society, Dalton Transactions, 2, 645-651.

García Calvo, J.L., Sánchez Moreno, M., Alonso Alonso, M.C., Hidalgo López A., García Olmo, J. (2013). Study of the Microstructure Evolution of Low-pH Cements Based on Ordinary Portland Cement (OPC) by Mid- and Near-Infrared Spectroscopy, and Their Influence on Corrosion of Steel Reinforcement. Materials, 6, 2508-2521.

Hartmann, A., Khakhutov, M., Buhl, J.-C. (2014). Hydrothermal synthesis of C-S-H-phases (tobermorite) under influence of Ca-formate. Materials Research Bulletin, 51, 389-396.

Nochaiya, T., Wongkeo, W., Chaipanich, A. (2010). Utilization of fly ash with silica fume and properties of Portland cement–fly ash–silica fume concrete. Fuel, 89(3), 768-774.

Sugiyama, D. and Fujita, T. (2006). A thermodynamic model of dissolution and precipitation of calcium silicate hydrates. Cement and Concrete Research, 36(2) 227-237.

Winter, N.B. (2012). Scanning electron microscopy of cement and concrete. WHD Microanalysis.

Yu, P., Kirkpatrick, R.J., Poe, B., McMillan, P.F., Cong, X. (1999). Structure of calcium silicate hydrate (C-S-H): Near-, Mid-, and Far-infrared spectroscopy. Journal of the American Ceramic Society, 82(3), 742-748.

Methodology to study the changes in microstructure and transport properties of the Boom Clay – concrete interface

Quoc Tri Phung[1*], Norbert Maes[1], Francis Claret[2], Stephane Gaboreau[2], Jef Leysen[3]

[1] SCK•CEN, Belgian Nuclear Research Centre (BE)
[2] BRGM, Bureau de Recherches Géologiques et Minières (FR)
[3] EURIDICE, European Underground Research Infrastructure for Disposal of nuclear waste in a Clay Environment (BE)

* Corresponding author: qphung@sckcen.be

Abstract

Cement-based materials have been increasingly used as part of engineered barrier system for radioactive waste disposal facilities. Under its service environment, these concrete structures undergo complex interactions with the geological matrices (e.g., clay, granite) because of the contrasting chemistry. Within this contribution, we present an experimental methodology to understand how the interface between Boom Clay and concrete alters due to chemical interaction. Experiments are performed on in-situ interfaces as well as newly made interfaces. A detailed procedure to sample the in-situ interface using resin anchor concept and new laboratory setups to accelerate the interface interaction are described. Changes in microstructure and mineralogy of concrete and clay will be characterised by using µ-tomography, SEM/EDX, FIB-SEM and TEM while changes in transport properties will be characterized by water permeability, diffusivity of dissolved gases and water sorptivity using newly developed methods.

Introduction

Cementitious materials (concrete, mortar, cement paste, grout) are omnipresent in civil and infrastructure constructions, from small components to huge buildings. Despite the development of novel/advanced materials, cement-based materials remain by far the most widely used materials in construction over the world. Besides their classical use in construction, these materials are also used for encapsulation of radioactive waste and as engineered barriers for disposal of radioactive waste, both in near-surface repositories for low/intermediate active waste and geological/deep repositories for high level long-lived radioactive waste (see Figure 1). The role of cement-based materials in these applications is not only focused on the retention of radionuclides (attributed to its low transport properties, high internal surface area) but also in helping creating beneficial conditions for the waste package integrity (steel corrosion) and limiting the solubility of many cationic species (Read et al., 2001) because of its high-pH buffering capacity for a very long period.

Under its service environment, these concrete structures undergo complex interactions with the geological matrices (e.g., clay, granite). The interaction processes are very slow but important for the long-term durability assessment (> 1 000 years). The interaction disturbs the equilibrium between the pore solution of the cementitious materials and the solid phases of the cement matrix which results in dissolution and/or precipitation of minerals. The alteration is even more pronounced at the host rock side. The interaction typically results in alteration of mineralogy and microstructure which are mostly followed by alteration of its transport properties and influences the long-term stability of an engineered structure.

The Boom Clay formation, which is a dark organic-rich and poorly indurated argillaceous clay with a high pyrite and glauconite content in the more silty layers, has been considered as a potential host rock for intermediate and high level radioactive waste in Belgium. This paper describes the sampling of concrete-clay interfaces with more than 13 year contact duration stemming from the underground research laboratory (HADES) located at 225 m in depth at Mol, which are representative for disposal conditions. The laboratory experiments to mimic the real interaction conditions are also described in this paper.

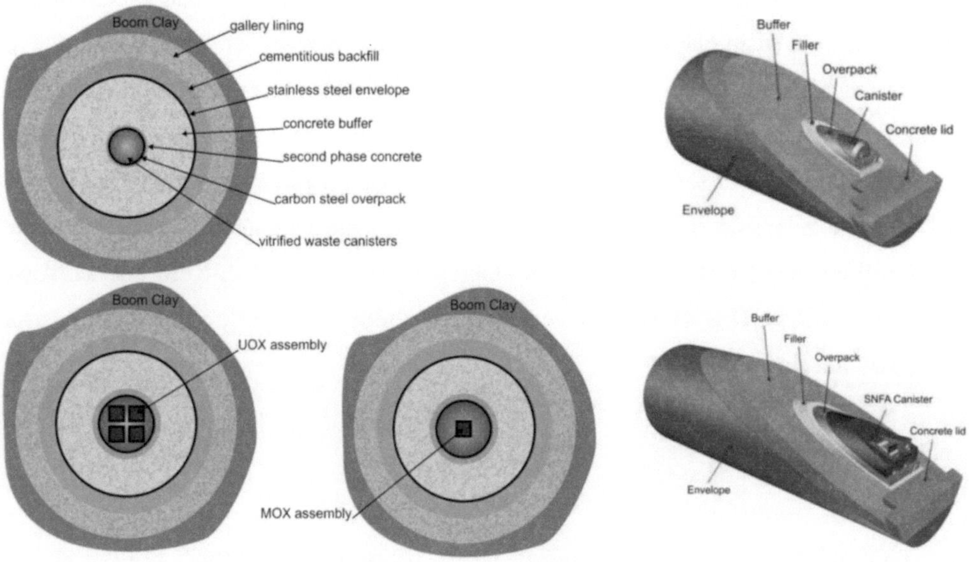

Figure 1: *Concrete used for disposal gallery lining, buffer, and backfill in high level waste facility – Belgian Supercontainer concept.*

Examined materials

In-situ concrete

The tests are performed on high strength concretes (wedge blocks) used as lining material in the Connecting Gallery of HADES URL. Age of interface concrete - Boom Clay host rock is about 13 years. The composition of concrete is shown in Table 1.

Table 1: *Mix compositions of in-situ concrete: SP – Superplasticizer.*

For 1 m^3 concrete	Cement (kg)	Fly ash (kg)	Coarse agg. (kg)	Fine agg. (kg)	Admixture		Water (L)
					SP (L)	µ-silica (kg)	
Connecting gallery	335 (CEM I)	115	1 252	540	4.5	90	135

Cement pastes for laboratory studies

With the purpose of accelerating the interaction processes, we intend to make cement pastes with high porosity for the laboratory experiments. The mixture compositions are for backfill materials, but without superplasticizer. In case concrete liners of disposal facilities (see Figure 1) are collapsed, the backfill materials will contact directly with the Boom clay. Note that the backfill compositions for Belgian radioactive waste facilities have not yet fixed. The typical composition composes water, Portland cement, limestone filler, hydrated lime and superplasticizer to improve the workability. We also investigate the effect of w/c ratio on the interaction processes on mixtures with three w/c ratios of 0.5, 0.7, and 0.9 as shown in Table 2. Type I ordinary Portland cement (CEM I 52.5 N) is used. This cement has a Blaine specific surface area of 435 m^2/kg and density of 3 100 kg/m^3. Note that experiments (Dehghanpoor Abyaneh et al., 2014) are performed on cement pastes as the main degradation processes occur in the paste matrix, and aggregates only play a limited role in these processes.

Table 2: *Mix compositions of cement pastes for laboratory experiments.*

Mixtures	Cement	Limestone filler/cement	Lime/cement	Water/cement
BF	CEM I 52.5 N	3/1	0.377	1.5
C05	CEM I 52.5 N	-	-	0.5
C07	CEM I 52.5 N	-	-	0.7
C09	CEM I 52.5 N	-	-	0.9

Boom Clay properties

The main minerals in Boom Clay are quartz and clays, in different proportions. Table 3 presents the Boom Clay mineralogical compositions (min – max ranges) at the Mol site (Honty and De Craen, 2012). Note that the same minerals are present over the whole Boom Clay formation although the proportions of the various minerals vary in the vertical profiles. The reference composition of Boom Clay pore water at the Mol site in Belgium is also described in Table 3 (De Craen et al., 2004). The carbonate concentration in pore solution is quite high (~ 0.014 mol/L $NaHCO_3$) which may accelerate the carbonation of the concrete side.

Table 3: *Mineralogical and chemical properties of Boom Clay and its pore water.*

Mineralogical composition Boom Clay (%)		**Boom clay pore water**	
Quartz	22 - 66	Na	359 mg/L
Na-plagioclase	0 - 6.3	K	7.2 mg/L
K-Feldspar	0.4 - 8	Ca	2.0 mg/L
Siderite	0 - 1.5	Mg	1.6 mg/L
Calcite	0 - 4.6	Fe	0.2 mg/L
Dolomite	0 - 1	Cl	26 mg/L
Apatite	0 -0.9	SO_4	2.2 mg/L
Pyrite	0.3 - 5	HCO_3	879 mg/L
Illite/muscovite	5 - 37	DOC	115 mg/L
Smectite and illite-smectite	6.8 - 37	Eh (mV)	-274 mV vs SHE
Kaolinite	2 - 14	pCO_2 (atm)	$10^{-2.62}$ atm
Chlorite	0.5- 4	pH	8.5

Experimental program

Interface sampling

Within this work, we aim at studying both existing interface and newly created interfaces. We perform experiment on both in-situ interface and newly made interface by acceleration in the laboratory.

In-situ study

Boom Clay-concrete interfaces are sampled in the HADES underground research facility (Mol, Belgium). Samples are taken from the Connecting Gallery, in which low-permeability high-pH concrete wedge-blocks are in contact with Boom Clay for 13 years.

Details of drilling procedure are described below:

We first drill a hole of 100 mm in diameter and 350 mm in depth in the concrete block (total thickness is 400 mm) cooling with water. This part is removed; the leftover is 50 mm of concrete at the outside of the lining, which is dried with compressed air. Note that we cannot drill the hole with the diameter larger than 100 mm to avoid reduction in mechanical strength of the gallery.

In order to reinforce the concrete-clay interface and ensure that the interface remains intact as one piece during caring, we install two anchors (Figure 2a and b) entering from concrete into the clay. Two holes of 18 mm in diameter are drilled through the 50 mm leftover concrete and entering 100 mm deep into the Boom Clay. The holes are drilled by dry drilling in order to avoid disturbing the contact between Boom Clay and concrete. A guiding system (Figure 2a) is used to keep the drilling bit parallel with the boreholes.

Pouring the resin (Figure 2c) into the 18 mm diameter holes. Note that the holes are inclined (45°) to the horizontal direction with which allows the resin staying in the holes after pouring. The choice of an efficient resin is crucial to avoid leakage around the side of the sample. The resin should have low viscosity, good contact with clay and concrete, high strength, and low heat generation during polymerization. As suggested from previous study (Phung, 2015), Sika® Injection-451 was selected as the most optimal choice.

After polymerization of the resin, we dry-drill with a 100 mm drilling bit (cooling with compressed air) through the 50 mm concrete leftover and into the Boom Clay for about 100 mm (Figure 2d).

Finally, the concrete-clay interface sample is retrieved from drilling bit using a hydraulic press and kept in aluminium foil to prevent any drying and reaction with the air.

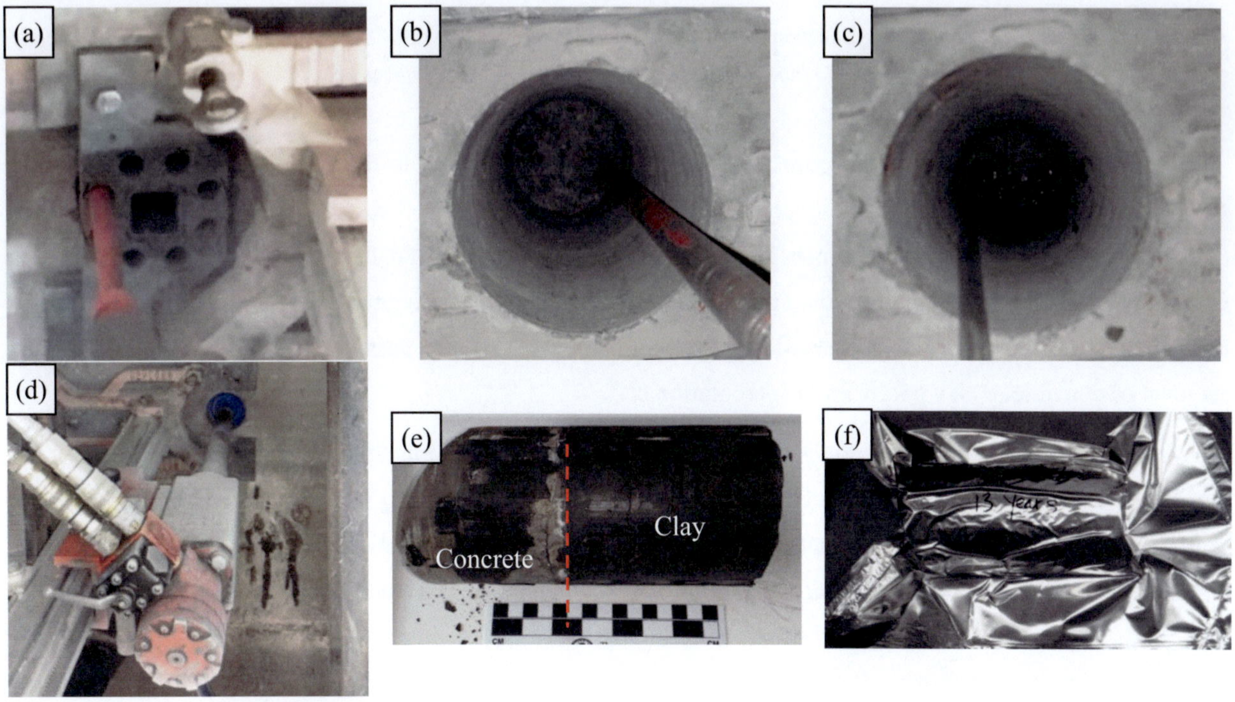

Figure 2: *In-situ drilling procedure:* ***(a)*** *Drilling small hole using guiding system;* ***(b)*** *Two small holes for resin filling;* ***(c)*** *Resin filling into two small holes to make two anchors after polymerization;* ***(d)*** *Dry drilling concrete-clay interface after reinforcing by anchors;* ***(e)*** *Concrete-clay interface sample after retrieving from drilling bit; and* ***(f)*** *Sample packaged in aluminium foil.*

Laboratory study

We perform percolation type experiments. The interface between Boom Clay-cement paste is created by putting Boom clay and hardened cement pastes in contact in a percolation cell which is a modified design of a permeability cell (Phung et al., 2013). In these experiments, the chemical conditions can be rigorously controlled. The conditions in the HADES underground research facility are mimicked by chemical compositions and advective flow of Boom Clay pore solution. A series of experiments are performed, which can be stopped as function of interaction time. For each interaction time, samples are used for measuring the changes in transport properties (i.e., permeability, diffusivity, sorptivity) and changes in chemical and microstructural properties.

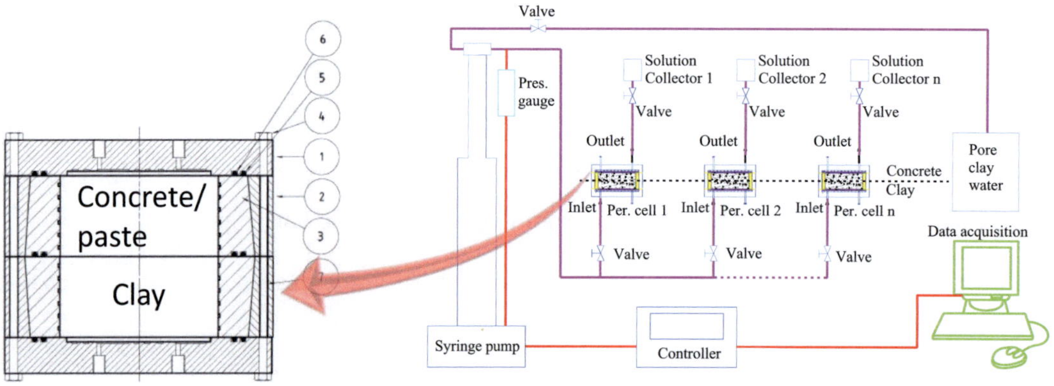

Figure 3: *Schematic test setup for percolation experiment – the percolation cell **(left)** composes (1) steel lid, (2) steel body, (3) inner polycarbonate body, (4) bolt, (5) inner O-ring, (6) outer O-ring, and (7) threaded rod.*

Figure 3 shows the percolation setup. The clay pore water is injected to the percolation cell from the clay side and solution is collected at the concrete side for chemical analysis. The pressure of the clay pore water (~ 22.5 bar) is controlled by a Syringe pump. The Syringe pump also enables to record the flow rate and cumulative pore water injected into the system. The percolation cell is designed in a special way which allows determination of the composite permeability of clay and concrete and also permeability of separate materials by removal of concrete or clay part from the percolation cell.

We also perform batch type experiments in which a cement disc is immersed into Boom Clay slurry. The experiments are conducted in closed system which is modified from leaching tests (Phung et al., 2015a; Phung et al., 2014) to prevent the environmental carbonation during the interactions. Several experiments are set up and the cement plug is removed for microstructural analysis regularly. The chemical composition of the slurry is also followed. These experiments are expected to induce more rapid cement degradation due to excess water compared to in-situ experiments.

All the samples need to be saturated following the saturation procedure proposed in (Phung et al., 2013). These saturated conditions simulate the real conditions in which concrete is getting saturated after a few decades in contact with the clay host formation.

Quantifying changes in transport properties

Water permeability

Prior to the percolation experiments, the permeability of cement paste and clay samples are determined separately using a controlled constant flow method as described in (Phung et al., 2013).The method is based on the application of a constant flow instead of a constant pressure in traditional methods, which overcomes the problem of measuring extremely low flow rate. A pressure gradient of around 5 to 10 bar is applied by controlling the pressure at both sides of a saturated sample embedded in a permeability cell (Figure 4). When the flow almost reaches steady state, the constant pressure mode is changed to the constant flow mode. The pressure is then measured until reaching the steady state in pressure, which is much more accurate than measuring flow rate. The pressure and water flow are controlled by precise Syringe pumps. A good contact between the samples (cement pastes, clay) and the permeability cell is obtained by using high strength and low viscosity resins. This method seems promising in terms of the required experimental time and the parameter control. It is also very flexible for further testing in which two permeability cells (one for clay and one for cement paste) can be reassembled to have a percolation cell (Figure 3) for percolation experiments. With this method, we are able to determine the permeability coefficient of cement paste/clay in a relative short time compared to other methods (within 1 week for permeability of 10^{-13} m/s).

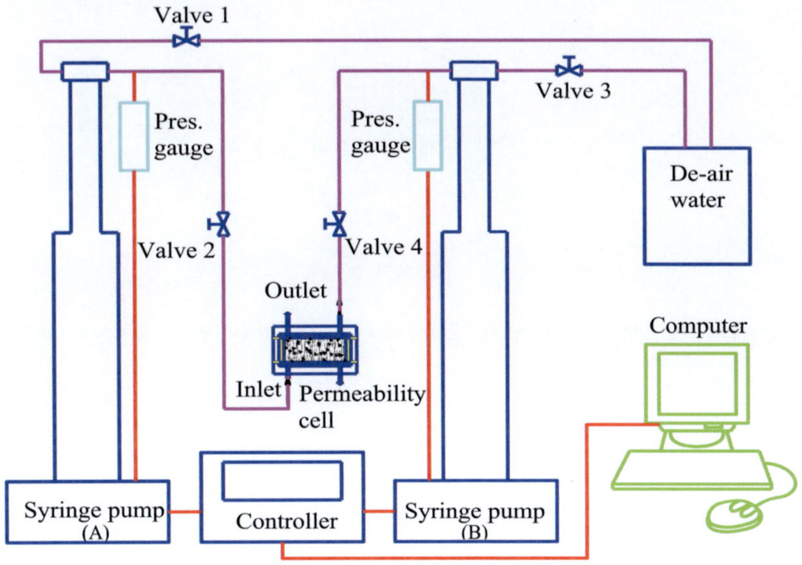

Figure 4: *Schematic test setup for permeability determination.*

The composite permeability of clay and cement paste samples can be calculated at any specific time during percolation experiments. At a certain interaction time, the percolation cell is decoupled (separate clay and cement paste samples) for determination of the permeability of each material after having interacted for a known period of time.

Dissolved gas diffusivity

Similar to permeability, we determine the diffusion coefficients of cement pastes and clay before and after a certain time of interaction. We use a through-diffusion methodology which allows simultaneous determination of diffusivities of two dissolved gases (He and CH_4) in a single experiment as described in (Jacops et al., 2013; Phung, 2015; Phung et al., 2015b). Note that thanks to its compatibility, the permeability/percolation cell can be easily disconnected and then reconnected to the diffusion setup.

Prior to the diffusion measurement, the sealing of the whole system is checked by applying a gas pressure of about 10 bar and following the pressure evaluation over time. Care must be taken to avoid testing gas contamination of each vessel (i.e., gas 1 present in vessel 2 and vice versa). A volume of 0.5 litre of degassed water is added to each vessel, whereafter they are pressurized by the respective gases to a similar pressure of about 10 bar to prevent advective transport. Gas samples are regularly (generally every 2 weeks) taken via external samplers until enough data points are collected to obtain diffusivity. The gas composition are analysed with a micro gas chromatograph. In order to interpret the experimental data, a 1-D diffusive transport model has been used (Jacops et al., 2013), which takes into account the drop of gas pressure during sampling.

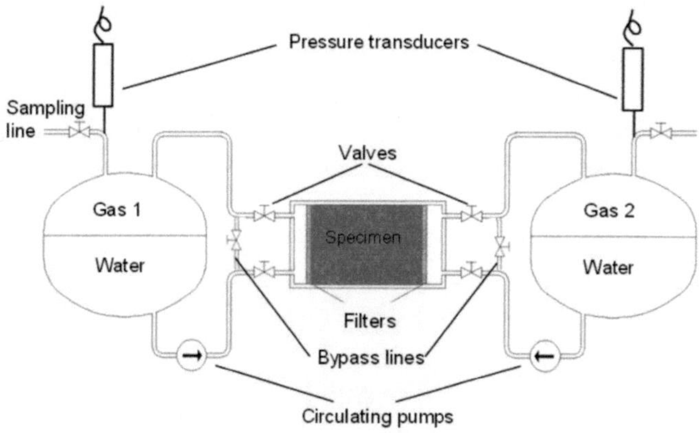

Figure 5: *Schematic experimental setup to measure diffusivities of dissolved gases (Jacops et al., 2013; Phung 2015; Phung et al., 2015b).*

Water sorptivity

Considering extremely slow concrete-clay interaction, the degraded depths, especially at the (high strength) concrete side, are expected to be limited. Our initial observations on few concrete samples taken from HADES underground research facility showed that the degraded depths were only few hundreds μm for 13 years of interaction, which are indeed not sufficient to perform permeability or diffusion measurements mentioned in previous sections. Therefore, we perform a water sorptivity test which is relevant to characterize the transport properties of an exposed surface of concrete or clay.

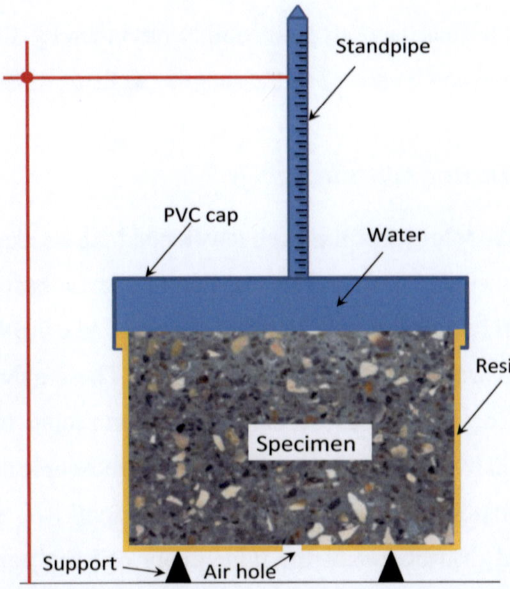

Figure 6: *Schematic experimental setup to measure water sorptivity.*

As the sorptivity is strongly affected by the water saturation of the sample, conditioning the samples is very important. Normally, the equilibrium relative humidity (RH) of the sample should be around 60% (ASTM, 2004). However, it could take a few months to years to reach equilibrium at such RH depending on the size of sample. In this study, we dry the sample in an oven at 50°C. This gentle temperature is chosen to, on one hand, reduce the conditioning duration and, on the other hand, prevent cracking to occur due to heating and decomposition of water in C-S-H phase. Furthermore, the required duration to perform water sorptivity tests on dried samples will be much shorter than on partially saturated samples.

The schematic sorptivity setup is shown in Figure 6. In this setup, water is on top of the concrete surface, gravity also will play a role in the water penetration. However, water ingress into the sample is mainly driven by capillary forces as it is much larger than water gravity (water head is about 0.5 m). The size of the sample is similar to the one for permeability/diffusion tests. The entire surface of the dry sample is covered by a resin except for the interface and a small area at the bottom to allow air to escape during sorption. In this way, sorption only occurs one-dimensionally, at the interaction surface. The volume of water absorbed by the sample is determined by following the level of the standpipe.

The sorption coefficient is defined as shown in the following Equation 1:

$$I = \frac{W}{A} = S\sqrt{t} + I_0 \qquad \textbf{\textit{Eq. 1}}$$

where I = cumulative water absorption on tested surface (mm^3/mm^2); W = volume of water absorption (mm^3); A= tested surface (mm^2); t = time variable (min); S = sorption coefficient ($mm/min^{1/2}$); and I_0 = initial water absorption (mm^3/mm^2). Water sorptivity is one of the independent transport indicators. However, the water sorptivity is strongly related to (unsaturated) permeability (and thereby diffusivity) and capillary pressure which depends on microstructure and water content of the sample (Kelham, 1988).

Quantifying changes in microstructure and mineralogy

A multi-scale investigation methodology that has been developed both on clay materials during the European FP7 CATCLAY project (Gaboreau et al., 2016) and on claystone - cement based material interfaces (Gaboreau et al., 2012; Gaboreau et al., 2011; Jenni et al., 2014; Lerouge et al., 2014) will be used to retrieve the μ-structural changes in association with the mineralogical changes. These quantifications will be done by the BRGM in France. The methodology integrates several bulk macroscopic characterization techniques and imaging methods to display quantitative data from macroscopic to nanoscopic scale. The association of several 2D/3D techniques (mineral cartography, autoradiography, μ-Tomography-RX, SEM, FIB-nT and TEM) allows to reach a quantitative and spatial distribution of the mineralogy and the pore network, from nanometer to micrometer. Therefore, the influence of the mineralogical and chemical changes on the microstructure will be obtained by: (i) 2D quantitative spatial distribution of the mineralogy and its evolution with the effect of the perturbation, (ii) 2D quantitative spatial distribution of the porosity to localize spatial heterogeneities and porosity evolution as regard to the interface, and (iii) focusing on localized areas to obtain a 3D pore network.

Outlook

This paper presents experimental methods to study the interface interactions between concrete and clay under real disposal conditions as well as under accelerated conditions. This work is a first stage in the development of a methodology which is suitable (confirmed so far by some preliminary experimental results) for characterization of the Boom Clay - concrete interface in terms of changes in transport properties and microstructure due to chemical interactions (combination of Ca-leaching and carbonation). The study focuses on changes occurring both on the concrete and clay side with the aim to answer some key questions: (i) whether clogging occurs in the clay or not; (ii) to which extent the permeability, transport properties and water sorptivity of the concrete are changed as a consequence of microstructural and mineralogical alterations; and (iii) whether the combined effect of carbonation and leaching could alter the microstructure and transport properties of the concrete in different manners compared to separate effect.

Acknowledgements

The research leading to these results has received funding from the European Union's European Atomic Energy Community's (Euratom) Horizon 2020 Programme (NFRP-2014/2015) under grant agreement, 662147 – Cebama. The first, second and last authors gratefully acknowledge partially financial support from NIRAS/ONDRAF.

References

ASTM (2004). ASTM C 1585-04: Standard Test Method for Measurement of Rate of Absorption of Water by Hydraulic-Cement Concretes: 6.

De Craen, M., Wang, L., Van Geet, M., Moors, H. (2004). Geochemistry of Boom Clay pore water at the Mol site. SCK• CEN Report, SCK• CEN-BLG-990.

Dehghanpoor Abyaneh, S., Wong, H.S., Buenfeld, N.R. (2014). Computational investigation of capillary absorption in concrete using a three-dimensional mesoscale approach. Computational Materials Science, 87, 54-64.

Gaboreau, S., Lerouge, C., Dewonck, S., Linard, Y., Bourbon, X., Fialips, C.I., Mazurier, A., Prêt, D., Borschneck, D., Montouillout, V., Gaucher, E.C., Claret, F. (2012). In-Situ Interaction of Cement Paste and Shotcrete with Claystones in a Deep Disposal Context. American Journal of Science, 312, 314-356.

Gaboreau, S., Prêt, D., Tinseau, E., Claret, F., Pellegrini, D., Stammose, D. (2011). 15 years of in situ cement–argillite interaction from Tournemire URL: Characterisation of the multi-scale spatial heterogeneities of pore space evolution. Applied Geochemistry, 26, 2159-2171.

Gaboreau, S., Robinet, J-C., Prêt, D. (2016). Optimization of pore-network characterization of a compacted clay material by TEM and FIB/SEM imaging. Microporous and Mesoporous Materials, 224, 116-128.

Honty, M. and De Craen, M. (2012). Boom Clay mineralogy–qualitative and quantitative aspects. Status 2011. SCK• CEN Report, SCK•CEN-ER-194.

Jacops, E., Volckaert, G., Maes, N., Weetjens, E., Govaerts, J. (2013). Determination of gas diffusion coefficients in saturated porous media: He and CH_4 diffusion in Boom Clay. Applied Clay Science, 83–84, 217-223.

Jenni, A., Mäder, U., Lerouge, C., Gaboreau, S., Schwyn, B. (2014). In situ interaction between different concretes and Opalinus Clay. Physics and Chemistry of the Earth, Parts A/B/C, 70–71, 71-83 .

Kelham, S. (1988). A water absorption test for concrete. Magazine of Concrete Research, 40, 106-110.

Lerougea, C., Clareta, F., Tournassata, C., Grangeona, S., Gaboreaua, S., Boyerb, B., Borschnekc, D., Linardd, Y. (2014). Constraints from sulfur isotopes on the origin of gypsum at concrete/claystone interfaces. Physics and Chemistry of the Earth, Parts A/B/C, 70–71, 84-95.

Phung, Q.T. (2015). Effects of Carbonation and Calcium Leaching on Microstructure and Transport Properties of Cement Pastes. PhD thesis, Ghent University.

Phung, Q.T., Maes, N., De Schutter, G., Jacques, D., Ye, G. (2013). Determination of water permeability of cementitious materials using a controlled constant flow method. Construction and Building Materials, 47, 1488-1496.

Phung, Q.T., Maes, N., Jacques, D., De Schutter, G., Ye, G. (2015a). Investigation of the changes in microstructure and transport properties of leached cement pastes accounting for mix composition. Cement and Concrete Research, 79, 217–234.

Phung, Q.T., Maes, N., Jacques, D., Jacop, E., Grade, A., Schutter, G.D., Ye, G. (2015b). Determination of diffusivities of dissolved gases in saturated cement-based materials. Paper presented at the International Conference on Concrete Repair, Rehabilitation and Retrofitting IV, Leipzig, Germany.

Phung, Q.T., Maes, N., Jacques, D., Schutter, G.D., Ye, G. (2014). Microstructural and permeability changes due to accelerated Ca leaching in ammonium nitrate solution. In: Grantham M, Basheer PAM, Magee B, Soutsos M (eds) Concrete Solutions - 5th International Conference on Concrete Repair. CRC Press, 431-438.

Read, D., Glasser, F.P., Ayora, C., Guardiola, M.T., Sneyers, A. (2001). Mineralogical and microstructural changes accompanying the interaction of Boom Clay with ordinary Portland cement. Advances in Cement Research, 13, 175-183.

Experiments on interface processes at the cement/Callovo-Oxfordian claystone interface and the impact on physical properties; initial findings

Robert Cuss[1*], Andrew Wiseall[1], Jon Harrington[1], Jean Talandier[2], Xavier Bourbon[2]

[1] BGS, British Geological Survey (UK)
[2] Andra, Agence nationale pour la gestion des déchets radioactifs (FR)

* Corresponding author: rjcu@bgs.ac.uk

Abstract

The evolution of flow and strength properties of the concrete/Callovo-Oxfordian Claystone (COx) interface is of importance to the long-term performance assessment of a geological disposal facility for radioactive waste. A series of shear/flow experiments using a bespoke Direct Shear Rig will examine the evolution of flow and strength as the cement ages. Two lithofacies of COx will be tested, the so-called repository and high carbonate varieties, both cored from the Meuse/Haute-Marne underground research laboratory. This paper outlines the agreed methodology and the current status of the experiments.

Introduction

The weakest part of any gallery or deposition hole seal is likely to be at the interface between the sealing components and the host rock. A single seal may comprise a number of elements reflecting different design criteria in order to address specific engineering challenges associated with changes in geochemistry and stress. The interaction of these components with the host rock, their evolution in terms of strength/bonding, cation exchange behaviour and interfacial permeability, and the sensitivity of these properties to an evolving geochemical and physical environment will be key factors in determining the long-term seal performance. In the French repository concept, low-alkali cement is proposed as a mechanical support for bentonite seals as they slowly hydrate to isolate sections of the repository system. The interaction between the cement, bentonite and host rock must be understood in order to perform performance assessment. As such, a matrix of bespoke tests using state-of-the-art experimental systems capable of simulating key repository components will be performed.

Particular emphasis will be placed on defining the temporal evolution of the interface to changes in geochemistry, mineralogy and stress to assess their impact on the development of hydraulic permeability and strength (shear strength). Diagnostic tests will be performed in a number of heavy duty shear rigs providing real-time data on the hydromechnical and transport behaviour of the interface as it evolves in time and space.

This information (including shear strength, hydraulic fracture transmissivity, volume change, normal and shear stresses etc) will be combined with post-test petrological and geochemical analyses to develop a conceptual model describing seal behaviour. This model will then be used by Andra (the French waste management organisation) to guide and verify future predictive numerical models aimed at repository performance assessment.

Experimental apparatus

Two Direct Shear Rigs will be used. These have been successfully used on several studies to date. The custom made experimental equipment (Figure 1) comprises 5 main components: (1) a rigid frame that had been designed to deform as little as possible during the experiment; (2) a normal load system comprising a rigid loading frame, a hydraulic ram, and a normal load thrust block; (3) a shear force actuator designed to drive shear as slow as 14 microns a day at a constant rate (equivalent to 1 mm in 69 days); (4) a pore pressure system that can deliver either water or gas through the centre of the top sample directly to the fracture surface; (5) a state-of-the-art custom designed data acquisition system using National Instruments LabVIEW™ software facilitating the remote monitoring and control of all experimental parameters.

a) 

b) 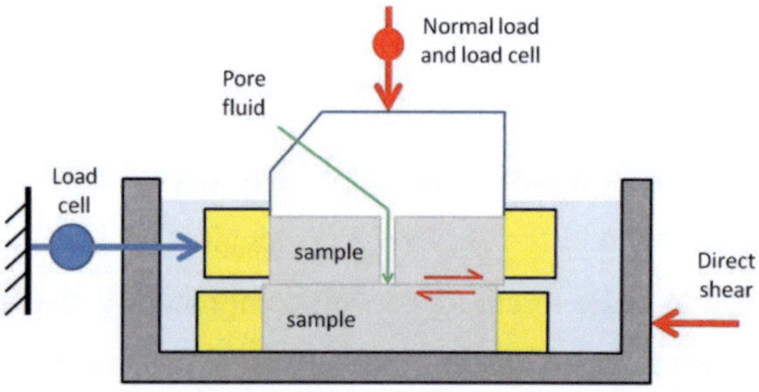

Figure 1: *The direct shear rig **(a)** photo of the rig, **(b)** schematic of the apparatus.*

The rig was designed to achieve 10 MPa normal stress (36 kN force) on a sample 60 mm × 60 mm. At this normal load, it is expected that shear force will not exceed 25 kN. Normal load is imposed by a hydraulic ram that is pressurised using an ISCO/Teledyne syringe pump. For the Cebama study the apparatus has been modified in order to use cylindrical samples of 60 mm diameter. This was deemed necessary as it makes sample preparation more straightforward. This also acts to increase the rig capacity to 12.75 MPa normal load.

The shear-force actuator is comprised of an ISCO/Teledyne 500 series D syringe pump, which has been dismantled and modified to provide lateral movement of the shear box. In the current setup, the barrel of the syringe pump has been removed and the drive-train directly connected to the sample assembly, itself mounted on a low friction bearing. A third ISCO/Teledyne 500 series D syringe pump can be used to deliver pore pressure, be it as water or gas. The shear-force actuator and pore pressure pumps are controlled from a single controller.

Test protocol

Samples of Callovo-Oxfordian Claystone (COx) will be provided by Andra from freshly cored boreholes at the Meuse/Haute-Marne URL at Bure, approximately 300 km east of Paris. Each core barrel will be sub-sampled into 75 mm long sections using a diamond tipped saw, from which a 60 mm diameter and 53 mm height sample is prepared using a lathe. At all times during sample preparation the "exposure" to air time is minimized to prevent oxidation and drying, with the samples stored in vacuum-packed plastic as much as is practicable. Off-cut material will be weighed and basic geotechnical properties will be determined. A total of 6 samples will be produced. In addition, 8 samples of 26.5 mm height will be produced. Table 1 summarises the tests to be conducted in the current test program.

Table 1: *List of tests to be conducted. Rough refers to fractures induced in the shear apparatus from cubic samples, whereas Planed refers to diamond milled smooth surfaces.*

			COx	Concrete		Test
Cebama_01		COx/Conc	Repository	T_L	Rough	Yr0
Cebama_02		COx/Conc	Repository	T_L	Rough	Yr1
Cebama_03		COx/Conc	Repository	T_L	Rough	Yr2
Cebama_04		COx/Conc	Repository	T_L	Rough	Yr3
Cebama_05		COx/Conc	Repository	T_L	Planed	Yr0
Cebama_06		COx/Conc	Repository	T_L	Planed	Yr1
Cebama_07		COx/Conc	Repository	T_L	Planed	Yr2
Cebama_08		COx/Conc	Repository	T_L	Planed	Yr3
Cebama_09		COx/Conc	High Carb	T_L	Rough	Yr0
Cebama_10		COx/Conc	High Carb	T_L	Rough	Yr1
Cebama_11	Shear test	COx/Conc	High Carb	T_L	Rough	Yr2
Cebama_12		COx/Conc	High Carb	T_L	Rough	Yr3
Cebama_13		COx/Conc	High Carb	T_L	Planed	Yr0
Cebama_14		COx/Conc	High Carb	T_L	Planed	Yr1
Cebama_15		COx/Conc	High Carb	T_L	Planed	Yr2
Cebama_16		COx/Conc	High Carb	T_L	Planed	Yr3
Cebama_17		Conc/Conc	/	T_L / T_L	Rough	Yr1
Cebama_18		Conc/Conc	/	T_L / T_L	Rough	Yr3
Cebama_19		Conc/Conc	/	T_L / T_L	Planed	Yr1
Cebama_20		Conc/Conc	/	T_L / T_L	Planed	Yr3
Cebama_21		COx/Conc	Repository	T_L	Rough	Yr1

			COx	Concrete		Test
Cebama_22		COx/Conc	Repository	T_L	Rough	Yr3
Cebama_23		COx/Conc	High Carb	T_L	Rough	Yr1
Cebama_24		COx/Conc	High Carb	T_L	Rough	Yr3
Cebama_25	Uniaxial test	Conc	/	T_L		Month 3
Cebama_26		Conc	/	T_L		Month 6
Cebama_27		Conc	/	T_L		Month 12
Cebama_28		Conc	/	T_L		Month 20
Cebama_29		Conc	/	T_L		Month 28
Cebama_30		Conc	/	T_L		Month 36

The samples of COx will be placed in the experimental apparatus and will initially be re-hydrated to bring up to full saturation under a normal stress of 10 MPa. The samples will be hydrated using a synthetic pore fluid under a pore pressure of 1 MPa, see Table 2 for the pore fluid geochemistry. It is anticipated that re-hydration will take approximately 3 weeks.

The samples will be sheared for 5 days at a rate of 1 mm per day. The data recorded will be pore-fluid flow, shear displacement, shear stress, normal stress, and vertical displacement. This will give; initial flow rate, shear yield stress, shear modulus (*G*), shear peak strength, shear residual strength, dilation/contraction during shear, and change in flow during shear. Each fracture surface will be scanned using a NextEngine 3D laser scanner. This data will be used to give: roughness average, root mean square roughness, peak-to-valley height, kurtosis, skewness, texture direction, texture aspect ratio, and texture direction index. All samples will be vacuum packed for medium-term storage.

Table 2: *Mixing recipe to produce synthetic pore fluid for Callovo-Oxfordian claystone.*

	g/L
NaCl	1.950
$NaHCO_3$	0.130
KCl	0.035
$CaSO_4$, $2H_2O$	0.630
$MgSO_4$, $7H_20$	1.020
$CaCl_2$, $2H_2O$	0.080
Na_2SO_4	0.700

***Table 3:** T_L variant concrete composition (to make 1 m³).*

		T_L (kg)		Supplier
	Super Plasticizer	5.7	Fluid Optima 175	CHRYSO
Binder	Cement	76.0	CEM I 52.5 PM EC CP2	Lafarge in Le Teil
Binder	Silica fume	123.5	Ref DM95	Condensil
Binder	Blast furnace slag	180.5		ECOCEM
	Sand	855	0/4 mm	
	Gravel	949	Gravels 4/12 mm	
	Effective water E_{eff}	152		
	E_{eff}/L	0.4		
	G/S	1.1		

The T_L variant of low pH cement will be prepared using the mixture shown in Table 3. The use of additional moulds will allow concrete to be cast giving a total sample height of 53 mm. A water layer will be used on top of the concrete to ensure that the concrete is fully hydrated. Additionally, 6 samples will be cast of 54 mm diameter and 108 mm length for uniaxial testing. Care will be taken to ensure that the COx does not go into free-swell and that water exchange at the interface will not influence the test results.

A series of COx/concrete samples will be produced, half with COx from the repository depth and half of the high carbonate variety. In addition, test samples of concrete/concrete (T_L/T_L) will also be produced, along with 6 T_L uniaxial test samples. The test program calls for 4 tests soon after casting at Yr 0; these tests will be repeated annually. The concrete/concrete tests will be tested at Yr1 and Yr3. Additional control tests will be conducted at Yr1 and Yr3. Uniaxial tests will be performed after 3, 6, 12, 20, 28, and 36 months after casting to determine how the strength properties change as the concrete ages.

Prior to shear testing, a 4 mm-diameter bore will be drilled through the COx to the depth of the interface between the COx and concrete. Normal load of 8 MPa with a target pore pressure of 1 MPa will initially be used, which is dependent on the overall permeability of the COx/concrete interface. Once stable flow has been achieved the sample will be sheared 5 mm at a rate of 0.25 mm/day in order to observe changes in flow during shear. Flow along the interface and the shear properties will be determined. The failure surfaces will be laser scanned to determine fracture roughness parameters. Fluorescent scanning will be conducted to identify flow paths of the injection water. Scanning Electron Microscopy (SEM) will be used to determine chemical aging of the concrete, geochemical reactions between the COx and cement and identify flow paths.

The uniaxial tests on concrete will be performed using an MTS Systems Corporation® apparatus and will determine; Young's modulus (E), Poisson's ratio (v), bulk modulus of compressibility (K), yield stress (σ_y), and the unconfined compressive strength (q_u). Repeat testing will show how these parameters vary as the concrete ages.

Current status

Experience from previous shear testing of shale (Cuss et al., 2011, 2014, in prep) have shown that the preparation of cubic samples can be highly problematic. This has necessitated the conversion of the two existing shear rigs that employed cubic samples to test cylindrical samples. These conversions have delayed the start of testing considerably due to manufacturing issues. Core material is also awaited following the latest round of drilling activity at the Meuse/Haute-Marne URL. Testing should now start in June 2016.

Acknowledgement

The research leading to these results has received funding from the European Union's Horizon 2020 Research and Training Programme of the European Atomic Energy Community (EURATOM) (H2020-NFRP-2014/2015) under grant agreement n° 662147 (CEBAMA).

References

Cuss, R.J., Harrington, J.F., Milodowski, A.E., Wiseall, A.C. (2014). Experimental study of gas flow along an induced fracture in Opalinus Clay. British Geological Survey Commissioned Report, CR/14/051.

Cuss, R.J., Milodowski, A., Harrington, J.F. (2011). Fracture transmissivity as a function of normal and shear stress: first results in Opalinus clay. Physics and Chemistry of the Earth, 36, 1960-1971.

Cuss, R.J., Harrington, J.F., Sathar, S., Norris, S., Talandier, J. (In Prep.). The role of the stress-path and importance of stress history on the flow of water along faults; an experimental study. Journal of Geophysical Research – Solid Earth.

Interaction concrete/FEBEX bentonite: outlining experimental conditions and characterization approaches at laboratory and in situ scale

María J. Turrero[1*], Elena Torres[1], Antonio Garralón[1], Paloma Gómez[1], Lorenzo Sánchez[1], Javier Peña[1], Belén Buil[1], A. Escribano[1], J.M. Durán[1], R. Domínguez[1]

[1] CIEMAT, Research Centre for Energy, Environment and technology (ES)

* Corresponding author: mj.turrero@ciemat.es

Abstract

The participation of CIEMAT in CEBAMA project for the first year was focused in the following tasks: 1) Contribution to the State of the Art Report, specifically in the aspects related to the role of groundwater and pore water in the deep geological repository, as an important issue to further understand geochemical processes in the system; 2) Sampling, preservation and testing of material coming from the long-term FEBEX *in situ* test, with the aim of studying the shotcrete/bentonite interaction processes at the repository scale; 3) Monitoring of the HB6 cell, which was designed to investigate the concrete/bentonite interaction in a long-term laboratory experiment; and 4) Definition of the characterization methods of the experiments, including some preliminary data obtained to check procedures. Some of these aspects have been reported as joint WP1 deliverables. This contribution provides a compilation of the work done, the experimental conditions of each experiment and the samples extracted from them to accomplish the studies on cement based materials/bentonite interaction at different temporal and spatial scales. The work was made in close collaboration with CSIC and UAM groups. Furthermore, modelling of HB6 laboratory experiment will be carried out by UDC.

Introduction

To demonstrate the technical feasibility of installing the engineered barriers (EBS) for a high level waste disposal facility in crystalline rock and to study their behaviour ENRESA initiated in 1995 the FEBEX Project. A part of the project consisted of a full-scale *in situ* test, performed at the Grimsel underground facility (Switzerland) which includes the Spanish reference bentonite and a plug constructed by the shotcreting technique (Huertas et al., 2006). Post-mortem samples of aged bentonite/shotcrete interface coming from the dismantling of the experiment will be studied by CIEMAT in the context of CEBAMA to analyse if materials maintain their technical requirements as barriers. In particular, data and interpretation will be based on the investigation of geochemical processes and transport phenomena at the interface and experiment scale and the extension of the altered zone.

Simultaneously, a series of six identical small-scale, long-term laboratory tests were initiated during EU project NF-PRO and continued during PEBS (Turrero et al., 2011; Torres et al., 2013). The tests were designed with the dual objective of complementing the information acquired from larger scale tests and having a temporal up-scaling by sequentially dismantling them. The dismantling and analysis of the last of those tests, after running during a little more than 10 years, will be part of CIEMAT work in CEBAMA. Its modelling will help the temporal up-scaling activities.

Groundwater and pore water in the DGR

Deep geological repository (DGR) of radioactive wastes includes cement-based materials and bentonite as materials to tunnel reinforcement, and to buffer, plug and seal main galleries, grouts and wells in the selected repositories. At the time scale of a repository these materials interacts each other and with the surrounding media, namely granite or clay rock plus groundwater and pore water, so its integrity in terms of barrier function isolating the wastes must be assessed.

Groundwater from the host rock and its interaction with the bentonite: Processes as water redistribution during desaturation and re-saturation periods and equilibration of EBS pore water with surrounding groundwater, as well as temperature gradients, will induce a series of physicochemical and mineralogical changes that will play an important role in the evolution of the DGR. Within this context, baseline composition of groundwater and pore water of the different constituents in the DGR and their evolution during the life time of the repository is a key issue in two aspects: (a) may condition the properties of the EBS materials, and (b) can provide valuable data to understand geochemical processes taking place in the repository. One of the Ciemat underway tasks is the integration of the hydro-geochemical results of the FEBEX *in situ* test at the experiment scale to establish the baseline composition and its evolution. This task is going on by compiling the results from sampling campaigns of boreholes surrounding the experiment.

Concrete/rock and concrete/bentonite interaction: The concrete pore water chemical composition depends on a number of parameters, namely the initial composition of the water in contact with the cement, the components of the material, the degradation of the solid phases in the system, etc. During the re-saturation stage, leaching will be a dominant process in the system, and the pore solution of concrete equilibrates with the surrounding water, or with the bentonite, reacting with them and affecting its physical and chemical properties. The main reactions described in the literature consist of ion exchange and mineral dissolution/precipitation, which may affect the sorption and transport properties of the whole system (see a more detailed explanation in Cuevas et al., this volume). Aged FEBEX *in situ* and HB laboratory experiments are being used to go in depth about soluble elements distribution (e.g., mass transport) and stable isotope analysis (e.g., carbonation) at the different scales (*in situ* and lab, from cm to meter scale) and to support modelling.

FEBEX in situ experiment

The FEBEX experiment at the Grimsel Test Site was a full-scale engineered barrier experiment performed under natural conditions and designed for the study of the near field of a repository of high level waste in crystalline rock following the Spanish reference concept in crystalline rock: canisters are placed horizontally in drifts surrounded by a clay barrier constructed from highly compacted bentonite blocks (Figure 1A). Two heaters were used to simulate the thermal effect of the wastes. FEBEX bentonite extracted from the Cortijo de Archidona deposit (Almería, Spain), compacted to 1.69 - 1.70 g/cm^3 dry density, was used as clay barrier. A concrete plug closed the gallery (ENRESA, 2000). Heating started in 1997 and since then a constant temperature of 100°C was maintained, while the bentonite buffer was slowly hydrating in a natural way. After five years the heater closer to the gallery entrance was switched off. In the following months the concrete plug, heater and all the bentonite and instruments preceding and surrounding it were extracted. The remaining part was sealed with a new concrete plug (Figure 1B) constructed by the shotcreting technique (ENRESA, 2006), whose composition is 430 kg/m^3 of CEM II A-L 32.5 R, 30 kg/m^3 of nanosilica MEYCO MS 660; 50 kg/m^3 of steel fibres Dramix ZP 306; 800 g/m^3 of polypropylene fibres; 1.5% of superplasticiser GLENIUM T803; W/C=0.40; 1 700 kg/m^3 of aggregate 0 - 8 mm; 1% of curing compound MEYCO TCC 735 and 6% accelerant MEYCO SA 160 E. The second operational phase started in 2002 until the definitive dismantling in 2015.

The hydration was natural and some boreholes were drilled radial, before installing the experiment, and parallel, after first dismantling, to the gallery and packed-off to sample and check the evolution of the groundwater (Figure 1C). The granite deep groundwater composition analysed from radial borehole samples around the FEBEX gallery is Na^+-Ca^{2+}-HCO_3^--type water, with relatively high concentration of F^- and SO_4^{2-}, with low electric conductivity (76 - 236 µS/cm) and pH neutral to alkaline.

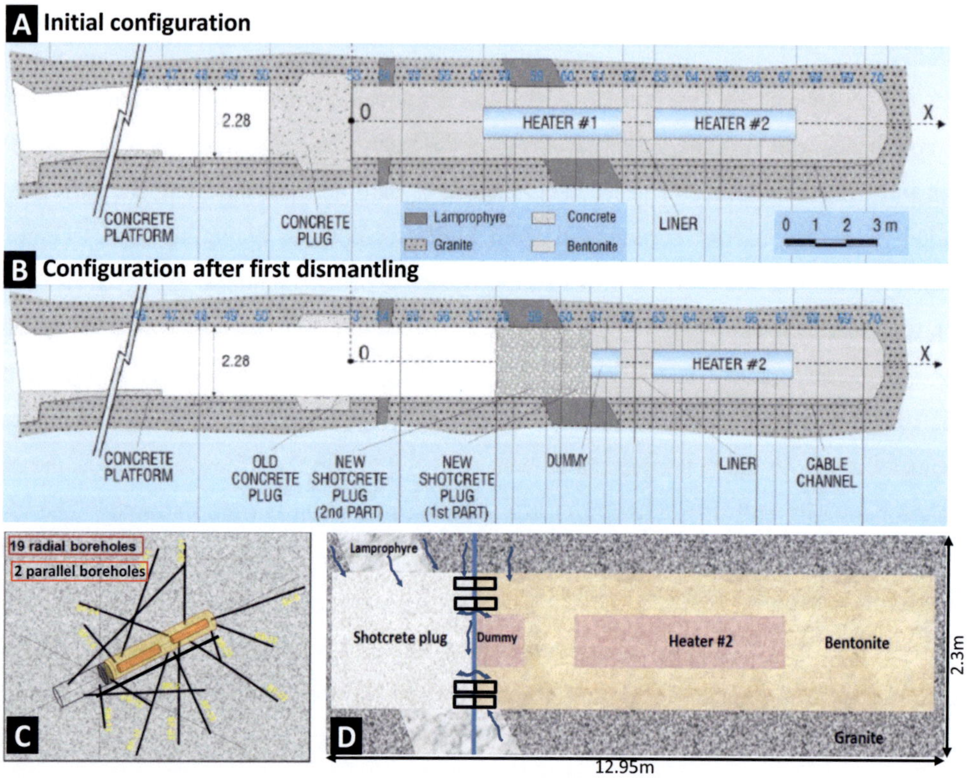

Figure 1: ***A.*** *Initial configuration of the FEBEX experiment. It was in operation for 5 years (1995-2002).* ***B.*** *Configuration of the FEBEX experiments after the first dismantling. It was in operation until 2015, when the experiment was completely dismantled.* ***C.*** *Location of boreholes for groundwater sampling around the FEBEX gallery.* ***D.*** *Simplified picture showing sampling position (rectangles) and water fluxes under consideration.*

Sampling of the shotcrete/bentonite interface and characterization procedures

Sampling took place in two stages: a) March 2015: 3 aged shotcrete/bentonite core samples were obtained by the overcoring technique (Urs Mäder Team, University of Bern). A quarter of these samples are shared and currently being analysed by CIEMAT and UAM teams (Figure 2A); the interfaces were well preserved and for the Ciemat analysis were dry cut in slices following the scheme in Figure 2A. b) May 2015: aged shotcrete/bentonite core samples were obtained by dry drilling with a diamond coring machine (AITEMIN, CIEMAT and UAM teams) (Figure 2B).

The interfaces were well preserved and dry cut in slices that were shared with UAM. The minimum size of sub-sampling to obtain reliable data according to the objectives (e.g., mass balance or porosity changes) had to be tested during this initial period of the project due to the small size of the cores. A minimum size of 1 - 1.5 cm is necessary to get analysis of main parameters committed. Then, slices of 1 - 1.5 cm were dry cut and an example of cation exchange capacity results is shown in Figure 1C. Consider that each characterization has replicates and the slices are being used for multiple determination, as stable isotopes, exchangeable cations, total exchange capacity, FTIR spectra, SEM-EDS, etc.

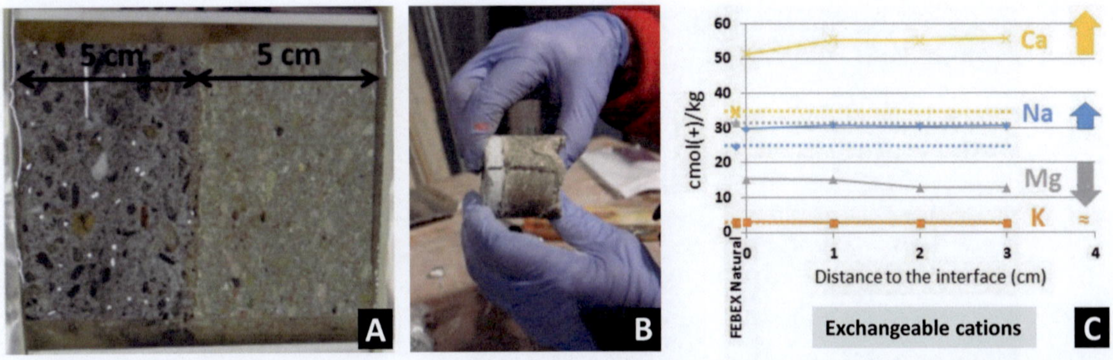

Figure 2: ***A.*** *Example of aged (13y) shotcrete/bentonite specimen obtained by overcoring (March'15).* ***B.*** *Example of aged (13y) shotcrete/bentonite specimen obtained by dry drilling with a diamond coring machine (May'15). Samples are shared by UAM and CIEMAT teams.* ***C.*** *Example of variations of cation exchange population in one of the dry drilled cores.*

HB laboratory experiment

A set of 6 identical tests (HB) was designed during EU NF-PRO Project to study the concrete/bentonite interaction in the EBS of a DGR and its evolution at different stages of the re-saturation phase of the repository. They are Teflon® cylindrical cells with an internal diameter of 7 cm and an inner length of 10 cm. The upper closing of the cells is a stainless steel plug with a deposit to water circulation at room temperature (25°C). The bottom part is a plane stainless steel heater set at a temperature of 100°C. In this way, a constant gradient between top and bottom of the sample is imposed. The cells contain a 7 cm-high FEBEX bentonite cylinder with its hygroscopic water content (14%) and a nominal dry density of 1.65 g/cm^3 and a 3 cm-high CEM I 42.5 R/SR cement cylinder with a dry density of 2.22 ± 0.01 g/cm^3 and initial water content 2.6 ± 0.8%. The hydration was made through the concrete blocks with synthetic Na^+-Ca^{2+}-SO_4^{2-}–type clay water from a stainless steel deposit pressurized to 12 bar. The deposit was periodically weighed to check the water intake. The cells were instrumented with capacitive-type sensors to measure temperature and relative humidity, placed inside the clay at two different levels. Five tests were dismantled after 6, 12, 18, 54 and 80 months. The last one will be dismantled within the CEBAMA Project. A picture showing the final assembling of the cells and the appearance of one of the dismantled cells is presented in Figure 3, in which a scheme of sampling to detailed studies of the concrete/bentonite interface is presented based on the experience gained from previous dismantling. Until dismantling by the end of 2016 monitoring of the experiment is made.

Some of the processes identified from cells already dismantled are pointed out in Cuevas et al. (these proceedings). In dismantled tests, cement leachates favoured the precipitation of carbonates (calcite, aragonite) and C(A)SH gels and sulfate-rich phases, ettringite in the 54-month test, gypsum after 82 months, were identified at the concrete/bentonite interface. The precipitation of these newly-formed phases at the interface resulted in a decrease in porosity and BET specific surface area in the first two millimetres of bentonite, fact predicted by reactive transport models. Aspects related to THC modelling of the HB series can be found in Samper et al. (these proceedings).

Figure 3: Left. *Picture of the medium cells used for the study of the interfaces in the laboratory.* ***Centre.*** *Picture showing the aspect of the material once one of the cells was dismantled.* ***Right.*** *Sample distribution after dismantling of the HB6 cell (10 y of operation).*

Summary and Future work

During the first year, the next tasks have been accomplished to answer to the CEBAMA programme: contributing to the state of the art report, mainly in those aspect related to the role of groundwater in a deep geological repository, as part of the further interpretation of samples coming from the FEBEX in situ experiment; contribution to the dismantling and conditioning of aged shotcrete/bentonite samples coming from the FEBEX in situ test; testing of minimum sub-sample size to have reliable results of parameters to be measured at the interface level; monitoring of the HB experiment; and contribution to meetings (general and with groups sharing samples and modelling) and deliverables.

Next steps are: going on with the characterization of aged shotcrete/bentonite interfaces coming from the FEBEX in situ experiment; advance in the interpretation of mass transport processes from the analysis of groundwater and pore water at experiment and interface scale; dismantling of HB6 cell.

Acknowledgement

The research leading to these results has received funding from the European Union's Horizon 2020 Reasearch and Training Programme of the European Atomic Energy Community (EURATOM) (H2020-NFRP-2014/2015) under grant agreement n° 662147 (CEBAMA).

FEBEX-DP consortium (http://www.grimsel.com/gts-phase-vi/febex-dp/febex-dp-introduction) provided shotcrete and bentonite samples for this study and made possible groundwater sampling.

References

ENRESA (2000). FEBEX Project. Full-scale engineered barriers experiment for a deep geological repository for high level radioactive waste in crystalline host rock. Final Report. Technical Publication ENRESA 1/2000.

Huertas, F., Fariña, P., Farias, J. [et al.] (2006). Febex Full-scale Engineered Barriers Experiment. Updated Final Report 1994-2004. ENRESA Publicación Técnica PT 05-0/2006.

Torres, E., Turrero, M.J., Escribano, A., Martín, P.L. (2013). Geochemical interactions at the concrete-bentonite interface of column experiments. PEBS EU Project. Deliverable D2.3-6-1.

Turrero, M.J., Villar, M.V., Torres, E., Escribano, A., Cuevas, J., Fernández, R., Ruiz, A.I., Vigil de la Villa, R., de Soto, I. (2011). Laboratory tests at the interfaces: First results on the dismantling of tests FB3 and HB4. PEBS EU Project. Deliverable D2.3-3-1.

Characterization of water penetration into concrete by means of neutron imaging

Andrea Sabău*[1], Yasin Yigittop[1], Denis Bykov[1], Jeroen Plomp[1], Lambert van Eijck[1], Jan Leen Kloosterman[1]

[1] Delft University of Technology (NL)

* Corresponding author: A.Sabau@tudelft.nl

Abstract

In the Netherlands, Boom Clay is considered a potential host rock for final disposal of radioactive waste. In most of the designs of radioactive waste repositories, cement and clay will be in direct contact. This interaction will lead to modifications across the interface, which consequently has repercussions on the structural and transport properties of both materials and may affect the performance of barrier systems. In permanent contact to water, cement-based materials may also undergo damage due to hydrolysis of hydration products which causes pores changes over time and affects the permeability at the interface. In this study was carried out the viability of neutron imaging as method that permits to visualize water penetration into concrete. Experiments using foamed concrete (CEM I SR 3 <3% C_3A) that is investigated as a backfill material in the geological disposal facility, were conducted for short term periods with the purpose to visualize water content and penetration through concrete. Another experiment using foamed concrete (CEM III/B 42.5 LH/SR) immersed in Milli-Q water, was performed as a function of time. The results on the mobility of D_2O and H_2O in concrete, observed with neutron imaging, indicate that this technique can be successfully applied for the permeability studies.

Introduction

The Rupelian Boom Clay formation in the Netherlands is one of the geological formations investigated to assess its suitability for future geological disposal of radioactive waste (low and intermediate level waste (LILW) and high level waste (HLW)). OPERA (OnderzoeksProgramma Eindberging Radioactief Afval), begun in 2011, is the third Dutch research programme to have been undertaken on the topic of geological disposal of radioactive waste and salt formations. In the Netherlands, an intermediate storage until 2130 is foreseen until enough waste is collected making the disposal facility economical viable. The existent options in the Dutch concept of geological disposal exploit cement materials having several functions: mechanical support of the galleries, backfill material and containment of waste. The pore water present in the natural clay is presumed to modify the composition of concrete and promote chemical reactions that will induce pore structure changes at clay/concrete interface. Despite the physico-chemical stress induced by the environmental impacts, the cement based materials should preserve their containment properties as much as possible.

Neutron imaging has been shown to be an adequate method for visualisation and quantification of water penetration into cementitious materials (McGlinn et al., 2010; Shafizadeh et al., 2015; Tremsin et al., 2015; Trtik et al., 2011; Zhang et al., 2010) and moisture penetration into building materials (Van Tittelboom et al., 2013). Thus, this technique is suited for the investigation of water movement in natural and engineered porous media. The surface of a porous material such as concrete in contact with a wetting liquid (D_2O/H_2O) will soak the liquid by capillary action.

The purpose of the presented work is to design an experimental set-up for performing neutron radiography studies of water penetration through the clay and concrete materials and to make the optimization of the measurement conditions through the preliminary permeability tests.

Materials and methods

Boom Clay

The inherent properties of Boom Clay in the Netherlands are also known to vary with both depth and location (Koenen and Griffioen, 2014). According to Arnold et al. (2015), the Belgian Boom Clay is a homogenous sediment with a varied mineralogy mostly composed by clay minerals (Kaolinite, Smectite and Illite) and non-clay minerals (Quartz, Kfeldspar, Na-plagioclase and Carbonates). A complete mineralogical characterization of the Dutch Boom clay drilled from various regions is presented by (Koenen and Griffioen, 2014). The samples consist mainly of quartz (16.3 - 86.3 wt%) and clay minerals (8.4 - 70.2 wt%, from bulk analysis). Plagioclase and K-feldspar are present in all samples and vary between 0.3 - 5.6 and 2.2 - 11.1 wt%, respectively. The carbonate content varies strongly between the samples and consists of calcite (up to 25.9 wt%), aragonite (up to 5.1 wt%) and occasionally small amounts of ankerite/dolomite and siderite. Pyrite and anatase are present in almost all samples with values up to 6.9 and 0.9 wt% respectively. The small amounts of gypsum in most samples can be assumed to have formed during drilling/storage as a result of calcium reaction (from calcite) with oxidized sulphur from pyrite. Also jarosite is assumed to be an artefact of pyrite oxidation. In the Netherlands the ground water has a high salinity comparable to sea water. pH variations between 6.8 and 8.1 were reported for various samplings area for groundwater (Griffioen, 2015).

The only Dutch available unoxidized Boom Clay samples are taken from a depth of 79 - 84 m near the shore. In this work we performed a characterization of the unoxidized Dutch Boom Clay using a combination of XRD, XRF and SEM-EDS, the results being in agreement with previous reported results (Koenen and Griffioen, 2014).

Concrete samples

In this project, we are focused on cementitious materials fulfilling three different functions in the prospective geological repository. The first one is foamed concrete CEM I SR3 < 3% C_3A investigated as a backfill material. The composition is detailed in Table 1.

***Table 1:** Composition enclosure emplaced waste (backfill – foamed concrete).*

Component	**Receipt for 1 m³ of Aercrete FC 1 200 to 1 600 kg/m³**	**Type for OPERA**	**1 600 (kg/m³)**
Cement	360 to 400 kg	CEM I SR 3 <3% C_3A	417
Water	140 to 160 kg	-	165
Fine aggregate	750 to 1 100 kg	Sand: 0 - 2 mm	1018
Foaming agent Synthetic surfactant	0.57 to 0.36 L	Foaming agent TM 80/23 Synthetic	1
Water	21.3 to 13.6 L	Water	
Air	434 to 277 L	Air	0

In the OPERA project, the Portland based backfill material is preferred because it may increase the lifetime of the waste package (the steel envelope) (Verhoef et al., 2014). A mortar is suggested in order to fill the smallest voids between the gallery lining and emplaced waste packages. The foamed concrete consists of cement, water, fine aggregate (quartz sand 0 - 2 mm) and foaming agent (TM 80/23). The dry density is ranged between 1 200 - 1 600 kg/m³.

CEM III/B 42.5 LH/SR (Table 2) is a type of concrete supposed to be used for the disposal of compacted waste in 200 litre drums.

***Table 2:** Concrete composition for the disposal of compacted waste in 200 litre drums.*

Component	**Type**		
Cement	CEM III/B 42.5 LH/SR	407 - 430	kg/m³
Water	-	175 - 185	kg/m³
Plasticiser	TM OFT-II B84/39 CON. 35% (BT-SPL)	3 - 5	kg/m³
Fine aggregate	Quartz sand : 0 - 4 mm	819 - 972	kg/m³
Coarse aggregate	Quartz gravel : 2 - 8 mm	891 - 763	kg/m³

The concrete has w/c = 0.45 and contains quartz and gravel together with the plasticiser TM OFT-II B84/39 CON. 35% (BT-SPL).

The first investigated sample, CEM I SR 3 < 3% C_3A, was a cylinder shape with a length of 5 cm and a diameter of 3 cm. The second sample, CEM III/B 42.5 LH/SR, is a concrete slice sawed from a cylinder having same dimensions with the previous one.

Neutron imaging

Neutron visualization of water penetration into concrete was performed at the Reactor Institute Delft (RID). The following parameters of the neutron beam were used in this study: neutron flux $1.3 \cdot 10^6$ $cm^{-2}\,s^{-1}$ and a collimation ratio L/D = 250. The neutron detector is a NeutronOptics camera which uses a converter of LiF and ZnS(Ag). The Li-6acts to convert the neutrons to gamma rays, the ZnS(Ag) is a scintillator which converts the gamma rays to visible light. The light is then reflected towards a CCD camera (Sony ICX85AL0) with a resolution of 1360 x 1024 and pixel size of 0.110 mm. Using a step motor, a tomography is performed over a rotation angle of 180° with a step of 0.225°. The studied concrete samples are kept in a specially designed closed aluminium sample container during the measurement. The sample preparation consists into drying the foamed concrete for 24 hours until the mass of sample remains constant. Then, the dried sample is immersed in D_2O or H_2O. Milli-Q and D_2O were selected for the different contrast in neutron scattering length densities and which permits to make the difference between the existent hydrogen atoms in the concrete.

Background noise and the blank neutron beam spot are subtracted from the images by processing the recorded dark images (closed shutter) and open beam measurements (no sample placed), respectively. The obtained raw data were processed using the program Octopus reconstruction (Inside Matters BVBA) yielding image dataset in the form of stacked slices with an intensity values per pixel. This value is identified as grey value, where one pixel corresponds to a length of 100 μm. The indication that D_2O/H_2O is penetrated in concrete is given by the similar grey values obtained in the immersed and dry area of the concrete. In Figure 2, the grey values corresponding to the D_2O/H_2O are highlighted in blue, whereas the pores are highlighted in red.

Time lapse studies of water penetration into concrete and Boom Clay were performed and analysed by neutron radiography. Because neutrons are highly sensitive to the presence of small amounts of water, neutron radiography was emphasized to be a non-destructive imaging technique to observe an increase of pore size at in the cement and clay samples (Trtik et al., 2011). We set up experiments where concrete specimens were introduced into D_2O in order to make a contrast difference with the existent water (H_2O) in the pores of concrete/clay. The difference in neutron scattering of D_2O and H_2O makes it possible to observe spatially distribution of D_2O soaked into concrete. The second type of experiments are made with the purpose to observe and quantify water mobility in concrete.

Results and discussion

Penetration of D_2O front across foamed concrete CEM I SR3 < 3% C3A is depicted in Figure 2. Data were processed using Avizo 9.0 software (FEI Corporate Overview).

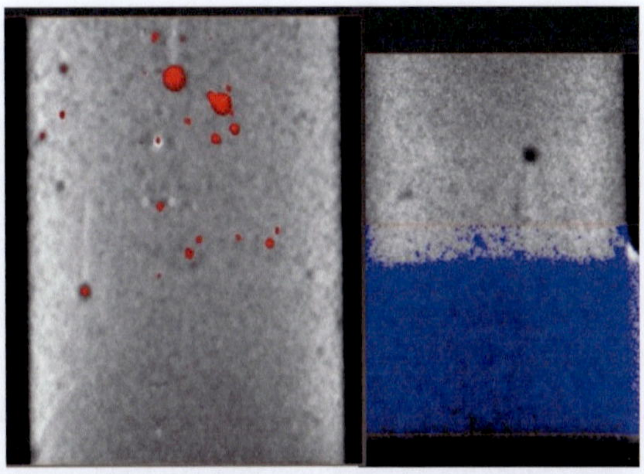

Figure 2: *Neutron imaging of dried foamed concrete (ϕ 30 × 50 mm) **-left-**(the red spots represent the voids in the concrete) and absorbed D_2O (blue) **-right-** into the foamed concrete.*

The left image in Figure 2 shows a cross section of a cylindrically shaped foamed concrete, where the red spots are the result of image processing to make it easier to recognize voids. The right image shows the cross section of the same type of concrete (different sample), immersed in D_2O. The blue marks are visualized to show matching grey value ranges between zones within the concrete and within the heavy water.

Figure 3 shows a kinetic sorption experiment of H_2O in concrete (CEM III/B 42.5 LH/SR) and provides a better representation of the water mobility within the sample.

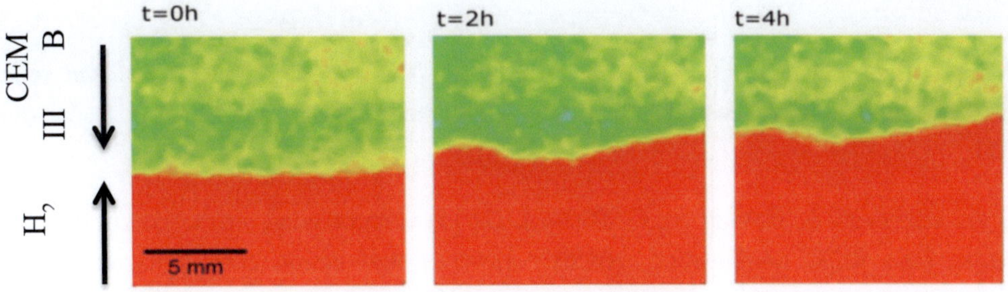

Figure 3: *Neutron imaging of foamed concrete CEM III B 42.5 LH/SR (yellow green) and absorbed H_2O (red).*

In this case the data were processed using Octopus reconstruction (Inside Matters BVBA) for filtering and empty beam subtraction, while AVIZO 9.0 software (FEI Corporate Overview) was used for colour-mapping. The yellow greenish colour in Figure 3 represents the hardened cement paste while the red colour is H_2O.

For Boom Clay, microstructural features are not observable using the presented neutron imaging setup. Neutron Imaging enables permeation recognition in concrete. However specific data to determine transport properties requires optimizations and improvements in the current neutron imaging setup.

Conclusions and Future work

The experiments presented herein showed D_2O and H_2O penetration into backfill foamed concrete, which allowed to make optimization of the setup for the permeability studies. Imaging of the samples will continue for more than 1 year when the samples will be cut and investigated in depth. More experiments are planned to be performed by immersing the concrete in Dutch Boom Clay pore water visible changes are expected to be seen after several months of contact time. Combined results from neutron imaging, positron annihilation studies, X-ray CT and SESANS (Spin Echo Small Angle Neutron Spectroscopy) are expected to provide adequate information to correlate porosity and mineralogical changes that are required for transport modelling. X-ray CT will be used to visualize changes in porosity at micrometric scale, while positron annihilation and SESANS are expected to provide information about changes of nanoporosity.

Acknowledgement

The research leading to these results has received funding from the European Union's European Atomic Energy Community's (Euratom) Horizon 2020 Programme (NFRP-2014/2015) under grant agreement, 662147 – Cebama.

References

Arnold, P., Vardon, J.P., Hicks, A.H., Fokkens, J., Fokker, A.P. (2015). A numerical and reliability-based investigation into the technical feasibility of a Dutch radioactive waste repository in Boom clay. OPERA-PU-TUD311.

Behrends, T., van der Veen, I., Hoving A., Griffioen J. (2015) Geochemical characterisation of Rupel (Boom) Clay material: pore water composition, reactive minerals and cation exchange capacity. OPERA-PU-UTR521.

FEI Corporate Overview http://www.fei.com/software/avizo-3d/ (consulted 28/04/2016).

Griffioen, J. (2015). The composition of deep groundwater in the Netherlands in relation to disposal of radioactive waste. OPERA-PU-TNO521-2.

Inside Matters BVBA https://octopusimaging.eu/ (consulted 02/05/2016).

Koenen, M. and Griffioen, J. (2014). Mineralogical and geochemical characterization of the Boom Clay in the Netherlands. OPERA-PU-TNO521-1.

McGlinn, J.P., de Beer, C.F., Aldridge, P.L., Radebe, J.M., Nshimirimana, R., Brew, R.M.D., Payne, E.T., Olufson, P.K. (2010). Appraisal of a cementitious material for waste disposal: Neutron imaging studies of pore structure and sorptivity. Cement and Concrete Research, 40, 1320-1326.

Shafizadeh, A., Gimmi, T., Van Loon, L., Kaestner, A., Lehmann, E., Maeder, U.K., Churakov, S.V. (2015). Quantification of Water Content Across a Cement-clay Interface Using High Resolution Neutron Radiography. Physics Procedia, 69, 516-523.

Tremsin, A.S., Lehmann, E.H., McPhate, J.B., Vallerga, J.V., Siegmund, O.H.W., White, B., White, P., Feller, W.B., De Beer, F.C., Kockelmann, W. (2015). Quantification of cement hydration through neutron radiography with scatter rejection. IEEE Transactions on Nuclear Science, 62, 1288-1294.

Trtik, P., Münch, B., Weiss, W.J., Kaestner, A., Jerjen, I., Josic, L., Lehmann, E., Lura, P. (2011). Release of internal curing water from lightweight aggregates in cement paste investigated by neutron and X-ray tomography. Nuclear Instruments and Methods in Physics Research, Section A: Accelerators, Spectrometers, Detectors and Associated Equipment, 651, 244-249.

Van Tittelboom, K., Snoeck, D., Vontobel, P., Wittmann, F.H., De Belie, N. (2013). Use of neutron radiography and tomography to visualize the autonomous crack sealing efficiency in cementitious materials. Materials and Structures/Materiaux et Constructions, 46, 105-121.

Verhoef, E.V., de Bruin, A.M.G., Wiegers, R.B., Neeft, E.A.C., Deissmann, G. (2014). Cementitious materials in OPERA disposal concept in Boom Clay. OPERA-PG-COV020.

Zhang, P., Wittmann, F.H., Zhao, T., Lehmann, E.H. (2010). Neutron imaging of water penetration into cracked steel reinforced concrete. Physica B: Condensed Matter, 405, 1866-1871.

Observation of growth of the altered zone regarding cement-bentonite interaction by using Ca-XAFS analysis

Hitoshi Owada[1*], Naoki Fujii[1], Daisuke Hayashi[1]

[1] RWMC, Radioactive Waste Management Funding and Research Center (JP)

* Corresponding author: owada@rwmc.or.jp

Abstract

To evaluate the thickness of altered zone of bentonite around C-B (cement-bentonite) interface, synchrotron Ca-XAFS (X-ray absorption fine structure) measurement of the cement-bentonite contact zone of a drilled core obtained from the "sandbox" of Grimsel test site had been carried out.

Dominant alterations around C-B interface are Ca dissolution in cementitious material and precipitation of C-S-H and also dissolution of montmorillonite in bentonitic material (Sakamoto, 2007). Former is easy to analyse by using such as EPMA (Electron probe X-ray micro-analyser). On the other hand, latter is hardly to analyse, especially to quantify, by usual microscopic or spectroscopic way because secondary C-S-H is an amorphous phase and its amount is quite small. Therefore, we tried to determine and quantify the secondary C-S-H in bentonite around C-B interface by synchrotron Ca-XAFS measurement.

C-S-H concentration of each point of the specimen was obtained by pattern fitting of XANES (X-ray absorption near edge) spectra with that of reference minerals.

The thickness of the altered zone was determined to be less than 5 mm from the results of the Ca-XAFS observation, in which the C-S-H content was slightly higher in the range of around 5 mm from the contact than in other domains.

Introduction

Because Japanese geological disposal is in the stage of site selection, investigation of disposal concept for the geological disposal is carried out for generic environments. Therefore bentonite is used as a buffer material to prevent water inflow in the generic hydraulic environment with relatively high groundwater flow. For TRU waste disposal, cementitious material will also be used as a filler of space in the container and a filler of space between containers (Japan Nuclear Cycle Development Institute and Federation of Electric Power Company, 2007).

Functions of a buffer material are to keep low groundwater flow in the disposal vault and to minimize the migration of radionuclides. Because the hydraulic properties of bentonitic material depend on the montmorillonite content and dry density (Japan Nuclear Cycle Development Institute, 2000), dissolution of the montmorillonite by the reaction with high pH cement leachate will be a cause of the deterioration of the performance of the EBS system. Therefore, prediction of the extent of the altered zone of bentonitic buffer material is important to estimate long term performance of EBS systems that include both cementitious and bentonitic materials.

Dominant alteration of bentonite around C-B interface are precipitation of C-S-H and also dissolution of montmorillonite. Therefore, we have tried to determine the altered zone by the increase of the quantity of secondary C-S-H.

We had developed a Ca-XAFS method to quantify C-S-H around cement-bentonite interface and obtain the thickness of the altered zone (Sakamoto, 2007). The thickness of the altered zone of cement-bentonite coupled specimens was about 6 mm (Kurosawa, 2009) for 4 - 5 years and 8 mm for 8 years (Negishi, 2013; Radioactive Waste Management Funding and Research Center, 2013).

In this study, we examined the cement-bentonite interface of a drilled core obtained from the "sandbox" of Grimsel Test Site. The sandbox was originally used as a test area for emplacement of sand/bentonite material in a concrete basin in preparation for the GMT Project, and it was preserved under water-saturated conditions.

Analyses of the cement-bentonite contact zone of boring core

Specimen

A core of the contact zone between bentonite sand mixture and concrete base was obtained from sandbox B1 of Grimsel Test Site. Concrete of this basin is made from OPC (W/C ratio = 0.50) which was cast more than 10 years before construction of the sandbox. Bentonitic material was a mixture of 20% Kunigel® V1 and 80% silica-sand of which saturation ratio was over 90%.

A specimen for the EPMA and Ca-XAFS analyses was cut out from the contact zone shown in Figure 1(C).

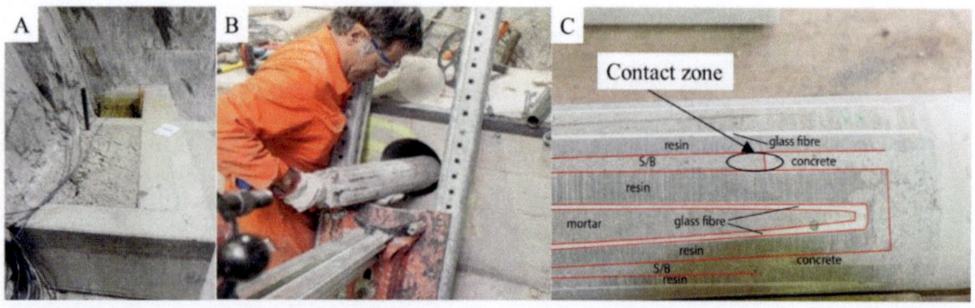

Figure 1: *Photographs of the appearance of sandbox B1* ***(A)****, Sampling* ***(B)*** *and the cross section of the boring core* ***(C)****.*

EPMA analysis to obtain the distribution of Ca concentration

Figure 2 shows the appearance of the specimen for the EPMA and Ca-XAFS analyses and the results of the 2-D mapping of the Ca concentration. In the cement side, decrease of the Ca content was observed in the 1to 2mm area from the contact. In this area disappearance of $Ca(OH)_2$ was observed by XRD analysis. No significant change of Ca concentration in bentonite was observed by EPMA analysis.

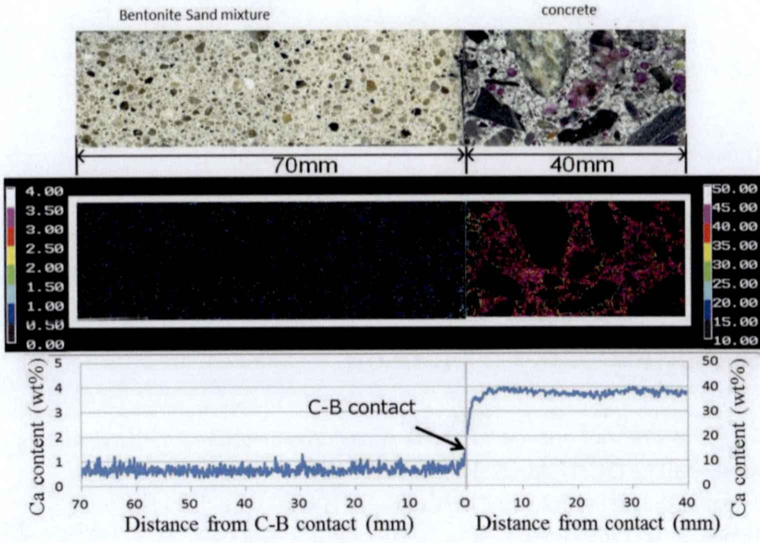

Figure 2: *Results of the 2-D mapping of Ca-concentration around cement-bentonite contact.*

Ca-XAFS analysis of bentonite side to determine the thickness of the altered zone

Figure 3 shows the X-ray absorption near edge (XANES) spectra of bentonite around the cement-bentonite contact. The absorbance near the contact was higher than that far from the contact. This means that Ca supplied from cement precipitated as secondary mineral near the contact. Because the dominant secondary mineral generated by cement-bentonite interaction was identified as C-S-H (Kurosawa, 2009) in our previous study, quantifying the C-S-H was done by a pattern fitting method. XANES spectra shown in Figure 3 were compared with those of standard minerals listed in Table 1.

Figure 4 shows the result of the quantification of Ca-containing minerals. C-S-H content within 5 mm from the contact is higher than that further away. C-S-H was detected even far from the contact, this might be caused by a trace amount of Ca containing silicate minerals such as tobermolite. The thickness of the secondary C-S-H precipitation, i.e., altered zone was about 5 mm. This is smaller than that of the OPC-bentonite coupled specimen immersed for 4 - 5 and 8 years. That specimen was obtained from the contact zone between basement concrete and bentonite, and the basement concrete was constructed 12 years earlier than the sandbox was constructed. Carbonation of the OPC before construction of the sandbox might be the reason of such retardation of cement-bentonite interaction.

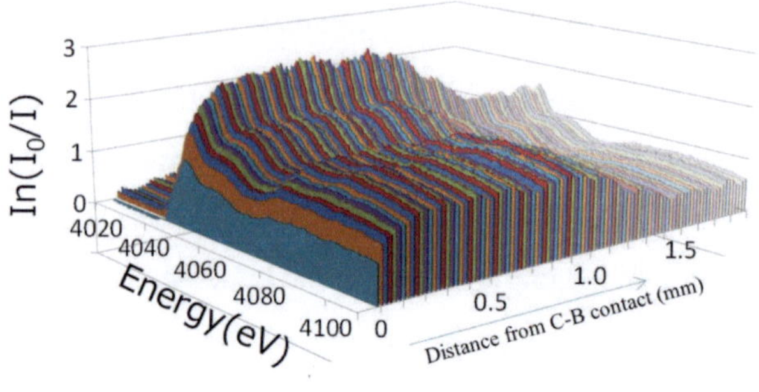

Figure 3: *XANES spectra of bentonite around the cement-bentonite contact.*

Table 1: *List of standard minerals for Ca-XAFS analyses.*

Kunigel-V1 (Na-type bentonite)
Ca compounds in cement: Quicklime (CaO), Calcite ($CaCO_3$), Portlandite ($Ca(OH)_2$) Hydrogarnet ($3CaO\cdot Al_2O_3\cdot 6H_2O$) Ettringite ($3CaO\cdot Al_2O_3\cdot 3CaSO_4\cdot 32H_2O$) C-S-H*
Ca compound that are considered to cause cement-bentonite interaction: Mono-sulfate ($3CaO\cdot Al_2O_3\cdot 3CaSO_4\cdot 12H_2O$) Mono-carbonate ($3CaO\cdot Al_2O_3\cdot 3CaSO_3\cdot 12H_2O$) Friedel's salt ($3CaO\cdot Al_2O_3\cdot 3CaCl_2\cdot 10H_2O$) C-S-H*
Ca-type bentonite (Ca-exchanged montmorillonite using a $CaCl_2$ solution)

*Four kinds of C-S-H are synthesized with different Ca/Si ratios (Ca/Si = 0.6, 0.83, 1.0, 1.4)

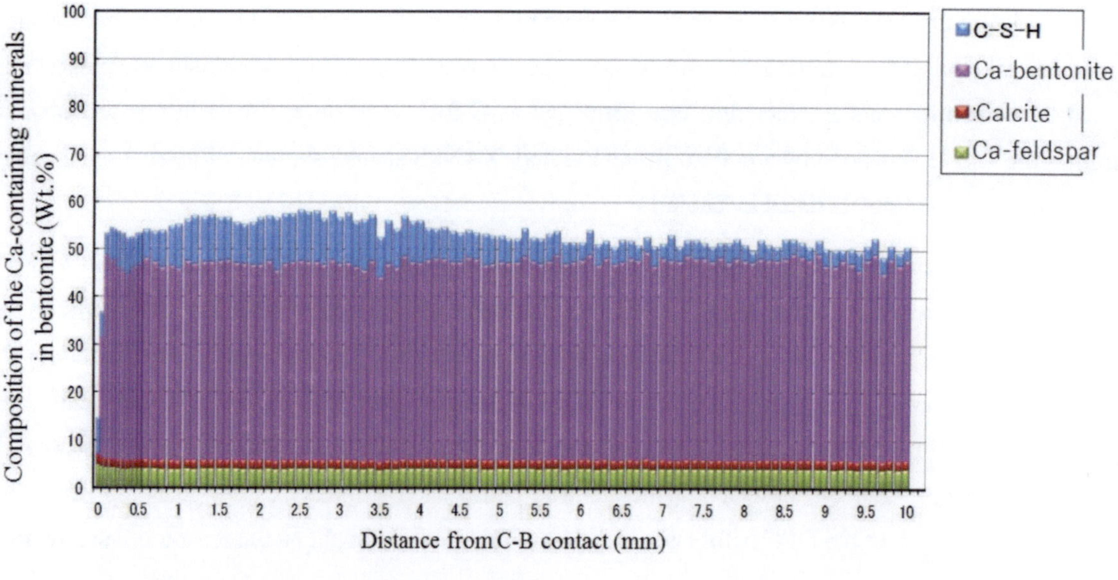

Figure 4: *Results of the quantification of Ca-containing minerals by pattern fitting method.*

Conclusions and Future work

The thickness of the altered zone of the core of the cement bentonite contact zone obtained from Grimsel Test Site was less than 5 mm after 12 years of contact. It was smaller than that of our previous work using the OPC-bentonite coupled specimen immersed for 4 - 5 years. Carbonation of cement before construction of the sandbox might be the cause of this retardation of cement-bentonite interaction compared to our laboratory tests.

Understanding of such generation and growth of altered zones will be important to predict the soundness and performance of cement-bentonite, or cement-clay systems. Our long term immersion test is still on going. We will obtain data of further generation and growth of the altered zone of cement bentonite systems.

Acknowledgement

This research is a part of "Advancement of Processing and Disposal Technique for the Geological disposal of TRU Waste" (FY2015) under a grant from Agency for Natural Resources and Energy (ANRE) in the Ministry of Economy, Trade and Industry (METI) of Japan

The research leading to these results has received funding from the European Union's Horizon 2020 Research and Training Programme of the European Atomic Energy Community (EURATOM) (H2020-NFRP-2014/2015) under grant agreement n° 662147 (CEBAMA).

References

Japan Nuclear Cycle Development Institute and Federation of Electric Power Company (2007). Second Progress Report on Research and Development for TRU Waste Disposal in Japan. JAEA-Review 2007-010, FEPC TRU-TR2-2007-01.

Japan Nuclear Cycle Development Institute (2000). H12: Project to Establish the Scientific and Technical Basis for HLW Disposal in Japan. JNC TN1410 2000-001.

Kurosawa, S., Sakamoto, H., Nitta, K., Numako, C., Haga, K., Shibata, M., Sato, T., Nakazawa, T., Owada, H. (2009). Effect of Calcium Silicate Hydrate Precipitates at Cementitious and Bentonite Material Interface on Long-Term Engineered Barrier System Performance in TRU Waste Disposal Facilities. MRS Proceedings, 1193, 489-504.

Negishi, K., Sakamoto, H., Ishii, T., Hayashi, D., Fujii, N., Owada, H., Nitani, H. (2013). XAFS analysis of C-S-H formed by cement-bentonite interaction. Proceedings of Goldschmidt 2013, 254.

Radioactive Waste Management Funding and Research Center (2013). FY2012 Annual Report of Development of the technique for the evaluation of long-term performance of EBS, Part 4/5, 4-29 (in Japanese).

Sakamoto, H., Shibata, M., Owada, H., Kaneko, M., Kuno, Y., Asano, H. (2007). Development of an analytical technique for the detection of alteration minerals formed in bentonite by reaction with alkaline solutions. Physics and Chemistry of the Earth, 32, 311-319.

Interaction between cement and Czech bentonite under temperature load and in in-situ conditions: an overview of experimental program

Petr Večerník[1*], Lucie Hausmannová[2], Radek Červinka[1], Radek Vašíček[2], Michal Roll[2], Jaroslav Hloušek[2], Václava Havlová[1]

[1] ÚJV Řež, a. s. (CZ)
[2] Centre of Experimental Geotechnics, Czech Technical University in Prague (CZ)

* Corresponding author: petr.vecernik@ujv.cz

Abstract

Two activities are planned concerning interaction between cement and Czech bentonite within our experimental collaborative programme: (i) long term laboratory (ageing procedures among Czech bentonite, cement CEM II, low pH binder and groundwater Josef under high temperature) and (ii) in-situ experiments in the Underground laboratory Josef (cartridges filled with compacted bentonite and cement paste, which are ongoing from 2010). The laboratory procedures started on 1st March 2016 and in periodic time intervals will be analysed. The main goal is to check changes in geochemical and geotechnical properties under temperature load and in-situ conditions, the comparison with previous results and transformation to arguments for safety analysis.

Introduction

The objective of both Czech partners (ÚJV Řež, a. s. and CEG CTU) is to contribute to experimental study on interaction between cement based materials and bentonite. Planned works include long term laboratory and in-situ experiments in the Underground laboratory Josef and subsequent laboratory analyses of cementitious and bentonite samples from these experiments. The effect of cement/bentonite/water interactions under defined temperature load will be studied. The main aims are study of changes in bentonite and cement properties due to mutual interaction under in-situ and laboratory conditions (diffusivity, porosity, hydraulic conductivity, fluid composition, uniaxial compressive strength etc.), study the influence of mutual interaction on safety function of both materials, continuation of research works done within the Czech project MPO TIP FR-TI1/362 (2009 - 2013), for more details see Vašíček et al. (2013), and at final to provide the input experimental data to WP3 dealing with modelling. One of the most important material property, influencing safety function of engineered barrier function, is diffusivity. Diffusivity can be easily quantified using radioactive tracer (^{3}H, ^{36}Cl).

Samples from in-situ and laboratory experiments will be used to determine diffusivity changes after mutual interactions. Changes in physical, chemical and geotechnical properties will be also studied on original unaffected and altered / interacted materials.

Obtained data could be used in safety assessment of deep geological repository (DGR). Assessing long-term performance of cement material for radionuclide encasement requires knowledge of the radionuclide-cement migration, interaction and mechanisms of retention (i.e., sorption or precipitation). This understanding will enable accurate prediction of radionuclide fate when the waste forms come in contact with groundwater. Long term physical, chemical and geotechnical changes of engineered barrier materials (cement, bentonite) are very slow processes so for the purpose of the project interaction and alteration experiments accelerated by increased temperature were designed. After the alteration process changes in main physical, chemical and geotechnical parameters will be studied and the comparison of non altered material properties and prediction of their behaviour could also be applied for safety assessment.

Materials and methods

Studied materials

Especially local Czech materials are studied in this project: bentonite "Bentonit 75" from 2010 (produced by Keramost, a.s., laboratory denoted as B75). For some bentonite characterization detail see Gondolli and Večerník (2014), chemical analysis of bentonite B75_2010 is shown in Table 1. Bentonite B75 was used in form of suspension (S:L ratio 1:5) with groundwater obtained from Underground laboratory Josef (denoted as GW Josef), cement material CEM II A-S 42.5R (denoted as CEM II, Lafarge Cement, a.s.). For international cooperation and interlaboratory comparison the low pH cement based material from VTT Finland (denoted as LPC) is used. From both cement materials the cylindrical samples 50 mm in diameter and height of 8.2 mm are prepared for laboratory procedures and testing.

Table 1: *Chemical composition of bentonite B75 from 2010.*

	SiO_2	Al_2O_3	TiO_2	Fe_2O_3	FeO	MnO	MgO	CaO	Na_2O	K_2O	P_2O_5	CO_2
wt %	51.91	15.52	2.28	8.89	2.95	0.11	2.22	4.60	1.21	1.27	0.40	5.15

List of methods used for characterization of investigated materials

Complete chemical analysis of liquid phase in water/cement/bentonite systems: standard methods of analytical chemistry are used. Especially evolution of Ca^{2+} and Mg^{2+} concentration and electrochemical parameters such as pH, conductivity and TDS are measured more systematically for water/cement systems.

Mineralogical changes: standard methods of x-ray diffraction are used.

Cation exchange capacity of bentonite: cation exchange capacity and exchangeable cations are measured by Cu(II)-trien method adopted from Meier and Kahr (1999).

Hydraulic conductivity and swelling pressure of bentonite: both parameters are measured together in the device with constant injection pressure of 1 MPa. The volume of the sample is constant during measurement. The hydraulic conductivity is evaluated according to Darcy´s law and swelling pressure is measured by force sensor installed above the sample.

Liquid limit of bentonite: cone penetration method described in Czech Standard (ČSN CEN ISO/ TS 17892-12).

Porosity of cement materials: for porosity determination two methods are routinely used and compared; mercury intrusion porosimetry (in external laboratory) and water immersion method based on Melnyk and Skeet (1986).

Diffusion properties of cement materials: for diffusion studies the through diffusion method is applied, ^{3}H and ^{36}Cl will be used as a tracer. The diffusion process is governed by Fick´s laws, reported elsewhere (Eriksen and Locklund, 1989; Ohlssons and Neretnieks, 1995; Bradbury and Green, 1986; Vilks et al., 2004; Löfgren and Neretnieks, 2006). ^{3}H and ^{36}Cl are commonly used as non-interacting tracers, but in case of cementitious materials, there are known C-S-H interactions, e.g., chloride interactions with C-S-H phases are dependent on H/S ratio, C/S ratio and nitrogen surface area (Beaudoin et al., 1990), tritium might also interact with cement (Ono et al., 1995; Furuichi et al., 2007).

Uniaxial compressive strength of cement materials: standard procedures (described in ČSN EN 196-1) for testing of strength properties of cementitious samples require samples with prescribed shape and low Surface Area/Volume ratio (e.g., cubes 40 x 40 x 40 mm). That means that samples are too large for our experimental purposes and effect of interaction would be very difficult to study on these samples. An alternative method rock mechanics using thin plates (Pauli and Holoušová, 1994) is good candidate but had to be validated for cementitious samples. CEG CTU verified the using of alternative method for evaluation of uniaxial compressive strength of thin samples (experimentally used for rock). The results were statistically evaluated and method was found applicable for cement paste samples (below it is described more in detail).

Description of Work to be Performed/Tasks

All experiment are planned for alteration and interaction studies on cementitious materials and bentonite as representatives of engineered barrier materials. Two types of experiments were designed: in-situ interactions and laboratory alteration procedures. Summary of procedures under in-situ conditions (cartridges and pressure vessels at 8 - 10°C) and in laboratory (pressure vessels at 95°C) is shown in the table below (Table 2) and documented there (Figure 1 and Figure 2).

Table 2: *Experimental program - new laboratory procedures and aged cartridges.*

	Ageing procedures	*Bentonite B75*	*GW Josef*	*Cement CEM II 42.5R*	*Low pH cement*	*Temperature 95°C*
New procedures: ageing 9/18/27 months	suspension + CEM II (LPC) + 95°C	x	x	x	x	x
	suspension + CEM II (LPC)	x	x	x	x	
	water + CEM II (LPC?) + 95°C		x	x	?	x
	water + CEM II (LPC?)		x	x	?	
	bentonite suspension	x	x			
	CEM II (LPC)			x	x	
Aged samples: 60 months	cartridge (bentonite)	x	x			
	cartridge (bentonite + CEM II)	x	x	x		

Cartridges filled with compacted bentonite and cement paste (ongoing in-situ test from 2010 in Underground laboratory Josef) will be dismantled, analysed during 2016 and results compared with data obtained in previous Czech project (MPO TIP FR-TI1/362). Material from laboratory alteration and interaction procedures will be analysed in time period of 9, 18 and 27 months to see dynamics of the process and changes of key parameters.

Figure 1: *Laboratory procedures: cementitious thin plates and bentonite suspension.*

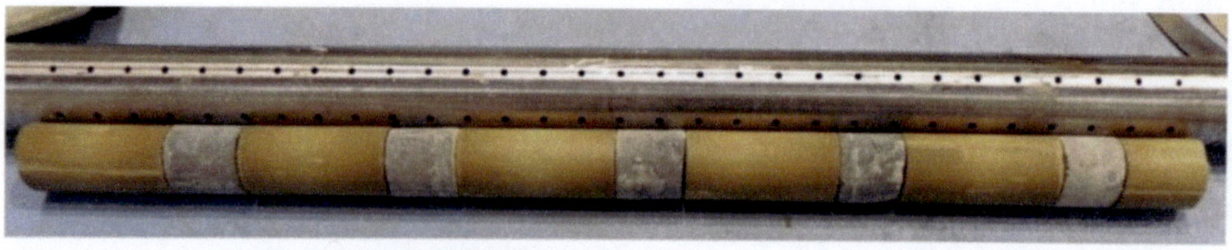

Figure 2: *Cartridges: bentonite and cementitious samples preparation for placing into cartridge for in-situ experiment (2010).*

Experiments and so far achieved results

Initial analyses for laboratory procedures

Firstly the initial analyses of GW Josef were performed. This groundwater was then used for preparation of bentonite B75 suspension (S:L ratio 1:5) and left to equilibrate for more than three months. Subsequently the porewater was squeezed out from the suspension using hydraulic press (MEGA 11-300, FORM+TEST Seidner+Co, Germany) with special squeezing cell containing Teflon frits under pressure 1.2 MPa for 46 hours. After the squeezing the rest of bentonite was analysed for CEC and interlayer cations. This suspension was used in laboratory experiments (pressure vessels). Data are summarised in Table 2 and Table 3. The qualitative phase analysis will be extended for qualitative representation of each mineral.

Table 2: *Composition of GW Josef and equilibrated GW Josef with bentonite B75.*

Species / concentration (mg/L)	GW Josef	Equilibrated GW Josef with bentonite B75 (squeezed porewater)
pH	7.7	8.8
Na^+	18.3	349.0
K^+	2.0	13.0
Ca^{2+}	76.1	5.0
Mg^{2+}	15.4	3.9
NO_3^-	0.1	9.7
Cl^-	14.7	21.4
F^-	0.7	1.2
SO_4^{2-}	106.2	159.5
HCO_3^-	199.6	604.1

The chemical composition of GW Josef is dominated by calcium and sulphate / carbonate ions. After the equilibration with bentonite B75, the porewater was highly enriched by sodium and sulphate / carbonate ions. The dissolution of impurities such as Na_2CO_3 in bentonite could increase concentrations some of these ions. In addition the exchange of sodium by calcium and magnesium in exchangeable complex slightly increased also the sodium concentration in porewater.

Table 3: *Phase composition and CEC measurement of bentonite B75.*

Bentonite B75 Phase (XRD analysis)	*Abundance*	*CEC and interlayer cations (meq/100g)*	*After equilibration with GW Josef*
Smectite (montmorillonite)	***	Na^+	27.20
K-micas (illite, biotite)	*	K^+	2.69
SiO_2 phases (quartz)	**	Mg^{2+}	17.07
Kaolinite	*	Ca^{2+}	10.97

Bentonite B75 Phase (XRD analysis)	***Abundance***	***CEC and interlayer cations (meq/100g)***	***After equilibration with GW Josef***
Calcite	*	CEC_{VIS}	48.44
Anatase	*	CEC_{SUM}	57.93
Siderite (Mg)	*		
Ankerite	*		

*Note: XRD analysis, ***major, **minor, *present, CEC_{VIS} is measured by UV/VIS, CEC_{SUM} is calculated as sum of exchangeable cations*

After the equilibration with GW Josef the bentonite is dominated by sodium and magnesium in exchangeable complex. The difference between CEC_{VIS} and CEC_{SUM} can be explained by dissolution of minerals / salts which overestimates CEC_{SUM}.

Preparation of cementitious samples (OPC)

Cylindrical samples with diameter of 50 mm and height of 8.2 mm were prepared from cementitious paste (CEM II) with water ratio 0.45. Material was casted into PVC tube formwork (inner diameter 50 mm and length about 200 mm) and stored in humid condition in 15°C for 1 week, afterwards the cylinders were removed from tubes and stored in the underground (humid condition, 10°C). After 28 days, cylinders were sliced into 8.2 mm thick plates and kept in the underground for another 16 days. The major part of the plates was placed into the pressure vessels after 44 days in total and 30 slices were used for testing of initial parameters. The uniaxial compressive strength of the samples was evaluated and reached the value of 32.6 MPa using alternative method. The recalculation to standard method gives value about 60 MPa.

Verification of Alternative Strength test method for thin samples

The size of the samples was chosen regarding to limited recommendations (diameter of 50 mm and thickness of 8.2 mm). Also standard shape samples (cubes 40 x 40 x 40 mm) were tested to validate the data (according to standard ČSN EN 196-1). Three types of cement with different strength properties were used (CEM II B-M 32.5 R 32.5, CEM I 42.5 R and super cement 52.5). Results are shown in Figure 3. The relation (equation) between these two methods is illustrated in Figure 4. In the CEBAMA project, the alternative method with following recalculation will be used for evaluation of uniaxial compressive strength of cementitious samples.

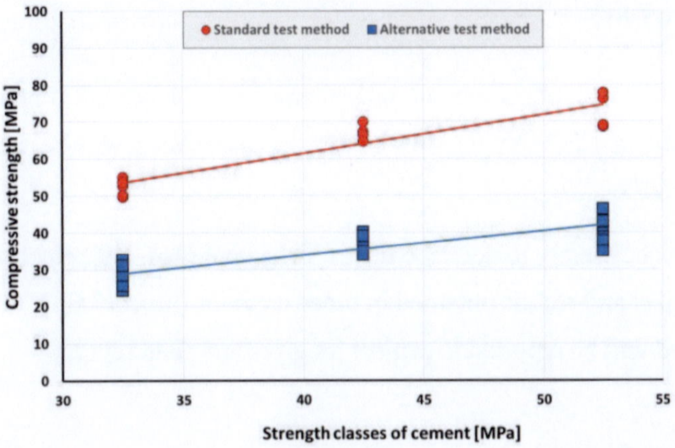

Figure 3: Comparison of two methods for testing of compressive strength of cement paste (w = 0.45): standard test method (cubes 40 x 40 x 40 mm) – red colour and alternative method - thin plates with diameter of 50 mm and height of 8.2 mm – blue colour.

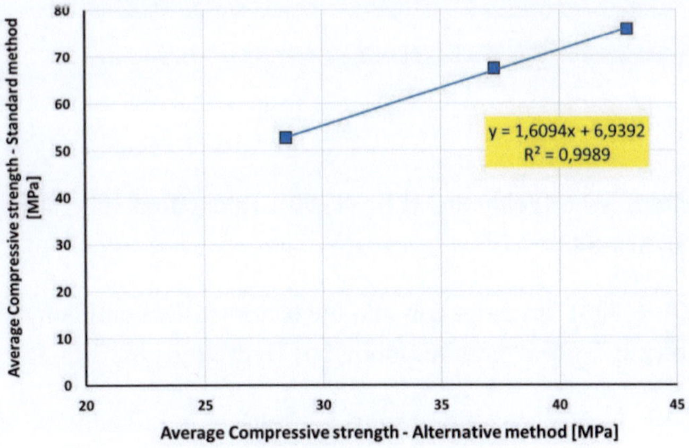

Figure 4: Relation between results of alternative and standard test method.

Ca^{2+} and Mg^{2+} concentration, TDS and pH evolution in the cement / water systems

Basic chemical analysis of groundwater in interaction with cement is performed in periodic time intervals (first one after one month of ageing and next ones every two weeks). For comparison the processes in pressure vessels are studied in two temperatures 10°C and 95°C. Note that the pressure vessels which are heated up to 95°C are weighed before sampling and evaporated water is refilled by distilled water. The reason of using distilled water for refilling is to preserve the sum of dissolved solids at constant value and not to increase the concentration of cations and anions by adding aditional amounts of GW Josef. TDS and pH of distilled water are controlled before every refilling. Reading from the first measurements showed, that in both cases distinct increase of pH values was observed, from pH~ 8 (GW Josef) to pH~ 13 (alkaline environment). Ca^{2+} concentration increased (from 30 mg/L to the range from 40 to 180 mg/L), similarly as TDS and conductivity. In contrary Mg^{2+} concentration decreased (from 27 mg/L to 7 - 10 mg/L). Generally, all parameters increase or decrease more rapidly for the samples which are heated. The measurements will continue until a steady state.

Conclusions and Future work

The overview of collaborative experimental programme was presented. Laboratory procedures already started and after preparation of LPC samples the laboratory procedures will be extended. Now, the data are collected for modelling in WP3.

This contribution is focused mainly to describe the experimental plan. Description and comparison of results achieved in CEBAMA project and results obtained in other previous project MPO TIP FR-TI1/362 is one of the main aims of the project and will be reported in project progress and final reports.

Acknowledgement

The research leading to these results has received funding from the European Union's European Atomic Energy Community's (Euratom) Horizon 2020 Programme (NFRP-2014/2015) under grant agreement, 662147 – Cebama.

References

Beaudoin, J.J., Ramachandran, V.S., Feldman, R.F. (1990). Interaction of chloride and CSH. Cement and Concrete Research, 20(6), 875-883.

Bradbury, M.J. and Green, A. (1986). Investigation into the factors influencing long range matrix diffusion rates and pore space accessibility at depth in granite. Journal of Hydrology, 89, 123-139.

ČSN CEN ISO/TS 17892-12. Geotechnický průzkum a zkoušení - Laboratorní zkoušky zemin - Část 12: Stanovení konzistenčních mezí. Praha: Český normalizační institut, 2005.

ČSN EN 196-1. Metody zkoušení cementu: Část 1: Stanovení pevnosti. Praha: Český normalizační institut, 2005.

Eriksen, E. and Locklund, B. (1989). Radionuclide sorption on crushed and intact granitic rock. Volum and surface effects. SKB Technical Report, TR 89-25.

Furuichi, K., Takata, H., Katayama, K., Takeishi, T., Nishikawa, M., Hayashi, T., Kobayashi, K., Namba, H. (2007). Evaluation of tritium behavior in concrete. Journal of Nuclear Materials, 367, 1243-1247.

Gondolli, J. and Večerník, P. (2014). The uncertainties associated with the application of through-diffusion, the steady-state method: a case study of strontium diffusion. Geological Society Special Publication, 400, 603-612.

Löfgren, M. and Neretnieks, I. (2006). Through electromigration: A new method of obtaining formation factors and investigating connectivity. Journal of Contaminant Hydrology, 87, 237-252.

Meier, L.P. and Kahr, G. (1999). Determination of the cation exchange capacity (CEC) of clay minerals using the complexes of copper(II) ion with triethylenetetramine and tetraetylenepentamine. Clays and Clay Minerals, 47, 386-388.

Melnyk, T.W. and Skeet, A. (1986). An improved technique for determination of rock porosity. Canadian Journal of Earth Sciences, 23, 1068-1074.

Ohlssons, Y. and Neretnieks, I. (1995). Literature survey of matrix diffusion theory and of experiments and data including natural analogues. SKB Technical Report, TR 95-12.

Ono, F., Tanaka, S., Yamawaki, M. (1995). Tritium sorption by cement and subsequent release. Fusion engineering and design, 28, 378-385.

Pauli, J. and Holoušová, T. (1994). Mechanika Hornin (Rock Mechanic textbook), České vysoké učení technické v Praze.

Vašíček, R., Levorová, M., Červinka, R., Hausmannová, L., Kaisr, Z., Venkrbec, Z. (2013). Výzkum vlastností materiálů pro bezpečné ukládání radioaktivních odpadů a vývoj postupů jejich hodnocení. PODETAPA 4.2 Laboratorní výzkum tlumících, výplňových a konstrukčních materiálů. Czech project MPO TIP FR-TI1/362 (2009-2013).

Vilks, P., Miller, N.H., Stanchell F.W. (2004). Phase II. In situ diffusion experiment. Rep. AECL No. 06819-REP-01200-10128-R00.

Characterisation of NRVB, a high-pH backfill cement

Rita Vasconcelos[1], Neil Hyatt[1], John Provis[1], Claire Corkhill[1*]

[1] NucleUS Immobilisation Science Laboratory, Department of Materials Science and Engineering, University of Sheffield (UK)

* Corresponding author: c.corkhill@sheffield.ac.uk

Abstract

NRVB (Nirex Reference Vault Backfill) is a high pH cementitious material considered for use as a backfill material in the illustrative concept for the geological disposal of intermediate level waste packages in a higher strength rock in the UK. Due to the required long-term performance of a backfill material, understanding ageing and alteration associated with groundwater interactions is of critical importance. In this study, preliminary results related to the characterisation of NRVB are presented. X-ray diffraction and scanning electron microscopy analysis were used to identify the main hydrate phases present at an early age. The results showed that the main hydrate phases present after 12 days of curing were calcite, portlandite, AFt and AFm phases. As future work, leaching experiments will be performed using three different types of groundwater, with characterisation of cement microstructure and chemistry as a function of time.

Introduction

In the UK, the current illustrative concepts for the disposal of intermediate level waste (ILW) packages are based on a multi-barrier Geological Disposal Facility (GDF) (NDA, 2010). In the concept for a higher strength rock, the vaults containing waste would be backfilled with a cementitious material (Francis et al., 1997; Crossland, 2001; Bamforth et al., 2012). An important function of the backfill material is to provide a chemical barrier to radionuclide release to the geological environment (Francis et al., 1997).

NRVB (Nirex Reference Vault Backfill) has been proposed as the backfill material in a higher strength rock. It comprises Portland cement, calcium hydroxide (hydrated lime) and calcium carbonate (limestone flour) (Francis et al., 1997). This composition contributes to some specific functions, including maintaining a high pH that will minimise the solubility and transport of many radionuclides, and also providing high sorption capacity and minimising gas pressure buildup through a high porosity microstructure (Francis et al., 1997).

The time over which the performance of a cementitious backfill is required is extremely long, and during such timescales the backfill material will age and alter due to interactions with the local environment, particularly with groundwater (NDA, 2010). In the UK, which does not yet have a GDF site, a number of different host rocks, hence a number of different ground water compositions, are possible (e.g., granitic and clay). It is important to understand how the properties of the cementitious materials will be affected by interactions with the groundwater. For example, coupled changes in chemistry and microstructure may significantly influence the pH buffering performance of the backfill, and also the permeability and radionuclide sorption capacity.

To understand these effects, it is first necessary to possess a thorough understanding of the chemistry and microstructure of NRVB. We present results of an ongoing ageing study, developed to identify the characteristics of NRVB, and discuss plans for future work.

Methodology

Materials and synthesis

Cement paste batches were formulated according to Table 1, with a water / solid ratio of 0.55. The Portland cement used is CEM I 52.5 N sourced from Hanson Cement Ltd, Ribblesdale works; $Ca(OH)_2$ ($\geq$ 95.0%) was sourced from Sigma-Aldrich; and $CaCO_3$ ($\geq$ 99.0%) was also sourced from Sigma-Aldrich. The components were mixed for 5 min in a Kenwood benchtop mixer. Subsequently, the cement paste was transferred into centrifuge tubes and cured at 20°C and 95% relative humidity. After 12 days of curing, the hydration was stopped using solvent exchange in acetone, and the sample analysed.

Table 1: *NRVB formulation.*

Material	Content (kg/m^3)
CEM I 52 5N	450
Calcium hydroxide	170
Calcium carbonate	495
Water	615

Characterisation

Samples were ground with a pestle and mortar to pass a 63 μm sieve prior to analysis by X-ray Diffraction (XRD), using a Bruker D2 Phaser diffractometer utilizing a Cu Kα source. Measurements were taken from 10° to 70° 2 θ with a step size of 0.02° and 2 s counting time per step. For SEM imaging and energy dispersive X-ray (EDX) analysis, NRVB samples were mounted in epoxy resin and polished to a 0.25 μm finish using diamond paste. Back scattered electron (BSE) images were recorded using a Hitachi TM3030 scanning electron microscope operating with an accelerating voltage of 15 kV. EDX analysis was performed using a Quantax 70 software and elemental maps were collected for 10 min.

Results

XRD analysis of NRVB after 12 days revealed the presence of calcite ($CaCO_3$) and portlandite ($Ca(OH)_2$) as the main phases in the hydrate product assemblage, as shown in Figure 1. It was not possible to identify C-S-H by XRD at this early age. Ettringite (AFt phase) and calcium monocarboaluminate hydrate (AFm phase) were also identified. This is in agreement with the results of Holland and Tearle (2003) who used modelling to predict the expected mineral assemblage of NRVB after 28 days. However, those authors also predicted that hydrogarnet would form, which is absent from our 12 days analysis. The degree of reaction of the cement clinker phases appears to be close to complete even at this early age, as the water/cement ratio of NRVB is high.

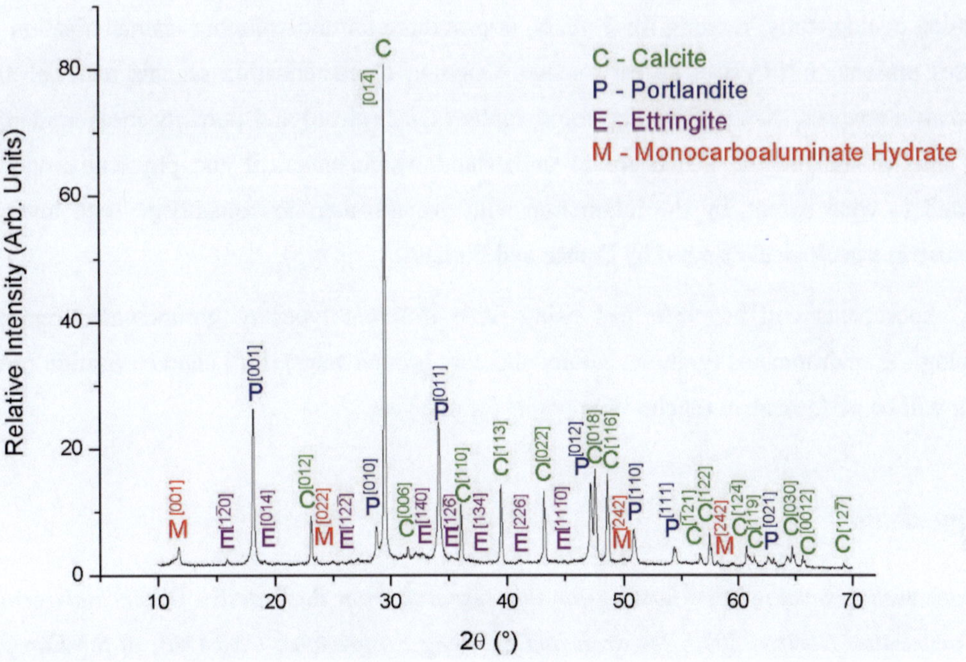

Figure 1: *XRD pattern for NRVB at 12 days of curing. Crystalline phases are labelled.*

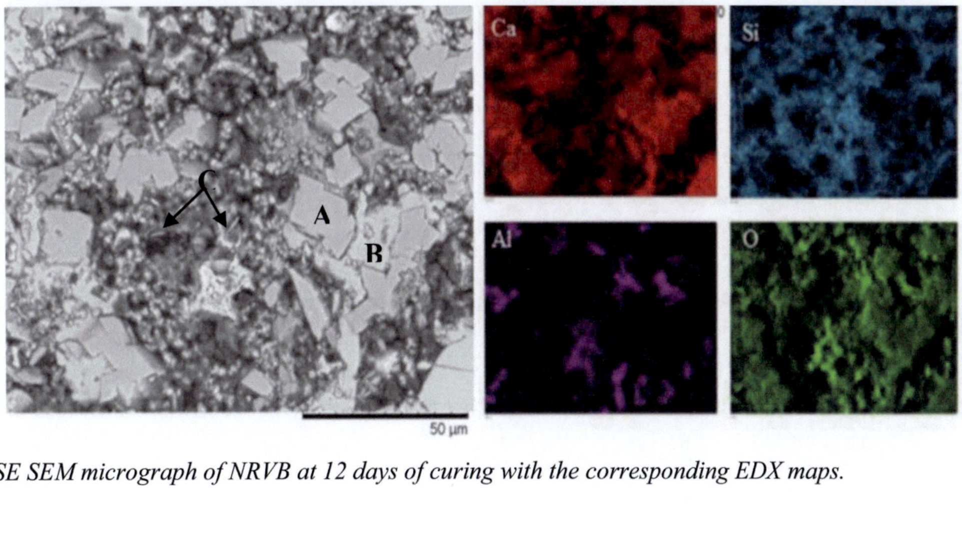

Figure 2: *BSE SEM micrograph of NRVB at 12 days of curing with the corresponding EDX maps.*

SEM imaging and EDX analysis confirmed the presence of the phases identified by XRD, as shown in Figure 2. Portlandite (Figure 2, labelled A) is identified as large Ca-containing rhombohedral crystals, indicating that these originate from the batch ingredients rather than resulting from cement hydration, which would be expected to yield finer crystals. The Ca-rich phase surrounding Portlandite crystals (Figure 2, labelled B) may be C-S-H. Areas containing higher concentrations of aluminium have a sheet-like morphology, suggestive of calcium monocarboaluminate hydrate. Small needle-like phases (Figure 2, labelled C) are associated with EDX maps of sulfur (data not shown), indicative of ettringite.

Conclusions and Future work

This ongoing ageing study, running for 3 years, is providing an underpinning characterisation of the main hydrate phases present in NRVB at an early stage. Ongoing characterisation on this material also includes thermogravimetric analysis, X-ray computed tomography (synchrotron) and porosity measurements. This will support the aims of the project, which are to understand which chemical and physical properties will be influenced, and to what extent, by the interaction with groundwater, in comparison with low pH cements proposed for use in a geological disposal by France and Finland.

Leaching experiments will be performed using three different types of groundwater, representative of different geological environments (granitic, saline, and clay groundwater). Full characterisation of the cements post-leaching will be performed, at regular time points for one year.

Acknowledgements

The authors wish to acknowledge funding for this research from Radioactive Waste Management and the European Commission Horizon 2020 Research and Training Programme, CEBAMA, of the European Atomic Energy Community (EURATOM), under grant agreement number 662147. Steve Williams (RWM) is thanked for his comments and input to this work.

References

Bamforth, P.B., Baston, G.M.N., Berry, J.A., Glasser, F.P., Heath, T.G., Jackson, C.P., Savage, D., Swanton, S.W. (2012). Cement materials for use as backfill , sealing and structural materials in geological disposal concepts. A review of current status. Serco Report SERCO/005125/001 Issue 3.

Crossland, I.G. and Vines, S.P. (2001). Why a cementitious repository. Nirex Report, N/034.

Francis, A.J., Cather, R., Crossland, I.G. (1997). Nirex Safety Assessment Research Programme: Development of the Nirex Reference Vault Backfill. Report on Current Status in 1994. Nirex Report, S/97/014.

Holland, T.R. and Tearle, W.M. (2003). Serco Assurance. A Review of NRVB Mineralogy. Serco Report, SERCO/ERRA-0455.

Nuclear Decommissioning Authority (2010). Geological Disposal: Near-Field Evolution Status Report. NDA Report, NDA/RWMD/033.

Kinetic parameters for assessing the pH –evolution of low-pH concrete

Tapio Vehmas[1*], Markku Leivo[1], Erika Holt[1]

[1] VTT Technical Research Centre of Finland Ltd. (FI)

* Corresponding author: tapio.vehmas@vtt.fi

Abstract

Ordinary Portland cement based materials play a major role in spent nuclear fuels repository systems. High pH of cementitious materials endangers the stability of specific clays in engineered barriers systems and low-pH mix designs have been developed to overcome the problem. Various modelling tools are used to understand and forecast the low-pH cementitious materials behaviour and interactions throughout the lifetime of the repository. The pH development of low-pH cementitious materials is primarily controlled by the propagation of pozzolanic reaction which transforms portlandite into calcium-silicate-hydrates (C-S-H) and later on transforms calcium-silicate-hydrates into low calcium C-S-H. Kinetic parameters are needed for time depended modelling. Arrhenius equation is widely utilized method to describe kinetics of a chemical process. We have determined the Arrhenius parameters for pozzolanic reaction by conduction calorimeter measurements. The activation energy for pozzolanic reaction was 67 ± 7 kJ/mol, which was in-line with the literature. According to results, Arrhenius equation is a solid method to incorporate the kinetic parameters for low-pH concrete pH development.

Introduction

Ordinary Portland Cement (OPC) based materials play a major role in spent nuclear fuel repository systems. Even in the KBS-3V Engineered Barrier System (EBS) (Palomäki et al,. 2012), where bentonite acts as a main barrier material and concrete acts only as a structural component for rock support and tunnel closure, OPC based materials are used in large amounts. OPC based materials have naturally high pH which endangers the stability of the bentonite buffer. Low pH mix designs have been developed that utilizes large amounts of pozzolanic materials to ensure the safety of the bentonite buffer.

The lifetime of long-term nuclear waste repositories (NWR) is extremely long. The radioactivity of the used nuclear fuel reaches the radiation level on natural uranium ore deposit in 250 000 years. Experimental research is incapable to study the long-term behaviour and interactions of materials in the NWR environment. Modelling is needed to understand and forecast the material behaviour and interactions throughout the lifetime of NWR.

Thermodynamic modelling and reactive transport modelling have been widely utilized tools to understand the long-term behaviour. Kinetic factors must be included in the models to predict the time dependent behaviour. In low-pH Portland cement-based materials, the time dependency is a very important aspect because the safety of the multi-barrier system depends on the pH–plume, caused by degradation of cementitious materials Koskinen (2014). The pH of cement-based materials is primarily controlled by the propagation of pozzolanic reaction which first transforms portlandite into calcium-silicate-hydrates (C-S-H) (Equation 1) and later on transforms C-S-H into low-calcium C-S-H Cau-Dit-Coumes (2006).

$$SiO_2 + Ca(OH)_2 \rightarrow CSH \qquad \textit{Eq. 1}$$

Implementation of kinetic parameters into modelling is not a straightforward task. Although the parameters are easily integrated into the modelling, their physical relevance is questionable. A good kinetic parameter can distinguish the various processes that have effect on kinetics. The kinetics can depend on the physical factors such as surface area and volume of the pozzolanic materials, but also the rate of the chemical reaction. Arrhenius equation is widely used equation to describe the kinetics of a chemical reaction. Arrhenius equation (Equation 2) is able to distinguish the rate of the chemical reaction and the physical factors Laidler (1984). Chemical reaction is characterized by activation energy ΔE. Physical parameters are characterized with frequency factor A. In this study, the applicability of Arrhenius equation to estimate kinetics of pozzolanic reaction was studied.

$$k = Ae^{-\Delta E/RT} \qquad \textit{Eq. 2}$$

Where k = rate of chemical reaction (s^{-1}), A = frequency factor (s^{-1}), ΔE = activation energy (J/mol), R = Gas constant (J/(K·mol)) and T = temperature (K).

Materials and Methods

Kinetic parameters for Arrhenius equation was determined from heat flow measured by conduction calorimeter (TAM-Air). Sample ampoules were filled with the mixtures described in the Table 1. Calcium hydroxide was purchased from Sigma Aldrich (ACS reagent ≥ 95.0%). Water used was ion-exchanged water. Pozzolanic material was amorphous silica, Silika Parmix purchased from Finnsementti. Silica was not pre-dispersed prior measurement. Samples were stirred manually for 5 minutes. Sample and reference ampoule were mixed respectively. Samples were introduced to calorimeter and the calorimeter was stabilized for four hours before the data acquisition. Four hours stabilization period ensured that the heat flow originated solely from the pozzolanic reaction. The reference ampoule had the exact composition of the studied sample but the silicon was in non-reactive, crystalline form (quartz). The heat capacities of the reference and the sample ampoules were identical which enabled good resolution for the calorimetric measurements. Two parallel samples (a and b) were measured and the experiment was conducted at temperatures 15ºC, 25ºC and 40ºC.

Table 1: *Sample compositions. Volume of the sample ampoule was 20 mL.*

Ampoule	m(SiO_2)	m(Quartz)	m($Ca(OH)_2$)	m(H_20)
Sample 1	0.5 g	-	2.0 g	2.0 g
Ref. 1	-	0.5 g	2.0 g	2.0 g
Sample 2	1.0 g	-	2.0 g	2.0 g
Ref. 2	-	1.0 g	2.0 g	2.0 g
Sample 3	2.0 g	-	2.0 g	2.0 g
Ref.3	-	1.0 g	2.0 g	2.0 g
Sample 4	3.0 g	-	2.0 g	2.0 g
Ref. 4	-	3.0 g	2.0 g	2.0 g

Data interpretation

Data acquisition started four hours after the samples were initially mixed. At that time the sample ampoules were inside the calorimeter in constant temperature (15ºC, 25ºC, or 40ºC). The heat flow of the sample was determined at various values of total heat. As the total heat corresponded to the reaction propagation, the selected method enabled the heat flow value determination at the various reaction degrees. Selected method enabled reliable determination of the kinetic parameters according to Arrhenius equation. Example of the data is presented in Figure 1.

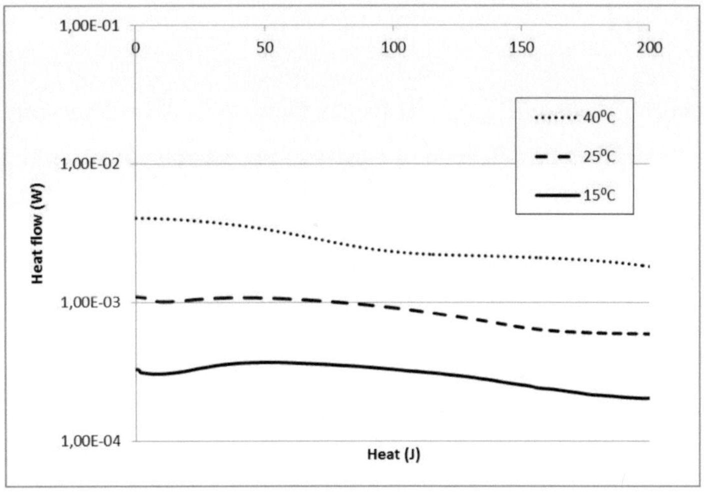

Figure 1: *Isothermal calorimetric results from Sample 4(b).*

Activation energy (ΔE/R) was determined with linear regression from the slope of ln(k) (k=heat flow (W)) and 1/T according to Equation 3. Example is presented in Figure 2.

$$ln(k) = -{}^{\Delta E}/_{R} * \frac{1}{T} + ln(A)$$ ***Eq. 3***

Where k = heat flow (W), ΔE =Activation energy (J/mol), R = Gas constant (J/(K mol)), T =Temperature (K) and A =frequency factor (s^{-1}).

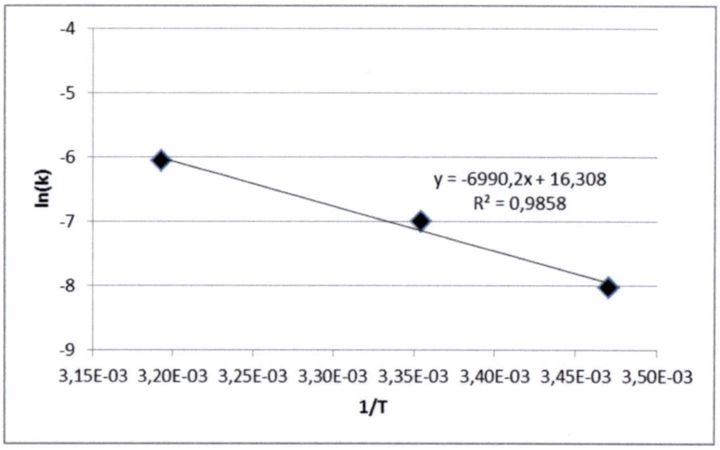

Figure 2: *Example of the activation energy calculation for sample 4(b) according to Equation 3.*

Frequency factor values were calculated according to Equation 4, from each individual heat flow value. Mean activation energy value (ΔE/R = - 8 000 K^{-1}) was used in the calculations which was previously determined according to Equation 3.

$$ln(A) = ln\left(\frac{k}{me^{\Delta E/_{RT}}}\right)$$ ***Eq. 4***

Where k = rate of chemical reaction (s^{-1}), A = frequency factor (s^{-1}), ΔE = activation energy (J/mol), R = Gas constant (J/(K·mol)) and T = temperature (K) and m = mass of the pozzolanic material (g).

Results

Activation energy

Calculated activation energy values are presented in Table 2. Table 2 also presents the heat flow values which were used to calculate the activation energies. Average activation energy (ΔE/R) was - 8 000 ± 900 K^{-1}.

Table 2: *Activation energies from various samples and the heat flow of the samples.*

Ampoule	Heat /J	Heat flow /W			ΔE/R
		15°C	25°C	40°C	
Sample 1	5	1.68E-05	5.43E-05	1.47E-04	-7 729.1
	10	1.58E-05	4.63E-05	1.21E-04	-7 264.1
	15	1.36E-05	4.03E-05	1.03E-04	-7 217.9
Sample 2(a)	10	1.36E-04	3.74E-04	1.52E-03	-8 700.9
	20	1.40E-04	4.04E-04	1.40E-03	-8 275.7
	30	1.38E-04	4.09E-04	1.23E-03	-7 831.7
	60	1.21E-04	3.39E-04	9.21E-04	-7 258.2
	90	1.04E-04	2.85E-04	7.72E-04	-7 172.1
Sample 2(b)	20	1.42E-04	3.90E-04	2.11E-03	-9 787.3
	40	1.44E-04	3.40E-04	1.93E-03	-9 448.3
	60	1.39E-04	2.94E-04	1.78E-03	-9 326.3
	90	1.15E-04	2.41E-04	1.63E-03	-9 712.5
Sample 3(a)	20	2.57E-04	7.43E-04	3.40E-03	-9 329.9
	40	2.95E-04	7.84E-04	3.06E-03	-8 445
	60	2.96E-04	7.26E-04	2.71E-03	-8 005
	80	2.76E-04	6.45E-04	2.32E-03	-7 701.4
	100	2.61E-04	5.70E-04	1.99E-03	-7 359.5
Sample 3(b)	20	2.66E-04	7.89E-04	2.73E-03	-8 363.1
	40	3.05E-04	8.36E-04	2.34E-03	-7 296.4
	60	3.03E-04	7.98E-04	1.96E-03	-6 668.4
	80	2.81E-04	7.29E-04	1.83E-03	-6 699.6
	100	2.62E-04	6.64E-04	1.77E-03	-6 846.8
Sample 4(a)	20	3.37E-04	1.07E-03	4.18E-03	-9 051.2
	40	3.86E-04	1.14E-03	3.85E-03	-8 257
	60	3.92E-04	1.12E-03	3.37E-03	-7 709.6
	80	3.70E-04	1.06E-03	2.86E-03	-7 307.7
	100	3.44E-04	9.74E-04	2.53E-03	-7 124.7
Sample 4(b)	20	3.20E-04	1.04E-03	3.90E-03	-8 977.9
	40	3.62E-04	1.08E-03	3.57E-03	-8 213
	60	3.64E-04	1.05E-03	3.11E-03	-7 677.5
	80	3.48E-04	9.93E-04	2.62E-03	-7 214.3
	100	3.26E-04	9.09E-04	2.31E-03	-6 990.2
				Average:	-8 000 ± 900

Frequency factor

Results from the frequency factor calculations are presented in Table 3. The average frequency factor was $\ln(A) = 18.6 \pm 0.47$.

Table 3: *Frequency factor values and the heat flow values of the samples.*

Ampoule	Heat /J	Heat flow /W			Ln(A)		
		15°C	25°C	40°C	15°C	25°C	40°C
Sample 1	5	1.68E-05	5.43E-05	1.47E-04	17.439	17.683	17.389
	10	1.58E-05	4.63E-05	1.21E-04	17.378	17.523	17.194
	15	1.36E-05	4.03E-05	1.03E-04	17.228	17.384	17.033
Sample 2(a)	10	1.36E-04	3.74E-04	1.52E-03	18.855	18.937	19.046
	20	1.40E-04	4.04E-04	1.40E-03	18.884	19.014	18.967
	30	1.38E-04	4.09E-04	1.23E-03	18.870	19.026	18.837
	60	1.21E-04	3.39E-04	9.21E-04	18.738	18.838	18.548
	90	1.04E-04	2.85E-04	7.72E-04	18.587	18.665	18.371
Sample 2(b)	20	1.42E-04	3.90E-04	2.11E-03	18.892	18.972	19.371
	40	1.44E-04	3.40E-04	1.93E-03	18.906	18.835	19.280
	60	1.39E-04	2.94E-04	1.78E-03	18.871	18.690	19.201
	90	1.15E-04	2.41E-04	1.63E-03	18.681	18.491	19.113
Sample 3 (a)	20	2.57E-04	7.43E-04	3.40E-03	18.780	18.932	19.162
	40	2.95E-04	7.84E-04	3.06E-03	18.938	18.985	19.057
	60	2.96E-04	7.26E-04	2.71E-03	18.941	18.908	18.936
	80	2.76E-04	6.45E-04	2.32E-03	18.871	18.790	18.780
	100	2.61E-04	5.70E-04	1.99E-03	18.815	18.666	18.627
Sample 3(b)	20	2.66E-04	7.89E-04	2.73E-03	18.833	18.990	18.941
	40	3.05E-04	8.36E-04	2.34E-03	18.970	19.048	18.787
	60	3.03E-04	7.98E-04	1.96E-03	18.963	19.001	18.610
	80	2.81E-04	7.29E-04	1.83E-03	18.888	18.911	18.541
	100	2.62E-04	6.64E-04	1.77E-03	18.818	18.817	18.508
Sample 4(a)	20	3.37E-04	1.07E-03	4.18E-03	18.665	18.891	18.963
	40	3.86E-04	1.14E-03	3.85E-03	18.801	18.954	18.881
	60	3.92E-04	1.12E-03	3.37E-03	18.817	18.936	18.748
	80	3.70E-04	1.06E-03	2.86E-03	18.759	18.881	18.584
	100	3.44E-04	9.74E-04	2.53E-03	18.686	18.797	18.461
Sample 4(b)	20	3.20E-04	1.04E-03	3.90E-03	18.614	18.856	18.894
	40	3.62E-04	1.08E-03	3.57E-03	18.737	18.902	18.807
	60	3.64E-04	1.05E-03	3.11E-03	18.743	18.876	18.668
	80	3.48E-04	9.93E-04	2.62E-03	18.698	18.816	18.497
	100	3.26E-04	9.09E-04	2.31E-03	18.632	18.728	18.370
				Average:		18.6 ± 0.47	

Discussion

Activation energy and amount of pozzolanic material in the sample are presented in Figure 3. According to Figure 3, the activation energy is fairly constant and irrespective to the total amount of pozzolanic material in the sample. Frequency factor and amount of pozzolanic material is presented in Figure 4. The frequency factor was calculated according to Equation 3, which depends on mass of pozzolanic material as a factor. This approach seems to be appropriate because the ln(A) value is fairly constant with various dosages of pozzolanic material. The overall reaction rate for pristine Parmix silica is presented in Equation 5.

$$k_{Granulated\ silicafume}^{SiO_2,Ca(OH)_2} = \gamma * (120 * 10^6)m * e^{\frac{-67kJ/mol}{RT}} \qquad \textbf{\textit{Eq. 5}}$$

Where k = rate of pozzolanic reaction (mol/s), γ = correlation coefficient of heat flow and reaction propagation (mol^{-1}), m = mass of the pozzolanic material (g), R = Gas constant (J/(K·mol)) and T = temperature (K).

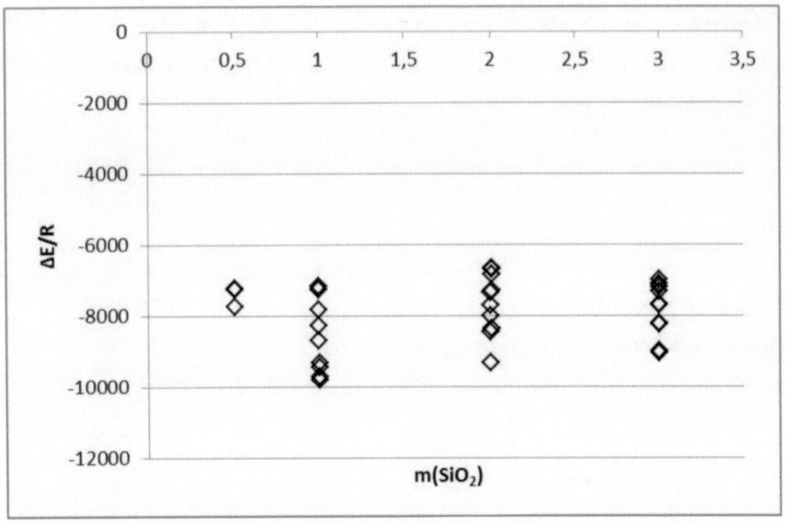

Figure 3: *Activation energies and the amount of granulated silica fume (Table 1) in the sample.*

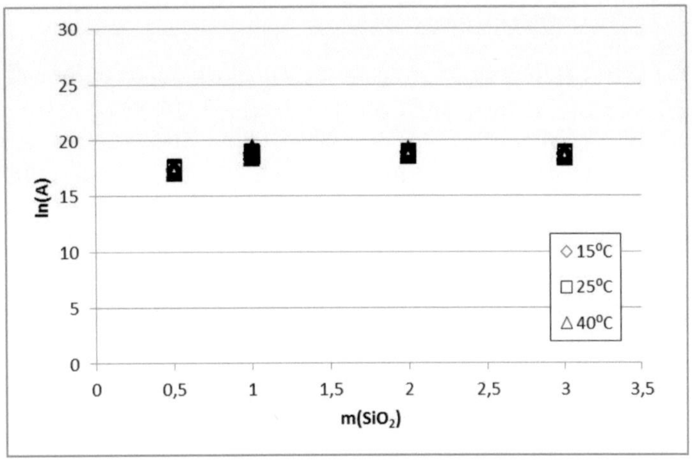

Figure 4: *Frequency factor and the amount of granulated silica fume (Table 1) in the sample.*

Conclusions and Future work

Arrhenius equation was proven suitable to model kinetics of pozzolanic reaction and transformations of calcium-silicate-hydrates. The parameters for chemical reaction and the physical factors were successfully identified. The activation energy for pozzolanic reaction of portlandite and amorphous silicon dioxide was 67 ± 7 kJ/mol, which is in-line with the literature Jensen (1999).

In the future, activation energy of C-S-H transformations must be defined and the physical parameters for various pozzolanic materials must be defined. Transport properties of the low-pH concrete potentially alter the reaction kinetics significantly. Also the chemical environment of Ordinary Portland Cement differs greatly from pure equilibrium compositions of calcium-silicate-hydrates. These two factors are being incorporated into kinetic models by variable parameters that describe the effect of chemical environment and transport properties of low-pH concrete.

Acknowledgements

The research leading to these results has received funding from The European Union's European Atomic Energy Community's (Euratom) Horizon 2020 programme (NFRP-2014/2015) under grant agreement 662147 – Cebama.

References

Cau-Dit-Coumes, C., Courtois, S., Nectoux, D., Leclercq, S., Bourbon, X. (2006). Formulating a low-alkalinity, high-resistance and low-heat concrete for radioactive waste repositories. Cement and Concrete Research, 36(12), 2152-2163.

Jensen, O. and Hansen, P. (1999). Influence of temperature on autogeneous deformation and relative humidity change in hardening cement paste. Cement and Concrete Research, 29, 567-575.

Koskinen, K. (2014). Effects of Cementitious Leachates on the EBS. Posiva Report, PORIVA 2013-04.

Laidler, K. (1984). The development of the Arrhenius Equation. Journal of Chemistry, 61(6), 494-498.

Palomäki, J. and Ristimäki, L. (2012). Facility description. Posiva Working Report, 2012-66.

Reference mix design and castings for low-pH concrete for nuclear waste repositories

Tapio Vehmas[1*], Anton Schnidler[2], Mia Löija[1], Markku Leivo[1], Erika Holt[1]

[1] VTT Technical Research Centre of Finland Ltd. (FI)
[2] Auburn University, Fullbright-VTT Scholar (US)

* Corresponding author: tapio.vehmas@vtt.fi

Abstract

A reference low-pH concrete for the Cebama Project was suggested at the WP1 workshop (London, 2.11.2015). A ternary binder composition was selected to be the reference mix design. For the reference mix design, ternary binder composition of Ordinary Portland cement, silica fume and blast furnace slag was selected. The basis of the Cebama reference low-pH concrete was a Posiva's ternary mix design. Posiva's ternary mix design has been developed for nuclear waste repository deposition tunnel end plugs. Workability of Posiva's ternary mix design and Cebama reference concrete was similar. Strength development of the reference concrete was faster than Posiva's ternary mix design. In the future, additional compression strength and pH measurements will be collected as the samples mature. These results will be shared within the Cebama Project community at a later stage.

Introduction

A reference low-pH concrete for the Cebama Project was suggested at the WP1 workshop (London, 2.11.2015). A reference concrete enables data comparison between the project's various studies. A ternary binder composition was selected to be the reference mix design. Ternary binders typically consist of Ordinary Portland cement (OPC), silica fume (SF), and either fly ash (FA) or blast furnace slag (BFS). For the Cebama reference mix design, a ternary binder composition of OPC, SF and BFS was selected. BFS based low-pH concretes are typically less studied than with FA–based and the Cebama Project offered a unique opportunity to develop a comprehensive understanding of BFS based low-pH concretes. BFS is also considered to be more homogeneous material than FA and it will be longer available than FA in the future.

The basis of the Cebama reference low-pH concrete was a Posiva's ternary mix design. Posiva's ternary mix design has been developed for nuclear waste repository deposition tunnel end plugs (Leivo et al., 2014; Holt et al., 2014). This ternary concrete mixture has been successfully utilized in a full-scale concrete plug demonstration in the DOPAS Project. Posiva's ternary mix design has been successfully used to cast a full-scale concrete plug structure with dimensions approximately 6 × 4.35 × 6.35 metres with a total volume of 160 m^3. These castings were placed 420 metres underground in Posiva's ONKALO research facility. The tunnel end plug is presented in Figure 1.

Posiva's ternary mix design was optimized with numerous batching trials. Due the limitations set by the concrete plug design, the water content of the concrete was minimized. Only with the minimal water content, the Posiva's ternary mix design was able to fulfil the plug design limitations (Holt and Koho, 2016; White et al., 2016). The effective water content of 120 L/m^3 was achieved by optimizing the particle packing of the mixture. Optimal particle packing with good flowability was only possible to achieve with optimal fine particle packing and superplasticizer.

The mix design of the Cebama reference concrete was similar to Posiva's ternary mix design with modified binder composition. As the FA was replaced with BFS, some modifications were made to account for this change. The FA could be replaced with the identical amount of BFS; however, the calcium content of BFS was significantly higher and the pH of the two mix designs would be significantly different. It was decided to use identical calcium/silica ratios for the two mix designs. This will allow the research team to compare the pH values of the Cebama reference mix design and Posiva's ternary mix designs over time.

In addition to the low-pH reference concrete, a reference paste was also developed. The paste was developed to offer an aggregate free option to study as requested by some of the Cebama Project's EU partners.

Figure 1: *Deposition tunnel end plug where Posiva's ternary mix design was successfully utilized.*

Materials and Methods

Materials

The Ordinary Portland cement used was Anläggningscement (CEM I 42.5 MH/SR/LA) from Cementa Ab. Anläggningscement is highly suitable cement for massive and underground low-pH structures due its moderate heat development, low alkalinity and high sulphate resistance. Silica fume was obtained from Finnsementti Oy (originating from Elkem). The silica was in granulated form and it is easy to transport and utilize. However, extra care must be taken to ensure the dispersion of the silica in the mixtures. Blast furnace slag was also purchased from Finnsementti Oy (product name "Masuunikuona KJ400").

Quartz filler from Sibelco Nordic Oy (product name "Nilsiä") was also used in the mixture. Crushed Finnish granitic aggregates were received from Rudus Oy. The particle size distributions of the fillers and aggregates are presented in Figure 2.

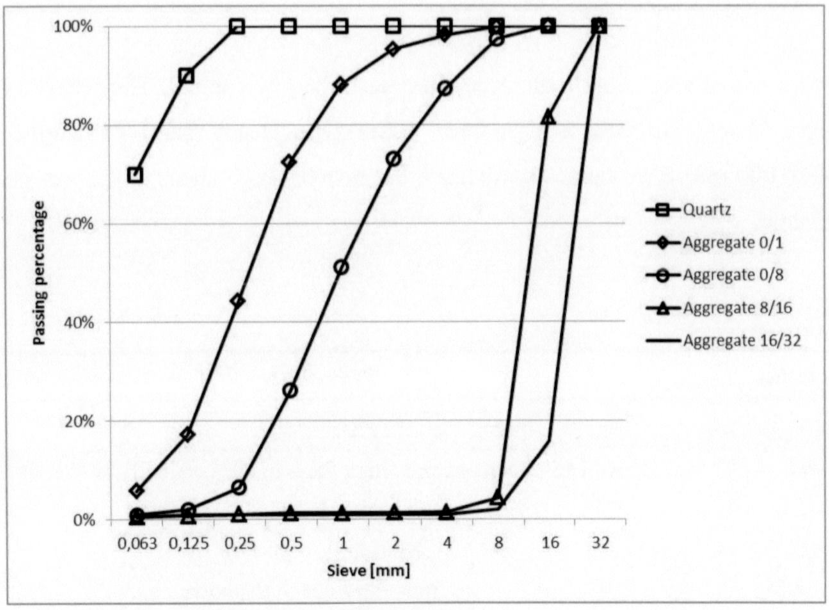

Figure 2: *Particle size distributions of quartz filler and aggregates.*

The superplasticizer used was a naphthalene-based Pantarhit LK (FM), supplied from Ha-Be Betonchemie GmbH. This specific superplasticizer was previously utilized in the full scale concrete plug demonstrations with good results. Typically utilized polycarboxylate–based superplasticizers, are not allowed in Finland's nuclear waste repository, though they have still been used by other such as SKB in Sweden Andersson et al. (2008).

Mixing procedure

Concrete batches were mixed with Zyklos –mixer (Figure 3) according to the procedure described in Table 1.

Paste samples were mixed according to the procedure described in Table 2. The planetary mixer used was a Hobart H400 (Figure 3) together with a high-shear mixer type Desoi AKM-70 (Figure 4) utilized at the maximum speed of 10 000 rpm. Two mixers were used, because the high-shear mixer was necessary to disperse the densified silica fume.

Table 1: *Concrete mixing procedure.*

Time (min)	Procedure
0 – 0.5	Add aggregate and 20% of water, mixing
0.5 – 3	Waiting time
3 – 5	Add cement, silica fume, fly ash, slag, quartz and 80% of water, mixing
5 – 7	Add plasticizer during mixing
8 – 10	Mix

Paste samples were mixed according to the procedure described in Table 2. The planetary mixer used was a Hobart H400 (Figure 3) together with a high-shear mixer type Desoi AKM-70 (Figure 4) utilized at the maximum speed of 10 000 rpm. Two mixers were used, because the high-shear mixer was necessary to disperse the densified silica fume.

Table 2: *Mixing of Cebama reference paste.*

Time (min)	Procedure
	Part 1: Planetary Mixer
0 – 0.5	Add 15 % cement and silica fume to 80% water, (low speed)
0.5 – 1.0	Add superplasticizer and 20% of water
1.0 – 1.75	Mixing (low speed)
1.75 – 2.25	Scrape the sides of the bowl
2.25 – 3.0	Mixing (low speed)
	Part 2: High-Speed Mixer
3.0 – 6.0	Mix at high speed (10 000 rpm)
	Part 3: Planetary Mixer
6.0 – 6.5	Add remaining 80% portland cement, slag and all quartz filler
6.5 – 7.0	Mixing (low speed)
7.0 – 7.5	Scrape the sides of the bowl
7.5 – 10.0	Mixing (high speed)

Figure 3: Left: *Zyklos mixer for concrete,* ***Right:*** *Hobart H400 mixer for paste.*

Figure 4: *High-speed mixer Desoi AKM-70.*

Measurements

Slump of the concrete was measured according to SFS-EN 12350-2. Air content was determined according to SFS-EN 12350-7. Compression strengths and densities of the concrete samples were measured according to EN 12390-3 and EN12390-9. Flow properties of the pastes were measured with Haegermann flow table according to DIN 1164. pH leachate from the samples were determined according to the method described in the SKB R-1202 report. pH of the leachate was determined by mixing 10 g of finely ground concrete sample with 10 mL of CO_2 -free ion exchanged water. The pH was measured with pH electrode.

Silica fume dispersion

Dispersion of the SF was measured with laser diffraction (Lecotrac LT-100). A small sample of the paste was introduced into laser diffraction equipment and three parallel measurements were performed. The measurements were performed with tap water at room temperature.

Results

Concrete mix design

The mix design of the Cebama reference concrete was based on the Posiva's ternary mix design. Posiva's ternary mix design is presented in Table 3. Aggregate and filler proportions of Posiva's ternary mix design were used as such. The binder composition was changed by replacing FA with BFS. An identical calcium/silica–ratio was maintained, as well as similar total binder and OPC content. By setting these factors constant, the proportions of SF and BFS were determined. The plasticizer content was adjusted in batching trials to get sufficient workability for the concrete. The mixture proportions for the reference concrete are presented in Table 4.

Table 3: *Posiva's ternary concrete mix design.*

Materials	kg/m^3
CEM I, 42,5	105
Silica fume	91
Fly Ash	84
Slag	-
Quartz	114
Aggregates 0/1	167
Aggregates 0/8	762
Aggregates 8/16	523
Aggregates 16/32	392
Plasticizer (from cement weight)	4.5%
Water (effective)	121

Table 4: *Cebama reference concrete mix design.*

Materials	kg/m^3
CEM I, 42.5	105
Silica fume	110
Fly Ash	-
Slag	65

Materials	kg/m^3
Quartz	116
Aggregates 0/1	168
Aggregates 0/8	770
Aggregates 8/16	532
Aggregates 16/32	396
Plasticizer (from total binder weight)	6.0%
Water (effective)	120

Paste mix design

Reference paste samples were designed keeping in mind that it very challenging to match the properties of the concrete and paste samples. Since the paste samples do not contain any aggregate, the lack of aggregate surface area will cause the paste to have much higher workability as compared to a concrete sample with the same water/binder ratio and superplasticiser dosage. Two options exist to proportion a paste from a concrete with known proportions: 1) keep the same water/binder ratio or 2) keep the same superplasticiser dosage. With option one, the superplasticizer content must be significantly reduced to prevent segregation of the samples. With a reduced superplasticizer dosage, the silica dispersion is not optimal and the porosity of the samples will be significantly higher than in the concrete samples. With option two, the water content must be significant reduced, because of the high workability imparted by the high dosage of superplasticiser. Also the SF is able to disperse completely which is the actual state of the SF in concrete mixes. With lower water content, the chemical composition of the hydration products might not correspond to the concrete samples. For low-pH concrete, option two is preferred, because it is essential to disperse the large amount of silica fume in this type of concrete.

In Table 5, compression strength results of the paste samples are presented. With the exact water content of the concrete, the paste sample had significantly lower strength and therefore inferior microstructure compared to the actual concrete samples. On the basis of the planned studies, it was decided to cast the paste samples by using Option two, which ensures that the silica fume is well dispersed. The mixture proportions of the Cebama reference paste is presented in Table 6.

Table 5: *Compression strengths of the paste samples at the age of 7 d. Water/binder ratio = 0.45 corresponds to identical water content paste and water/binder ratio = 0.25 to identical superplasticizer content.*

Water/Binder Ratio	Compression strength (7 d)
0.45	19.5 MPa
0.25	69.0 MPa

Table 6: *Cebama reference paste mix design.*

Materials	kg/m^3
CEM I, 42.5	468
Silica fume	491
Fly Ash	-
Slag	290
Quartz	517
Aggregates 0/1	-
Aggregates 0/8	-
Aggregates 8/16	-
Aggregates 16/32	-
Plasticizer (from total binder weight)	6.0%
Water (effective)	312

Silica fume dispersion

During the first paste casting the high-speed mixer was not used and low compressive strength was experienced due to poor silica dispersion. Poor dispersion of SF was confirmed with laser diffraction measurements (Figure 5). Mixing the paste only with a planetary mixer was not sufficient to disperse the SF. This problem was not encountered in concrete samples. Granulated silica fume was highly usable in concrete samples. In the paste samples, pre-dispersion of silica was necessary. Good dispersion of SF was observed when pre-dispersion of SF was performed using a high-speed mixer. However, for the superplasticizer to become effective, the addition of a small amount of OPC was necessary to increase the pH of the batch. The results of using different mixer on the SF dispersion are presented in Figure 5. Another option naturally, was to use pre-dispersed commercial silica slurries. However, that would not allow the comparison of paste and concrete samples.

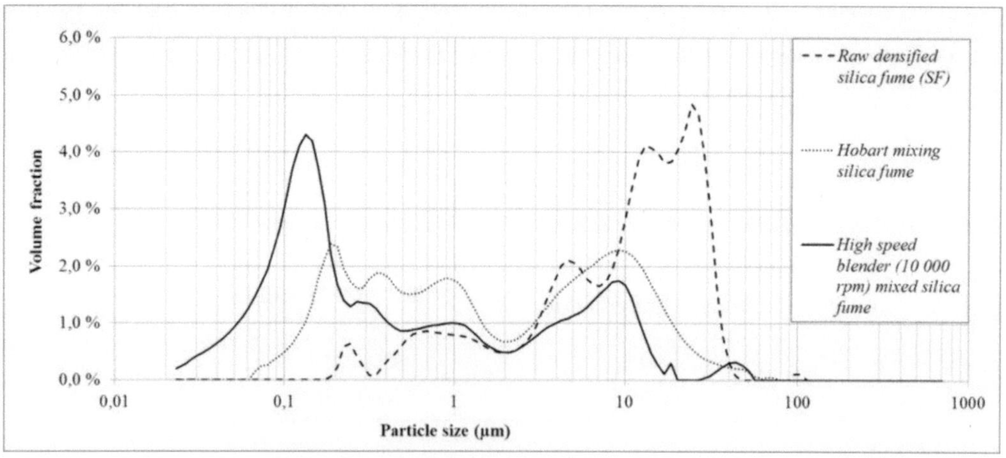

Figure 5: *Dispersion state of SF in is densified state, after planetary mixing, and after high-speed mixing.*

Fresh stage properties

Fresh stage properties of the concrete and the paste samples were measured. The slump of the concrete was 180 mm, which was similar to the Posiva's ternary mix design. Also the air content was similar. The fresh state properties of the Cebama reference concrete and Posiva's ternary mix design are presented in Table 7.

Table 7: *Fresh stage properties of the concretes.*

Concrete	Slump (mm)	Air content
Posiva's ternary mix design	120	0.9%
Cebama reference mix design	180	0.9%

Fresh stage properties of the reference paste were measured using a Haegermann flow table. Haegermann flow of the paste was 190 mm. The paste was extremely sticky, indicating high viscosity.

Hardened properties

Compression strength results of samples continuously cured in saturated Potassium hydroxide solution are presented in Figure 6. The pH of leachates is presented in Figure 7.

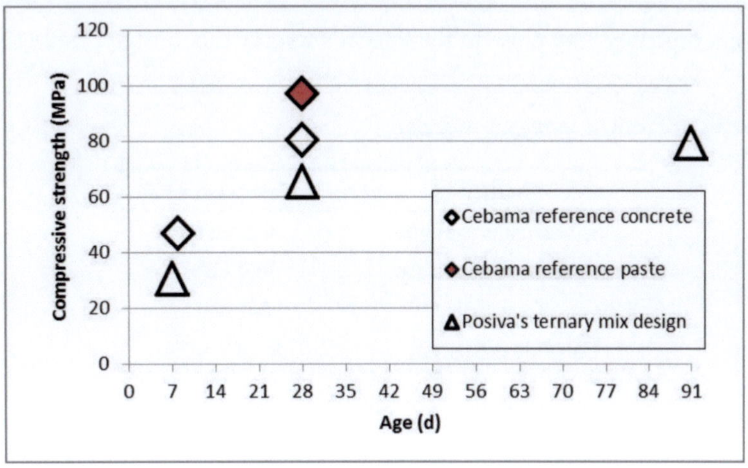

Figure 6: *Compression strengths average of three concrete and paste samples. Standard deviation of the parallel samples were within 1 MPa.*

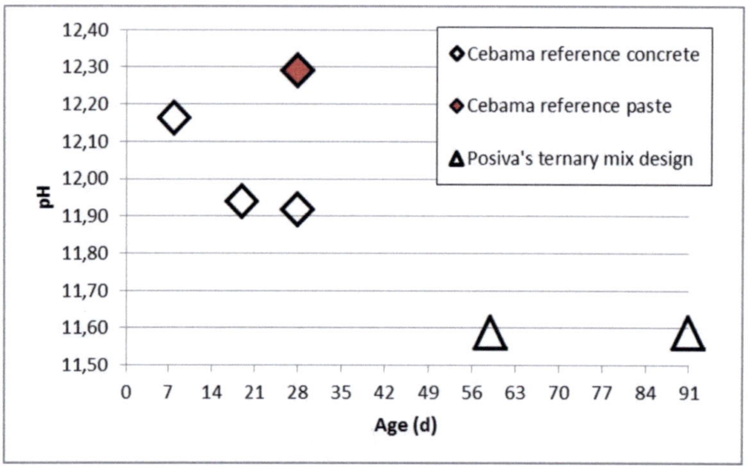

***Figure** 7: pH of leachates of concrete and paste samples in ion-exchanged water. Standard deviation of the parallel samples were within 0.05.*

Discussion

Workability of Posiva's ternary mix design and Cebama reference concrete was similar. It can be concluded that the workability of the concretes is not highly dependent on the binder composition. The workability is probably controlled by the large content of quartz filler and superplasticizer dosage. Mechanical properties of the reference concrete are probably very close to the Posiva's ternary mix design (Table 8) (Vehmas, 2015).

***Table 8:** Reported properties of Posiva's ternary mix design.*

Quality	Slump (mm)
Split tensile strength	4.5 MPa
Modulus of elasticity	34.2 GPa
Water front	5.0 mm
Chloride diffusivity	$2.8 \cdot 10^{-12}$ m^2/s

Conclusions and Future work

Reference mix designs for concrete and paste were successfully developed. It is important to vigorously mix paste samples containing high amounts of densified silica fume to make sure these particles are effectively dispersed. Fresh stage properties of the reference concrete were close to Posiva's ternary mix design, which was previously successfully utilized in full-scale deposition tunnel end plug castings. Strength development of the reference concrete was faster than that of Posiva's ternary mix design.

The compression strength and pH measurements will be continued at the future. These results will be provided to the Cebama partners.

Acknowledgement

The research leading to these results has received funding from The European Union's European Atomic Energy Community's (Euratom) Horizon 2020 programme (NFRP-2014/2015) under grant agreement 662147–Cebama.

References

Alonso, M.C., García Calvo, J.L., Walker, C., Naito, M., Petterson, S., Puigdomenech, I., Cunado, M.A., Vuorio, M., Weber, H., Ueda, H., Fujisaki, K. (2012). Development of an accurate pH measurement methodology for the pore fluids of low pH cementitious materials. SKB Report, R-12-02.

Andersson, M., Ervanne, H., Glaus, M., Holgersson, S., Karttunen, P., Laine, H., Lothenbach, B., Puigdomènech, I., Schwyn, B., Snellman, M., Ueda, H., Vuorio, M., Wieland, E., Yamamoto, T. (2008). Develompent of Methodology for Evaluation of Long-Term Safety Aspects of Organic Cement Paste Components. Posiva Work Report, 2008-28.

Holt, E. and Koho, P. (2016). DOPAS Deliverable D4.5: POPLU Experimental Summary Report.

Holt, E., Leivo, M., Vehmas, T. (2014). Low-pH concrete developed for tunnel and plugs used in nuclear waste containment. Concrete Innovation Conference 2014, CIC 2014 11-13 June 2014 Oslo, Norway.

White, Doudou, Bosgiraud, Foi, Czaikowski, Gentles, Grahm, Holt, Jobmann, Koho, Svoboda. (2016) DOPAS Deliverable D4.4: WP4 Integrated Report: Summary of Progress on, and Performance Evaluation of, Design, Construction and Monitoring of Plugs and Seals.

Leivo, M., Vehmas, T., Holt, E., (2014). Developing low pH concrete for tunnel plugging structures in nuclear waste containment. Proceedings of the Nordic Concrete Research 2014, Reykjavik, Iceland.

SFS-EN 12350-2:en (2009). Testing of fresh concrete, Part 2: Slump test.

SFS-EN 12350-7:en (2009). Testing of fresh concrete, Part 7: Pressure methods.

SFS-EN 12390-3 (2009). Testing of hardened concree, Part 3: Compression strength.

SFS-EN 12390-8 Testing of hardened concrete Part 8: depth of penetration of water under pressure.

Vehmas, T. and Leivo, M. (2015). POPLU Concrete mix design and performance study. VTT Research Report, VTT-R-0057-15.

PET/CT during degradation processes at the cement-clay interface and derivation of process parameters

Johannes Kulenkampff[1*], Urs Mäder[2], Marion Gründig[1], Sebastian Eichelbaum[3], Johanna Lippmann-Pipke[1]

[1] HZDR, Institute of Resource Ecology (DE)
[2] University of Bern, Institute of Geological Sciences (CH)
[3] Nemtics Visualization (DE)

* Corresponding author: j.kulenkampff@hzdr.de

Abstract

Degradation at the cement-clay interface is supposed to be a heterogeneous process, with mutual interactions between structural alterations and transport process properties. As complementary method to µCT, which is the standard method for structural characterization, we apply PET (positron emission tomography) as means for direct process observation by quantitative spatio-temporal visualization of radiotracer concentration. We benefit from the special features of both methods, high spatial resolution of µCT for structural imaging and high sensitivity of PET for process visualization.

We present applications of PET for observation of advectional and diffusional transport in the mobile phase in geological materials that is comparable to materials considered in CEBAMA. The characteristics of PET imaging require special algorithms for process parameter extraction from the spatiotemporal data.

The experimental setup for the planned degradation experiments at the clay-cement interface is shown and issues of the parameter derivation algorithms are discussed.

Introduction

Observation of degradation processes is complicated by the heterogeneous nature of the process. Typically, preferential transport pathways are observed in complex media, like the interface between cement and host rock, and thus dissolution and precipitation also occur preferentially along these pathways. Better understanding of such a heterogeneous process requires methods with spatio-temporal resolution for both monitoring the fate of chemical species (mobile phase) and structural alterations (stationary phase).

In the past decade, we empowered positron emission tomography (PET) for quantitative transport visualization in geological media – GeoPET (Figure 1) (Richter et al., 2000; Gründig et al., 2007; Kulenkampff et al., 2008; Kulenkampff et al., 2015). It has an unrivalled sensitivity and robustness for quantitative, non-destructive, spatiotemporal concentration measurements, $c_{PET}(x,y,z,t)$. Our latest achievements are detailed correction procedures to bring the GeoPET images into sharp focus (Kulenkampff et al., 2016a, b), and its benchmarking for geoscientific applications (Lippmann-Pipke et al., 2016). Thus, PET has become a significant method particularly for process observations related to the mobile phase. µCT adds structural information of the stationary phase (solid material).

Process understanding of geochemical transport takes advantage from the fundamental differences of these two imaging modalities. Being knowledgeable about the particularities of the two different methods allows extracting information from PET/CT overlays that is more than the sum of the two images:

- The molecular sensitivity of PET enables to operate with tracer concentrations in the picomolar range. The application of radiotracers avoids retroaction on the transport process.
- The detected amount of tracer is allocated to PET voxels with ~ 1 mm edge length. This spatial resolution is low compared to that of µCT. Nevertheless, when the resolution of µCT is underrun, voxels containing significant tracer concentration in an effective pathway will be detected as active and provide process information which is not available from µCT-imaging. However, the localization of that transport pathway underlies an uncertainty principle and thus is limited to anywhere within that voxel.
- For PET the observable c_{PET}, i.e., the number of tracer nuclides per voxel, is the product of fluid tracer concentration c_f and the local porosity n, $c_{PET} = c_f \cdot n$. So the detection limit depends on both, the concentration of the fluid, c_f, and the local porosity n.
- Sometimes gaps appear along anticipated flow lines (gaps in $c_{PET}(x,y,z,t)$ fields). They can be related to abruptly reduced and re-increased transport effective porosity. In such gap-zones c_{PET} does not exceed the noise threshold/detection limit. With e.g., c_f and the detection limit for c_{PET} known, an upper limit of the volume of the transport effective transport path, n, can be determined. This apparent pore size can even be smaller than the µCT resolution.
- We noticed in numerous case studies that PET-based process observations on the drill core scale and corresponding µCT-based simulated transport results differ frequently: The deduction of effective transport pathways from µCT imaging easily results in significant over- or underestimation (Figure 2 and Figure 3). This is because the small pore sizes below the resolution threshold of µCT generally control connectivity of the pore space and internal surface area. Therefore, models of alteration processes, acting at the fluid-solid interface, rely on a-priori assumptions about small pore classes or on higher resolving experimental methods, which then miss effects from the larger scale. This missing information shall be provided by PET process observations.

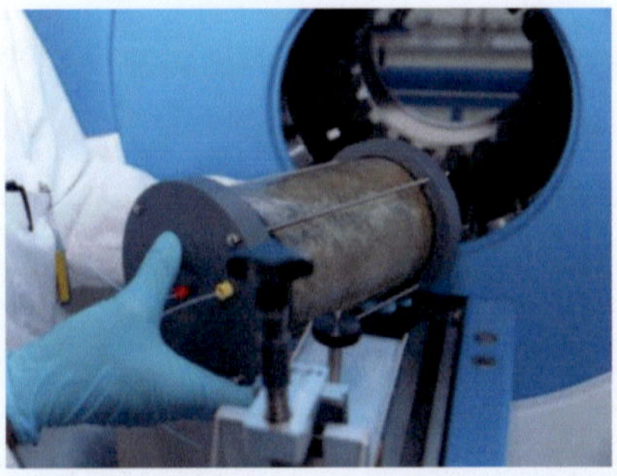

Figure 1: *Typical sample for visualization of tracer flow with GeoPET.*

Method

Examples

In the past, many GeoPET-experiments have been conducted at the HZDR in order to observe

- advection (flow) of conservative tracers,
- transport of reactive tracers and particles,
- diffusion

in fractured and porous materials. An overview of such experiments has been published in (Kulenkampff et al., 2016a).

As examples we show:

- in Figures 2 and 3: flow of brine, labelled with ^{18}F, through a fractured halite drill core with a diameter of 100 mm,
- in Figure 4: diffusion of ^{22}Na in synthetic Opalinus Clay pore water from an axial drill hole in a Opalinus Clay drill core with a diameter of 100 mm (Kulenkampff et al., 2015; Kulenkampff et al., 2016b).

Like in Figure 2, it is obvious from the overlay of PET and CT images in Figure 3 that only a small portion (about 10%) of the voids effectively takes part in the transport process. This effective pore volume also determines the effective surface area that is exposed to the mobile portion of the fluid. We also observe a gap in the tracer propagation pattern, which is a typical pattern in regions where the tracer concentration per voxel falls below the very low detection threshold of PET. This effect is a challenge for inverse transport parameter derivation algorithms. It is comparable to the effect of the limited, yet very high-resolution of μCT, which impedes the segmentation of very small pores.

In contrast to flow processes, diffusion processes can occur in smallest pore structures that are not detectable with μCT-methods. Highest resolving modalities for structural imaging have to be applied, which are working on millimetre-sized samples and thus miss heterogeneities on the larger scale. In this case, PET has a unique potential to visualize propagating species on the relevant scale of drill cores. Figure 4 shows pure diffusion in an Opalinus drill core, without driving pressure. Apart from anisotropic effects, identifiable as evolving elliptically shaped tracer distribution there are indications for preferential propagation within a layered structure. Such effects of heterogeneous diffusion would decrease the effective volume significantly, though less pronounced than in the advection example.

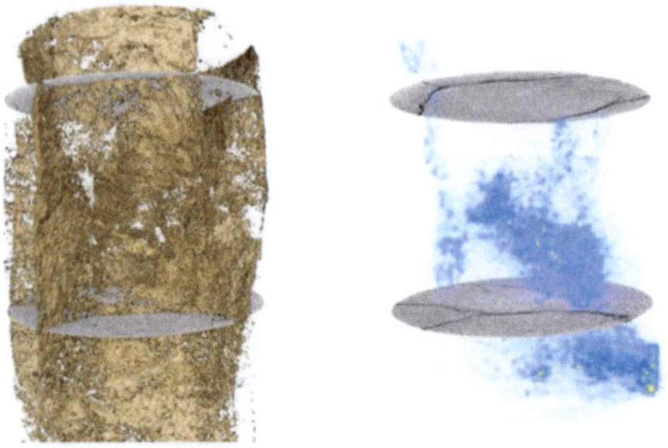

Figure 2: *Different information provided by µCT and PET in a sample showing preferential pathways.* ***Right****: fracture surface from µCT-imaging,* ***left****: tracer concentration $c_{PET}(x,y,z)$ from one PET-frame during flow experiment.*

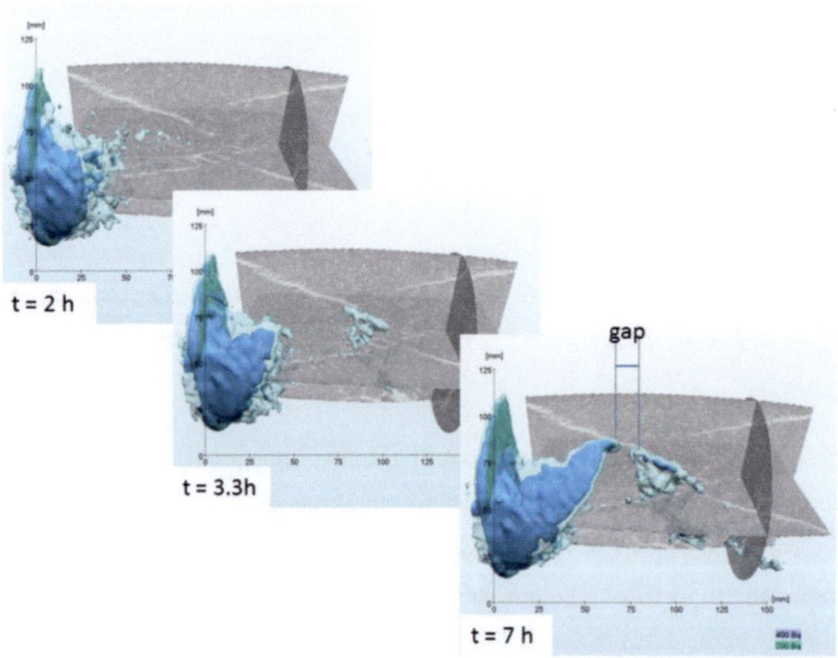

Figure 3: *Perpendicular µCT-slices (grey) of the fractured halite core from Figure 2, overlaid with three PET time frames from a flow experiment with $[^{18}F]$-salt solution. The continuous activity data is represented as iso-surfaces at 200 (turquoise), resp. 400 (blue) Bq/voxel. The gap between 60 and 70 mm persists throughout the whole time of the experiment. In this zone the activity concentration per voxel never exceeded the noise threshold: $c_{PET}(x,y,z) < c_{PET\_threshold}(x,y,z)$. This allows to determine an upper limit to the effective transport path crossing the gap, $n < c_{PET\_threshold}(x,y,z)/ c_f$. This size of this pathway can be smaller than µCT resolution.*

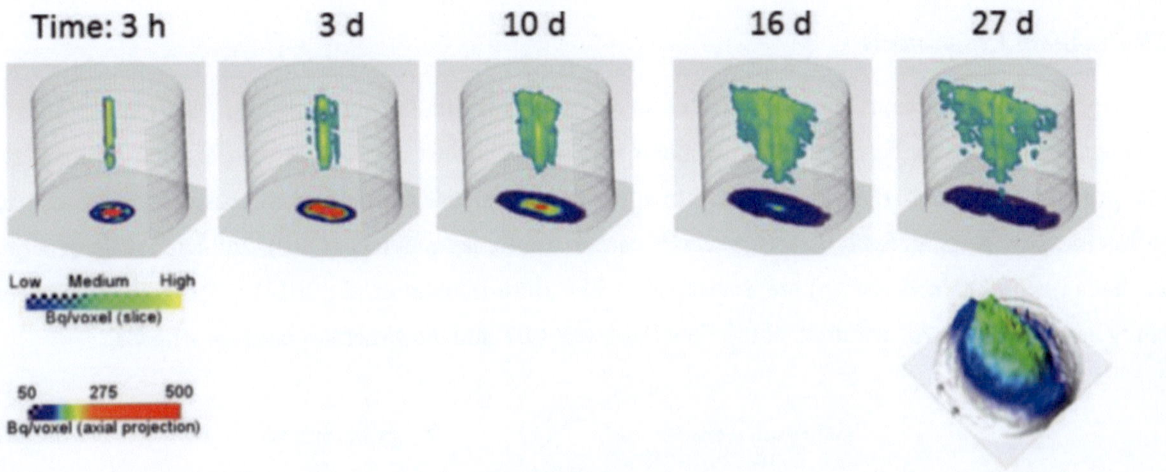

Figure 4: *PET of diffusion of ^{22}Na from a central drill hole in a 100 mm diameter Opalinus Clay core (from Kulenkampff et al,. 2015). The central drill hole was filled with synthetic Opalinus Clay pore water, labelled with ^{22}Na. Time frames: 3 h, 10 d, 16 d, 27 d after injection. The experiment continued over a period of more than 100 d.* ***Below:*** *axial projection showing anisotropy and heterogeneous effects of the last frame.*

Challenges: Parameter Derivation

In principal the effective pore volume and local velocities of the tracer transport pattern can be derived from GeoPET imaging. However, detection limits exist for any imaging modalities, also for PET and µCT. For example, considering structural imaging with µCT, detection limits are due to contrast and resolution thresholds. These thresholds delimit the validity of structural models based on µCT-imaging, because the large portion of small structures below the resolution threshold is not considered. In contrast, PET could detect tracer in smallest structures far below the spatial resolution even of µCT, as long as the mean tracer concentration within one PET voxel is above the tracer concentration detection threshold (Kulenkampff et al., 2016a). In case for PET, when we observe gaps in the tracer concentration propagation pattern (Figure 3), this could be caused both by tracer passage through very narrow, singular fissures with a high local velocity (bottle neck effect) and/or spreading of the tracer into a large rock volume with low porosity.

Currently, a parameter extraction algorithm is developed to include causality and continuity assumptions, which is based on graph theory and which spans these gaps artificially with a maximum likelihood function. The publication of this algorithm is in preparation.

Starting on such GeoPET image-based target parameter sets (velocity field and effective porosity field) it is possible to run a new type of intermediate-scale reactive transport simulation models where only equations for reactions have to be added while it runs in an unstructured (1 mm)^3 grid. We suggest these simulation models as realistic up-scaled supplementation to high-resolution pore-scale modelling based upon µCT-imaging.

Degradation Experiments

Degradation experiments are planned in close collaboration with the partners from SCK·CEN and Uni Bern. Realistic conditions for degradation experiments can be established in pressure vessels rated for simulated in-situ pressure conditions. Both PET and CT imaging requires radiation-transparent pressure vessels constructed of low-density material. Such vessels, made of carbon reinforced polymer (CRP), and applicable experimental methods (self-contained system) are developed at Uni Bern (Dolder et al., 2014) (Figure 5). This enables to study both the long-term evolution of the flow field with PET and the structural changes with CT.

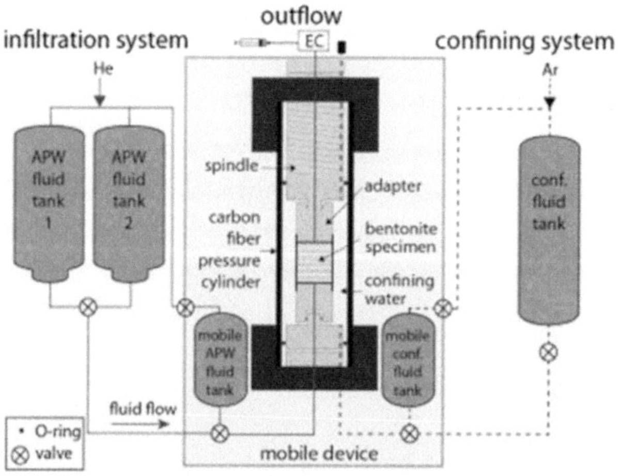

Figure 7: *Schematic of the core infiltration device consisting of an infiltration system and a confining system (from Dolder et al., 2014).*

Expected Benefits

Together with destructive post-experimental investigations of alterations these experiments will yield comprehensive experimental data sets of the degradation process.

In particular, PET will provide quantitative space and time resolved data of process variables, like tracer concentration, effective transport pathways, and residence times. These are not obtainable with other non-destructive methods.

Geochemical process simulations based on structural pore-scale models from μCT-images can be validated with the obtained experimental data. Additionally, a comparison of relevant results from both, the intermediate-scale model and from that based on the μCT geometrical structure is envisaged. Advantages of experimentally based intermediate-scaled models are reductions of computational demands and model complexity, as well as decreased challenges of scale transfer from lab observations to the relevant field scale.

Acknowledgement

This paper is produced within the project received funding from the European Union's Horizon 2020 Research and Training Programme of the European Atomic Energy Community (EURATOM) (H2020-NFRP-2014/2015) under grant agreement n° 662147 (CEBAMA).

References

Dolder, F., Mäder, U., Jenni, A., Schwendener, N. (2014). Experimental characterization of cement–bentonite interaction using core infiltration techniques and 4D computed tomography. Physics and Chemistry of the Earth, Parts A/B/C, 70-71, 104-113.

Gründig, M., Richter, M., Seese, A., Sabri, O. (2007). Tomographic radiotracer studies of the spatial distribution of heterogeneous geochemical transport processes. Applied Geochemistry, 22, 2334-2343.

Kulenkampff, J., Gründig, M., Richter, M., Enzmann, F. (2008). Evaluation of positron-emission-tomography for visualisation of migration processes in geomaterials. Physics and Chemistry of the Earth, Parts A/B/C, 33(14-16), 937-942.

Kulenkampff, J., Gründig, M., Zakhnini, A., Gerasch, R., Lippmann-Pipke, J. (2015). Process tomography of diffusion, using PET, to evaluate anisotropy and heterogeneity. Clay Minerals, 50(3), 369-375.

Kulenkampff, J., Gründig, M., Zakhnini, A., Lippmann-Pipke, J. (2016a). Geoscientific process monitoring with positron emission tomography (GeoPET). Solid Earth, 7, 1207-2015.

Kulenkampff, J., Zakhnini, A., Gründig, M., Lippmann-Pipke, J. (2016b). Quantitative experimental monitoring of molecular diffusion in clay with positron emission tomography. Solid Earth, 7, 2017-2031.

Lippmann-Pipke, J., Gerasch, R., Schikora, J., Kulenkampff, J. (2016). Benchmarking PET for geoscientific applications: 3D quantitative diffusion coefficient determination in clay rock. Computers and Geosciences (in review).

Richter, M. and Gründig, M. (2000). Positron emission tomography for studies of water flow in soil columns. In Rammlmaier, D. (Ed.), Applied mineralogy in research, economy, technology, ecology and culture. Proceedings of the Sixth International Congress on Applied Mineralogy ICAM 2000, Göttingen, Germany.

Definition of sampling and characterization of in-situ FEBEX-OPC concrete contact and design of new experiments studying surface interface reactivity

Jaime Cuevas[1*], Raúl Fernández[1], Ana Isabel Ruiz[1], Almudena Ortega[1], Javier González Yélamos[1], Daniel González Santamaría[1]

[1] UAM, Universidad Autónoma de Madrid (ES)

* Corresponding author: jaime.cuevas@uam.es

Abstract

During the first year of the Cebama project, the activities performed by the research team working at Universidad Autónoma de Madrid (UAM) have been: (1) The preparation of the state of the art document focused to concrete-bentonite interaction in a granite deep geological repository (DGR), (2) the setup of handling and testing procedures for in-situ taken samples at the FEBEX tunnel at Grimsel test site (GTS) and (3) the implementation of a small scale (laboratory temperature) bentonite-concrete experiments in order to determine surface reactivity at this interface by using several cement paste-mortars, FEBEX compacted bentonite and GTS groundwater. Cement pastes included CEM-I (high-pH), CEM-I+silica fume (low pH) and CEM-II-a-L (similar to the used for the shotcrete plug emplaced to seal the FEBEX tunnel after a first dismantling operation in 2002.

Introduction

The UAM contribution to CEBAMA project is to be developed by the achievement of two main objectives: (1) to perform the geochemical study of high pH concrete (OPC)-FEBEX-bentonite at in-situ and long-term experiments, and (2) to determine the characteristic surface interface reactivity of compacted FEBEX bentonite induced by OPC based concrete pore water (high and low pH).

Geochemistry in long-term experiments

The first objective it is covered in two aspects. First, it is being performed the characterization of aged (13 years) concrete-bentonite interfaces taken during the dismantling of the FEBEX tunnel at the GTS; and second, it is planned to study the material to be sampled in a long term (10-years) cell. Here, a concrete disc (high pH OPC) is hydrated towards a compacted bentonite column, which face a hot steel plate maintained at

100ºC. Then, a thermal gradient was induced between the hydration and the hot end. Other similar cells, referenced as HB, were studied during NF-PRO (Torres et al., 2009) and PEBS (Cuevas et al., 2014) projects. All of them were implemented by the CIEMAT group during the project NF-PRO. These experiments are being modelled for geochemical aspects by the UDC team as part of their contribution to WP3. For the GTS samples, in a first step, we handled in-situ taken interfaces (sampling performed by CIEMAT and UAM teams) by short length drillings practised when all the shotcrete was virtually excavated (May 2015). The shotcrete-bentonite contacts were reasonably well preserved and allowed us to gain experience prior to the sampling and characterization of high quality over-cored interface samples obtained by University of Bern team during February - March 2015.

Geochemistry in short-term small-scale experiments

The second objective is devoted to carry out groundwater infiltration experiments through new casted cement mortars (high pH and low pH) put in contact with compacted bentonite including a PTFE porous membrane separation (1.6 g/cm^3 dry density), both materials confined in small steel- methacrylate cells. For the design of mortars, we had the assessment of the CSIC (IETCC). The experiments were assembled to be easily dismantled, allowing the separation and study of the interface surface by means of different surface analysis techniques. The purpose of these different scale experiments is to tie short-term evidences of reactivity to that observed in long-term experiments or in-situ sampled real shotcrete-bentonite interfaces. This report contain also the contribution made by UAM to the state of the art deliverable (Vehmas and Holt, 2016), and the description of the set-up of new experiments, materials and methods to be used for the different tasks.

State-of-the-Art and defining concepts (up-scaling)

In a generic concept of DGR bentonite clay is used as hydraulic seal and physical-chemical buffer material, while concrete is used as a structural support or in the form of plug seals for constructing drifts or isolating galleries, respectively. The interactions between clay and concrete materials generate a complex and evolutionary system whose reaction pathways lead to a varied mineralogy depending on materials considered (Dauzeres et al., 2010). In crystalline rock (Neall, 2008) concrete plugs are used in order to tap the backfill and buffer clay materials. Thermal impact in the concrete plug-bentonite should be lower than in the clay rock-concrete annulus-bentonite interaction when a clay host rock is considered. The environmental conditions of concrete-bentonite interaction in granite deposits can be restricted to low-moderate temperatures 15 - 35ºC (Falth and Gatter, 2009; Mader et al., 2015). Then, the study of concrete-bentonite in granite-like repositories focus to local geochemical perturbations of the system which can be significant in describing the long-term evolution of the engineered barrier system (EBS) interfaces.

Local geochemistry in the concrete-bentonite interface surrounded by a granitic environment is driven by granite groundwater chemistry, concrete chemistry (high pH, low pH, mineralogy, ageing), and bentonite chemistry (exchangeable cations, smectite crystal-chemistry, secondary minerals d-p reactions, and porewater chemistry). The groundwater chemistry in the GTS has very low salinity (TDS < 10 mg/L) and it is composed of Na-Ca-HCO_3-Cl-SO_4; pH near 9 (Buil et al., 2010). These solutions are going to hydrate concrete and bentonite through the surface galleries contact. Then, the major chemical gradients in the EBS system are related to concrete interfaces and bentonite interfaces. In fact, bentonite porewater is saline (~ 0.3 M NaCl; pH ~ 8 type (Na-Mg-Ca-Cl-SO_4-HCO_3) in the case of FEBEX (Fernández et al., 2004), compared to GTS groundwater. Taking in mind the pH gradient (> 13 - 12 to 11 - 10 in high pH to low pH concrete) at the concrete-bentonite interface, geochemical processes, most of them related to previously observed reactions (Cuevas et al., 2014) at this interface are depicted in Figure 1.

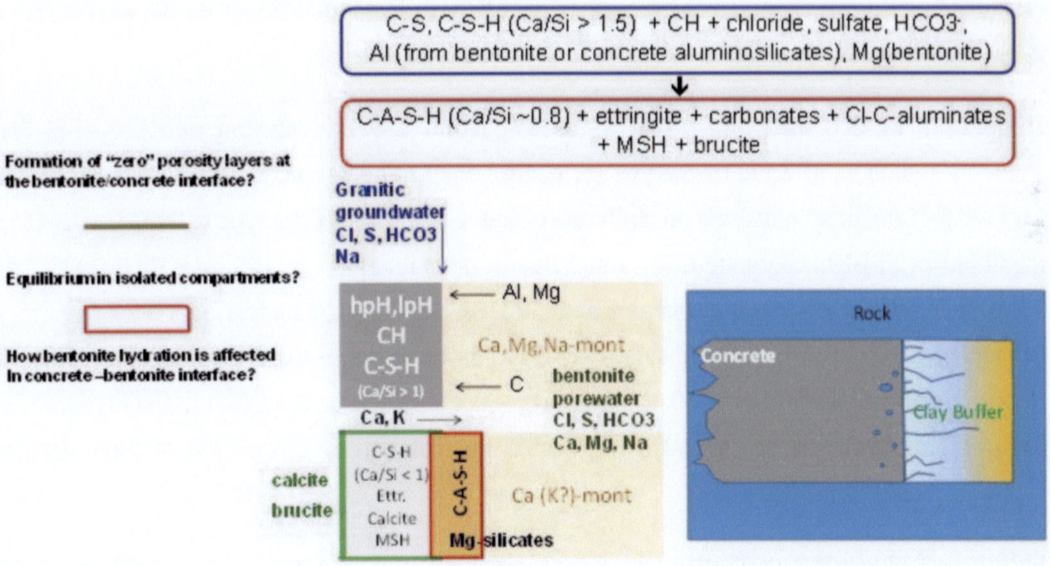

Figure 1: *Scheme of geochemical interactions in a concrete-bentonite interface within a concrete plug in a granite gallery context. hpH, lpH (high pH and low pH concrete).*

The pH gradient produces an alkaline front with the capacity to alter both the mineralogy of the bentonite and obviously of the concrete. This fact has been demonstrated by Dauzères et al. (2014), Jenni et al. (2014) and reviewed by Bildstein and Claret (2015) in clay rock contacts. Studies taking into account the effects of real volumes and composition of solutions migrating through the porosities of the involved materials, and the impacts on the concrete mineralogy and porosity are scarce. Environments dominated by $Ca(OH)_2$ (pH 12.5) or C-S-H phases (pH < 12) chemistry, with the contribution of carbonates, sulfates and chlorides interaction as have been commonly identified (i.e., Dauzeres et al., 2016). Consequently, a variety of processes can be summarized for the overall granitic groundwater bentonite-concrete scenario:

- **Carbonation of concrete and concrete-bentonite interface:** the initial high pH interface will favor the dissolution of CH (portlandite) and the precipitation of calcite rims in hydrated high pH concrete surfaces. Calcite, aragonite and gypsum have been observed to precipitate either in the concrete hydration surface or in the bentonite concrete interface in CEM-I concrete disc – FEBEX compacted bentonite experiments. Dissolution of CH and later high Ca/Si C-S-H compete with calcite formation. Moreover, Ca-Al-sulfates and carboaluminates are predicted to form as intermediates to yield finally calcite, gypsum, gibbsite and silica phases.

- **Cation exchange in montomorillonite**: high pH cements are initially a source for K and Ca during the migration of concrete porewater. Montmorillonite selective dissolution and K ion exchange can drive to an apparent illitization process, followed by precipitation of secondary phases as zeolites or C-(A)-S-H phases. Direct contact of compacted bentonite and OPC concrete at long term is characterised by the decrease of Mg and the increase of Ca concentrations both in pore waters and in the exchangeable cations population. Mg is exchanged and transferred to the alkaline media where it is precipitated as brucite (direct contact groundwater-concrete) or M-S-H phases depending on the availability of silica, released during montmorillonite or amorphous silica dissolution.
- **Dissolution of CH, development of a Ca-rich front, C-S-H evolution and C-A-S-H formation:** Concrete evolution, affected by the low pH environment of bentonite, is driven by portlandite and high Ca/Si C-S-H dissolution and the precipitation of low Ca/Si tobermorite type C-S-H. C-A-S-H have been identified in FEBEX bentonite-CEM-I and lime mortar interfaces at different time scales (months to 7 years; Cuevas et al., 2016; Fernández et al., 2016); detected at concrete interfaces forming part of a calcium rich rim in the bentonite side. C-A-S-H phases found in the experiments are consistent with narrow compositional types 0.2 - 0.3 Al/(Si+Al), ~ 0.8 Ca/Si ratios described in the literature. Moreover, ongoing structural analyses revealed a potential intercalation or association of montmorillonite and C-A-S-H phases at the pore scale.
- **Sulfate and chloride reactions in concrete**: Cement minerals can react with chloride in different ways to form chloroaluminates, Friedel's salt, AFm-like phases, or being adsorbed in C-S-H minerals (Baur et al., 2004; Yuan et al., 2012), thus, they can retain the chloride, which can be diffused from the compacted bentonite. Dissolved sulfates can react also during CH dissolution from concrete. Al dissolution from montmorillonite favour to form secondary ettringite-like (AFt) needle aggregates.

All these mentioned processes are globally in agreement with the observations regarding realistic conditions of reaction in terms of temperature (low) and transport (mainly diffusive in concrete-clay rock interfaces): carbonation of the interface, CH dissolution, low Ca/Si C-S-H formation, and ettringite precipitation. Low pH and high pH concretes exhibited different reactivity. The structure and chemistry of some of the forming phases is not precisely established in the case of M-S-H or C-A-S-H minerals. The proposed experiments or in-situ aged interfaces characterisation are focused then to upgrade our understanding of the concrete-bentonite interface dynamics and to support confidence in the predictive modelling capabilities.

Analytical methods for sample characterization: Sampling and conditioning of in situ large scale at OPC interfaces (FEBEX): (1-6 months)

Cylinder shaped samples (5 cm in diameter) were taken with a diamond coring tool at the contact of the shotcrete and compacted bentonite, when the demolition operation was approximately 10 cm before to cut the interface. The coring was performed under dry conditions by extraction of pieces not more than 4 cm in length. The core sample was fractured prior to extraction avoiding rotation of the shotcrete at the bentonite contact. Some of the interfaces were well preserved. The samples were cut with a diamond wire tool in small slices (< 5 mm), then immersed in liquid nitrogen and desiccated in vacuum (10^{-3} to 10^{-4} mm Hg). It was possible to polish the samples surface with no resin embedment in order to obtain chemical mappings and profiles. In any case, resin embedments are being performed in order to better preserve and visualize the bentonite-concrete contact. Figure 2 show an image sequence of the sample preparation and Figure 3 an example of mapping in a 3.5 cm length concrete-bentonite slice. This procedure is being actually applied to overcored interface samples obtained by the University of Bern team.

In addition to SEM-EDX measurements, bulk and detail (mm slice sampling) powder X-ray diffraction, BET N_2 surface, stable isotopes, TG-ATD and punctual ^{29}Si NMR measurements are being determined.

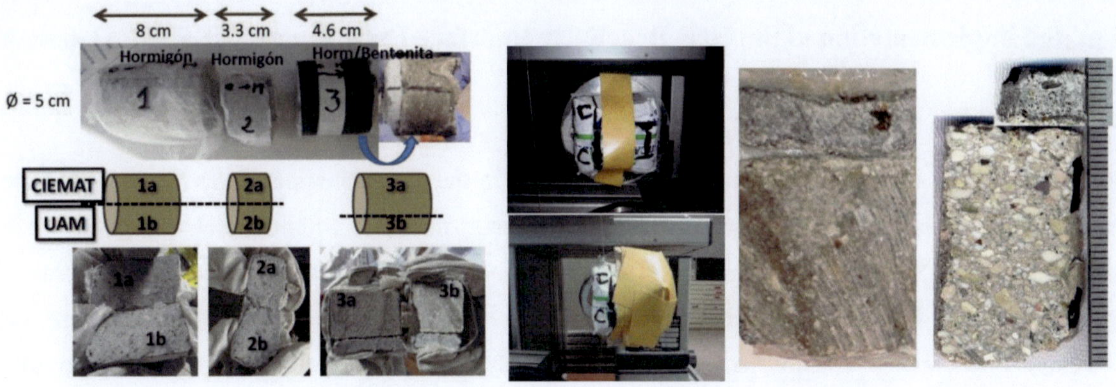

Figure 2: *Sequence of cutting and preparation of in-situ interface taken samples shared with the CIEMAT team.*

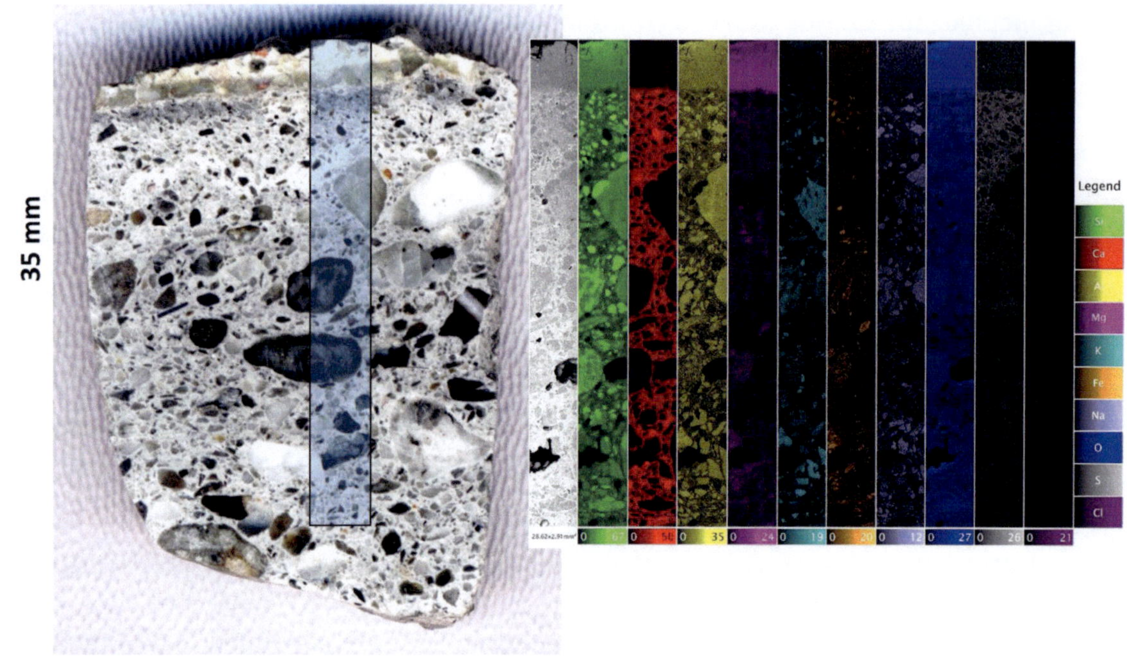

Figure 3: *Example of chemical mapping built by normalizing pixels in more than 15 SEM-EDX measure areas.*

Design and implementation of SurfacE Reactivity Interface Experiments (SERIE): (1-6 months)

Experiment design

The SERIE small scale tests have been designed to help in the discrimination of the reaction processes and its kinetics for representative conditions of concrete/bentonite interface within a real storage scenario. The overall objectives of SERIE experiments are (1) to determine the characteristic surface interface reactivity of compacted FEBEX bentonite induced by OPCs based concretes pore water resulting from Grimsel water-concrete-bentonite interaction; and (2) to conceptually Integrate chemical interactions at nm-µm scale with physical properties; i.e., measured hydraulic conductivity and porosity changes indicators. The identification of new forming phases from the first stages of alteration and the comparison of results of low pH versus conventional OPC high pH concrete, are expected to bring useful and relevant data for the repository design.

The interface concrete-bentonite (C-B) (9 mm length C and 9 mm length B; 20 mm diameter) are placed between two porous steel plates and separated by a PTFE filter membrane. Infiltration of granitic groundwater at 1 MPa through OPC (high pH) and OPC-Silica fume low pH)/ FEBEX bentonite interfaces is performed by automated control piston pumps. Percolated water will be collected and analysed for major ions chemistry. Average 60 days hydraulic conductivity of the concrete/bentonite in the actual running test is $1.5\ (\pm\ 0.5)\cdot 10^{-13}$ m/s. Figure 4 show the details of the experiment.

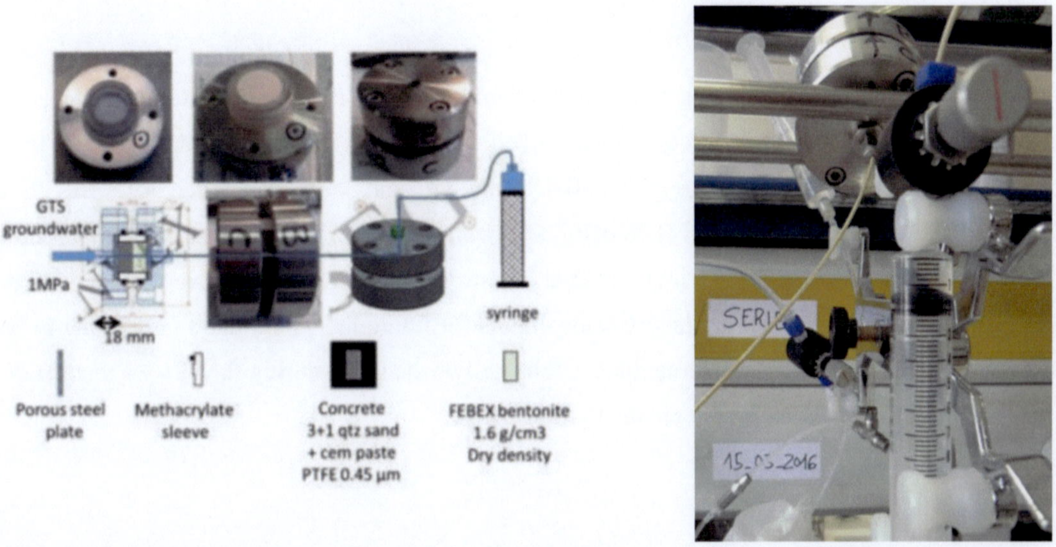

Figure 4: Scheme and details of the SERIE experiments.

Materials and initial characterization

The materials to be included in the SERIE experiments are several cement paste mortars: High pH: CEM-I; High pH: CEM II A-L SR and low pH: CEM-I (60%) + SILICA FUME (40%). They were mixed with quartz sand (< 0.5 mm grain size) in a 1+3 proportion with a water/cement paste ratio of 0.7. The bentonite is the FEBEX-bentonite grinded to < 0.5 mm, composed mainly by high charge montmorillonite and used compacted at a dry density of 1.6 g/cm^3. Grimsel groundwater (BO-ADUS Borehole); (Na-Ca-Cl-SO_4-HCO_3; pH 8 - 9; TDS < 10 mg/L) is used as aqueous solution in the infiltration experiment.

Methods and Work plan

Interface will be analysed for micro-structural, mineralogical and chemical properties at different scales (mm-µm-nm). Percolated water will be collected and analysed depending on its availability. An initially proposed plexiglass cell was substituted by a more robust steel confining design due to early breaking of the plastic cases (Figure 4). The PTFE protected flat interface is going to be analysed by means of XRD-thin film techniques (GI-XRD) to characterize µm-scale surface deposits. The whole specimen will be analysed by XRD in a detailed sampling by scraping the surfaces and cutting thin slices (1 mm). Part of the specimen will be freeze-dried in order to produce interface slice samples (either resin embedded or not) for electron microscopy observation in order to complement XRD phase characterization and to perform chemical profiling (µm scale depth). BET specific surface of mm sliced samples will be also measured. Specific geochemical modelling at Interface scale will be used to support the interpretation.

The plan schedule is to use the three cement paste materials (duplicated) running for 2 and 6 months, a total of 12 tests.

Conclusions and Future work

Main concepts revision for studying concrete–bentonite reactivity and implementation of the characterization plans have been completed. The on-going scheduled work is: (1) the characterization of in-situ FEBEX interfaces (6 - 42 months); (2), the sequential sampling and characterization of SERIE experiments (6 - 30 months); (3) the dismantling of HB6 cell (10 years), sub-sampling and characterization (12 - 38 month); and later on: (4), the conceptual modeling and data interpretation in order to achieve the integration for local to global (SERIE-HB6-in-SITU) geochemistry scale. Inputs for future model validation are going to be fulfilled in the last 38 - 46 months. However, intermediate results will be available during the 18 - 24 months of the project. There are no critical deviations for the planned project tasks

Acknowledgement

The research leading to these results has received funding from the European Union's Horizon 2020 Research and Training Programme of the European Atomic Energy Community (EURATOM) (H2020-NFRP-2014/2015) under grant agreement n° 662147 (CEBAMA).

References

Baur, I., Keller, P., Mavrocordatos, D., Wehrli, B., Johnson, C.A. (2004). Dissolution-precipitation behaviour of ettringite, monosulfate, and calcium silicate hydrate. Cement and Concrete Research, 34, 341-348.

Bildstein, O. and Claret, F. (2015). Chapter 5 - Stability of Clay Barriers Under Chemical Perturbations, In Developments in Clay Science, C. I. S. I. C. B. Christophe Tournassat B. Faiza Eds. Elsevier, 6, 155-188.

Buil, B., Gómez, P., Peña, J., Garralón, A., Turrero, M.J., Escribano, A., Sánchez, L., Durán, L.M. (2010). Modelling of bentonite-granite solutes transfer from an in situ full-scale experiment to simulate a deep geological repository (Grimsel Test Site, Switzerland). Applied Geochemistry, 25, 1797-1804.

Cuevas, J., Ruiz, A.I., Fernández, R., Torres, E., Escribano, A., Regadío, M., Turrero, M.J. (2016). Lime mortar-compacted bentonite–magnetite interfaces: An experimental study focused on the understanding of the EBS long-term performance for high-level nuclear waste isolation DGR concept. Applied Clay Science, 124-125, 79-93.

Cuevas, J., Samper, J., Turreo, M.J., Wieczorek, K. (2014). Impact of the Geochemical Evolution of Bentonite Barriers on Repository Safety Functions - PEBS Case 4. In: A. Schäfers & S. Fahland (Eds.): Proceedings International Conference on the Performance of Engineered Barriers: Backfill, Plugs and Seals. BGR, 35-42.

Dauzeres, A., Achiedo, G., Nied, D., Bernard, E., Alahrache, S., Lothenbach, B. (2016). Magnesium perturbation in low-pH concretes placed in clayey environment: solid characterizations and modeling. Cement and Concrete Research, 79, 137-150.

Dauzères, A., Le Bescop, P., Cau-Dit-Coumes, C., Brunet, F., Bourbon, X., Timonen, J., Voutilainen, M., Chomat, L., Sardini, P. (2014). On the physico-chemical evolution of low-pH and CEM I cement pastes interacting with Callovo-Oxfordian pore water under its in situ CO_2 partial pressure. Cement and Concrete Research, 58, 76-88.

Dauzères, A., Le Bescop, P., Sardini, P., Cau-Dit-Coumes, C. (2010). Physico-chemical investigation of clayey/cement-based materials interaction in the context of geological waste disposal: Experimental approach and results. Cement and Concrete Research, 40(8), 1327-1340.

Fälth, B. and Gatter, P. (2009). Mechanical and thermo-mechanical analyses of the tapered plug for plugging of deposition tunnels. A feasibility study. SKB Report, R-09-33.

Fernández, R., Ruiz, A.I., Cuevas, J. (2016). Insight into the formation of C-A-S-H phases as a consequence of the interaction between concrete or cement and bentonite. Clay Minerals (in press).

Fernández, A.M., Baeyens, B., Bradbury, M., Rivas, P. (2004). Analysis of the porewater chemical composition of a Spanish compacted bentonite used in an engineered barrier. Physics and Chemistry of the Earth, 29, 105-118.

Jenni, A., Mäder, U., Lerouge, C., Gaboreau, S., Schwyn, B. (2014). In situ interaction between different concretes and Opalinus Clay. Physics and Chemistry of the Earth, Parts A/B/C, 70-71, 71-83.

Mäder, U., Kober, F., Abplanalp, H., Baer, T., Detzner, K. (2015). Shotcrete-Bentonite Interface Sampling. FEBEX-DP Partner Meeting. Stockholm, September 15/16, 2015.

Neall, F. (2008). Safety assessment of a KBS-3H spent nuclear fuel repository at Olkiluoto Complementary evaluations of safety. SKB Report, R-08-35.

Torres, E., Escribano, A., Turrero, M.J., Martín P.L. (2009). Temporal evolution of the concrete-bentonite system under repository conditions. MRS Proceedings, 1124, 295-300.

Vehmas, T. and Holt, E. (2016). Cebama project deliverable D 1.03, WP1 Experimental studies – State of the art literature review (M09 - Feb 2016).

Yuan, Q., Shi, C., de Schutter, G., Audenaert, K., Deng, D. (2012). Chloride binding of cement-based materials subjected to external chloride environment - A review. Construction and Building Materials, 23, 1-13.

Characterisation of concrete aging due to interaction with groundwaters in contact with different engineering barrier system (EBS)

María Cruz Alonso[1*], José Luis García Calvo[1], Virtudes Flor Laguna[1]

[1] CSIC, Institute of Construction Science Eduardo Torroja (ES)

* Corresponding author: mcalonso@ietcc.csic.es

Abstract

CSIC focuses its studies on the concrete aging in contact with different EBS and the surrounding groundwater. Thus, the modifications promoted in concrete by its interaction with bentonite waters or granite and clayey groundwaters are being evaluated. Experimental studies of high and low pH concretes are under consideration. Characterisation of long-term interaction of concrete with groundwaters in real exposure conditions (concrete/Febex bentonite, 13 years, and HB6 cell, 10 years) and short-term interactions with groundwaters using aged and new concretes by means of percolation tests is underway. All this work is being carried out in close collaboration with CIEMAT and UAM.

During the first 6 months of CEBAMA project CSIC has contributed to the definition of testing methods for water-concrete interaction characterisation and of the experimental boundary conditions for experimental studies: 1) SoA on testing methods for water-concrete interaction characterization, 2) SoA on evaluation of the concrete groundwaters interactions in EBS, and 3) Definition of the materials and test methods to be evaluated by CSIC and sampling identification of Concrete/Febex bentonite.

State of the art literature review: Experimental studies

CSIC was involved in the writing of the Deliverable D 1.03 considering three aspects:

1) Chapter 2: SoA on the background on the Spanish Disposal Concept (CIEMAT, CSIC, UAM) and reviewed by ENRESA, so it is not explained in the present S+T report.
2) Chapter 3.1: SoA of testing methods for water-concrete interaction characterization.
3) Chapter 5.7: SoA of the geochemical interaction of the groundwater (GW)-concrete (C)-bentonite (B) system and its impact in the near field and EDZ. This last issue was addressed together with UAM and CIEMAT, and the CSIC. Efforts were focused on Chapter 5.7.2, dealing with SoA on concrete-groundwaters interactions in EBS.

The analysis of existing knowledge carried out by CSIC in relation to groundwater interaction with concrete has been used to identify the type of concrete leaching scenario and testing method to be employed for evaluating the concrete aging in contact with groundwater. The main considerations of the study are:

- Numerous types of leaching tests have been developed in order to verify the stability and confinement of nuclear and hazardous wastes in construction materials. While some tests methods are focused on the confinement ability of the cement paste to retain the waste, others consider the leachability of cementitious materials in aqueous media, focusing on studying the stability of hydrated solid phases to know their integrity under water contact. The appropriate selection of a testing method for characterisation of concrete performance under different waters implies to know two main aspects: the type of scenario to simulate and the reproducibility of the leaching mechanisms involved.

  Several testing protocols exist and have demonstrated their capacity; however, the classification of leaching tests is a complex matter due to the wide number of parameters that influence the leaching process. CEN/TC 292 adopted a distinction of leaching methods in relation to the level of understanding needed in practice: a) basic characterisation tests, b) compliance tests, and 3) on-site verification tests. Another classification for laboratory leaching tests follow a division into three groups based on the type of the specimen (Barth et al., 2001).

  There is no agreement between the leaching methods used concerning the regime of leachant/concrete contact but also no agreement exists in testing methodologies with the selection of leachant composition. For instance, tests using acid media as leachant agent are been employed, also deionised water, or water with chlorides, or even bubbling of CO_2 through the demineralised water, that affects the leaching of Ca (Adenot et al., 1992; Andac et al., 1999; Hidalgo et al., 2000; Alonso et al., 2006; García Calvo et al., 2010). In any event, the most important aspect is to reproduce as well as possible the boundary conditions that will exist in the real concrete site.

- The degradation processes expected in cementitious materials based on Portland cement (PC) in contact with groundwaters considering the most relevant studies related to the use of concretes in radioactive waste repositories. Most of the related studies deal with conventional Portland concretes but there are also some studies concerning low-pH cementitious materials. Although concrete is stable in high humid environments, the direct contact with water, stagnant, percolation, flowing, produces a diffusion of the pore solution and also alteration of the solid phases. Degradation of concrete due to leaching occurs when the hydrates in cementitious materials dissolve into the surrounding water and the precipitation of new phases takes place. Moreover, carbonation reactions can occur and different aggressive agents dissolved in water can attacked the existing hydrates (e.g., sulphates attack). All these phenomena can cause a loss of strength. Although the expected degradation rate is slow, its evaluation is very important for structures near field Deep Geological Repository (DGR), where extremely long-term stability is needed.

In realistic conditions of DGR, concrete durability is based on its interaction with clays and/or granite groundwaters. In general, groundwaters are mineralised solutions; however their exact composition in site must be known for evaluating their interaction with concretes. Salinity of groundwater is very variable depending on the emplacement of the repository, while in south of Europe the more expected groundwaters have low salinity, in the North it would be expected water with high saline content, around 50 g/L (AA.VV., 2010).

Regarding conventional PC materials, most of the published results refer to tests made in laboratory conditions. It is well known that the solid hydrates of cement paste are more persistent at pH above 12 - 13, but at lower pHs the hydrated phases no longer remain stable and dissolve. The pore solution of a typical Portland cement paste is highly alkaline, so that the leaching process starts by removing alkalis, followed by dissolution of portlandite and subsequently by the leaching of calcium from silicates, e.g., C–S–H (Taylor, 1997). Aluminate phases are also affected, dissolution/precipitation processes of AFm, ettringite and calcite are observed, and decreases in the C/S ratio and/or incorporation of aluminium in the C–S–H gel were observed (Faucon et al., 1997; Alonso et al., 2006). Silica gel formation has also been detected in the outer layer of cement pastes exposed to leaching in deionized water. All the reported phenomena are going to promote modifications in the porosity profiles of the used concretes. However, according to the published studies, the main detrimental reactions of concrete are taking place at small depths from the exposure surface, which is typically relatively small part of the whole structure and, especially, of the total multiple technical barriers.

Regarding low-pH cementitious materials, results from the scarce leaching tests made, show, in general, a good resistance of the low-pH cementitious materials against water aggression, although an altered front can be also observed from the surface in all the tested samples. From leaching analyses using deionised water, real groundwater (from the Äspö Site) or simulated fresh groundwater, a Ca^{2+} released in the leaching solution has been described, as well as a decalcification process governed by diffusion being the Ca^{2+} flux not only balanced by the release of OH^- but also by that of sulphate ions (Codina et al., 2008; García Calvo et al., 2010). Apart from a very low CaO/SiO_2 ratio (0.3 to 0.4 for all pastes), a disappearance of ettringite and enrichment in a hydrotalcite-like phase have been reported near the leached surfaces of low-pH cement pastes. After the test periods, low-pH concretes show a small altered front that can be observed from the surface. In this altered front, decalcification of the C-S-H gels is observed but its intensity varies in the different published studies although, in general, it seems that the degradation front is higher when clayey environments are considered and lower when aggressive granitic waters are taken into account. Therefore, it is essential to respect the in-situ conditions for understanding cement paste degradation mechanisms. This C-S-H decalcification is followed by the incorporation of magnesium ions from ground water into them (even forming M-S-H phases) and into the anhydrous phases (as “magnesia nodules”). In fact, recent results have demonstrated that Mg-perturbation is systematically observed for low-pH cementitious materials placed in clayey, or granitic environment, or in any kind of environment containing at least 3 mmol/L of magnesium in its pore solution (Dauzeres et al., 2016).

Definition of the materials to be evaluated by CSIC within CEBAMA project

The testing protocols, samples preparation requirements and testing methodology for percolation tests have been defined accordingly with analysis of literature review and existing knowledge. CSIC have identified different "systems" to be evaluated. The work conducted by CSIC will study the modifications promoted in high pH and low pH concretes by its interaction with two types of groundwaters. In this sense, CSIC will consider 8 types of systems, five of them refer to existing aged concretes and the other three will consider new concrete samples. Moreover, 2 systems refer to long-term in-situ concrete-groundwater interactions and 6 to percolation laboratory tests that will last for two years (within Cebama project). The 6 type of concretes to be evaluated are:

Existing aged concretes:

1. FEBEX shotcrete plug (high pH): aged concrete cores extracted from a full-scale experiment in the natural conditions of the Grimsel test site in contact with granitic groundwaters (BOADUS). Operation time: 13 years.
2. HB6 (high pH) concrete: from a small-scale laboratory cell experiment from CIEMAT: aged concrete exposed to clayey groundwater from the concrete face and heat (100ºC) from the bentonite. Operation time: 10 years. The dismantling of the test is expected to be performed by the end of this year (2016).
3. High pH concrete: cores from non-altered front of FEBEX shotcrete plug. 13 years concrete.
4. Low pH concrete based on PC CEM I + 40% silica fume. Cores extracted from a low-pH shotcrete plug fabricated in Äspö (Sweden). The cores were extracted in 2007 and have been cured in humidity chamber (98% RH, 20ºC) up to now. 9 years low pH concrete.

New concretes:

5. High pH concrete: with similar composition to HB6 concrete.
6. Low pH concrete: based on PC CEM I+Silica Fume+Slag, one of the Cebama benchmarks. The samples have been fabricated by VTT.

Moreover, two different groundwaters will be considered for percolation tests of GroundWater/Concrete (GW/C) interaction:

- Simulated granitic groundwater from Grimsel site (BOADUS).
- Simulated clayey groundwater.

A summary of the systems to be evaluated by CSIC is showed in Table 1.

Table 1: *Concrete-groundwater systems to be evaluated by CSIC within CEBAMA project.*

Exposure conditions	Ground water	1) Febex concrete plug	2) HB6 concrete	3) Febex concrete plug Not altered	4) Low-pH concrete Äspö plug	5) HB6 concrete new	6) low-pH concrete VTT
In-situ	Granitic	X					
	Clayey		X				
Lab Percolation test	Granitic			X	X		X
	Clayey				X	X	X

In addition to the laboratory tests, the characterisation of concretes from FEBEX concrete plug is being carried out. This concrete has been during 13 years in direct contact with bentonite. Location and definition of test characterisation of concrete damage have been already decided.

Samples to be analysed are shown in Figure 1.

Figure 1: *Concrete samples for aging characterisation of degradation.* ***Up-left)*** *FEBEX concrete 13y age bentonite-waters interaction,* ***Up-right)*** *Low pH Concrete from Äspö site (9 y age),* ***Down)*** *FEBEX concrete plug 13 y age.*

Summary and Future work planning

- The CSIC contributions in CEBAMA project during first 6 months work have allowed identifying the testing methodologies for concrete water interactions to select the most suitable method based on GW/C percolation test. Besides the analysis of existing experience with GW-C interactions and main degradation processes in function of WG and C composition has allowed selecting the concrete and GW for further evaluation of aging of concrete.
- CSIC has been also contributed to Deliverables: D1.02, D1.03 and is contributing to D1.04 and D1.05.
- Next further step will concentrate in the characterization of aging of the shotcrete plug of FEBEX due to the interaction with Grimsel granitic and bentonite waters. In addition the preparation and initiation of concrete samples for lab water interaction (percolation tests) is underway.

Acknowledgement

This paper is part of the work receiving funding from the European Union's Horizon 2020 Research and Training Programme of the European Atomic Energy Community (EURATOM) (H2020-NFRP-2014/2015) under grant agreement n° 662147 (CEBAMA). Also acknowledge to the consortium of FEBEX-DP for the sampling supply.

References

AA.VV. (2010). Geological repository systems for safe disposal of spent nuclear fuels and radioactive waste. Ed. Joonhong Ahn & Michael J. Apted, Woodhead Publishing Limited.

Adenot, F. (1992). Durabilité du béton: Caractérisation et modélisation des processus physiques et chimiques de dégradation du cément. Doctoral Theses, Univ. D´Orleans.

Andac, M. and Glasser, F.P. (1999). Long-term leaching mechanism of Portland cement- stabilized municipal solid waste fly ash in carbonated water. Cement and Concrete Research, 29, 179-186.

Alonso, C., Castellote, M., Llorente, I., Andrade, C. (2006). Ground water leaching resistance of high and ultrahigh performance concretes in relation to the testing convection regime. Cement and Concrete Research, 36, 1583-1594.

Barth, E.F., Percin, P., Arozarena, M.M. (1990). Stabilization and solidification of hazardous wastes. U.S. Environmental Protection Agency, 5, 35-46.

Codina, M., Cau-dit-Coumes, C., Le Bescop, P., Verdier, J., Ollivier, J.P. (2008). Design and characterization of low-heat and low-alkalinity cements. Cement and Concrete Research, 38, 437-448.

Dauzeres, A., Achiedo, G., Nied, D., Bernard, E., Alahrache, S., Lothenbach, B. (2016). Magnesium perturbation in low-pH concretes placed in clayey environment-solid characterizations and modelling. Cement and Concrete Research, 79, 137-150.

Faucon, P., Adenot, F., Jordá, M., Cabrilac, R. (1997). Behaviour of crystallised phases of Portland cement upon water attack. Materials and Structures, 30, 480-485.

García-Calvo, J.L., Hidalgo, A., Alonso, C., Fernández-Luco, L. (2010). Development of low-pH cementitious materials for HLRW repositories. Resistance against ground water aggression. Cement and Concrete Research, 40, 1290-1297.

Hidalgo, A., Andrade, C., Alonso, C. (2000). An accelerated leaching test to evaluate the long term behaviour of concrete in waste disposal. 5th CANMET/ACI Inter. Conf. on Durability of Concrete. Barcelona June, 373-381.

Taylor, H.F.W. (1997). Cement Chemistry. Chap 9, Concrete Chemistry, 380-383.

Physical and chemical behaviour of Low Hydration Heat/Low pH concretes

Xavier Bourbon[1], Yannick Linard[1], Nicolas Gilardi[1], Sylvie Delepine-Lesoille[1] and Isabelle Munier[1]

[1] Andra R&D Division (FR)

* Corresponding author: xavier.bourbon@andra.fr

Abstract

Tests proposed by Andra in the CEBAMA project context deal with the chemical and physical behaviour of Low Hydration Heat/Low pH Concretes. Two reference formulations, designed for plugs and seals, will be followed over several years: short term (from day to month scale): preparation, casting, hydration phase; medium term (from month to year scale): evolution in operating conditions.

This experiment is an in situ test performed in the Andra's underground research laboratory in Bure (Meuse/Haute-Marne laboratory). Two types of concretes are studied. These materials are ternary blends based on CEM I/Silica Fume/fly ash and CEM I/silica fume/blast furnace slag mixes.

The short term behaviour is assessed through the evolution of the physical properties during setting and hydration (temperature and shrinkage evolution with time). The medium term evolution is studied assessing the physical deformations of cubic meters samples.

To ensure the constant following of the physical properties, elements are monitored. Specific sensors to measure temperature, shrinkage and the water saturation evolutions with time have been used. To fulfill our knowledge of the chemical evolution in the same boundary conditions, samples will be taken time to time to assess the evolution of sound material compared to those at the interfaces with the atmosphere and with the clay host rock.

Introduction

The CEBAMA project (CEment BAsed MAterials, properties, evolution and barrier functions) started in June 2015. In this context, Andra proposed an in situ test to assess the physical and chemical behaviour of Low Hydration Heat/Low pH concretes in operating conditions at the concrete/host rock interface. Data acquisition, in situ measurements and modelling are planned over a 4 year program.

This test is performed in the French Underground Research Laboratory (URL) in Bure (Meuse/Haute-Marne, France), at 500 m depth in the Callovo-Oxfordian claystone layer, to be representative of the operating conditions of a geological disposal facility and to assess the evolution of the concrete/host rock interface.

Targeted data to be measured on two different formulations are:

- Within a short term time scale: hydration heat, evolution of the physical properties with respect to hydration (mechanical properties, shrinkage),
- Within a long time scale: physical evolution in atmospheric conditions (i.e., impact of drying on shrinkage and cracks, impact of atmospheric carbonation); chemical reactivity at concrete/host rock interface and physical consequences of the chemical reactivity with the geological medium.

The cementitious materials to be studied are "Low Hydration Heat/Low pH concretes", specifically designed for plugs and seals. To assess their physical and chemical behaviour at a representative scale, large elements have been built. To follow the evolution of their physical properties with time, the concrete elements are monitored. Temperature and geometrical evolution will be followed with time. To assess the chemical behaviour of such cementitious materials, samples will be taken to analyse both interfaces with the atmosphere and the geological medium.

In situ test design

This in situ test has two main goals:

1. To assess the short timescale physical behaviour of the low hydration heat/low pH concrete formulations at a large scale to validate a physical model including the hydration of such blended cements;
2. To assess the physical consequences of the evolution at the interface with the host rock and the interface with the atmosphere in the underground facilities.

The cementitious materials to be studied are those formulated for seals, with specifications in connection to the long term evolution of large/broad elements in contact with the host rock. Such concretes must have a low hydration heat to prevent temperature increase during setting and a low pH (pH of the hydrated cement pore water) to prevent significant chemical evolution of clay materials (the Callovo-Oxfordian claystone, as well as the swelling clay of the seal core). To fulfil these requirements, two types of "hydraulic binders" have been designed and used.

Formulations

Two formulations are studied, both recipes are dedicated to plugs and seals in the French concept design of the underground nuclear waste disposal facility. Both are based on a ternary blend with Clinker, Silica Fume and Fly Ash or Blast Furnace Slag. Compositions of the two concretes, respectively called T_{CV} and T_L are reported in Table 1.

Table 1: *Compositions of the ternary blends.*

Compounds (kg/m³)	TCV	TL
CEM I 52.5 PM-ES	140.6	76
Silica Fume	121.9	123.5
Fly Ash	112.5	-
Blast Furnace Slag	-	180.5
superplasticizer (Chryso – Optima 175)	5.63	5.70
Sand (0/2 mm – 0/4 mm)	248.6 - 581.3	253.8 - 593.5
Gravels (5/12 mm)	930.5	940.4
Total water	150	152

The first goal of this study (Thermo-Hydro-Mechanical assessment) leads to the design of a large element to be representative of a structural element of the disposal design. The second goal (chemical assessment) requires the concrete to be cast directly in contact with the host rock. Therefore this test has been designed as a wall cast on the clay host-rock in the French-URL, thick and large enough to reach a volume close to 1 m^3. As two formulations have been planned to be tested, elements have been duplicated.

<u>*Concrete walls*</u>

To assess the chemical evolution of the concrete with time, we plan to drill some samples. To prevent from disturbing the physical measurements, two more thin walls have been dedicated to that purpose. Four specimens were prepared -two concrete elements for each formulation- so that there was the opportunity to measure the physical evolution and the chemical evolution with time of the two tested formulations. Thick walls are widely instrumented (see below) and thin walls are poorly instrumented. These four elements are placed between steel sliding arches in the GAN3 gallery of the URL. The theoretical sizes of each element are reported in the following table (Table 2).

Prior to casting the concretes, the concrete liner on the gallery walls (shotcrete), has been removed and the host rock surface has been cleaned to have a mechanically stable surface. The total volume of concrete required was higher than first anticipated. A 3D-scan has been performed to give an estimation of the total volume required for each element (Figure 1, Figure 2 and Figure 3).

Thick walls have a total volume between 1.1 and 1.3 m^3 and thin walls have a volume between 0.6 and 0.7 m^3. This requires a total volume close to 2 m^3 for each formulation.

***Table 2:** Theoretical size of the tested concrete walls.*

	Height (m)	Width (m)	Thickness (m)
Thick walls	1.50	0.83	0.83
Thin walls	1.50	0.63	0.20

***Figure 1:** Experimental location in the GAN3 gallery of the concrete elements.*

***Figure 2:** 3D-Scan of the concrete elements in the GAN3 gallery.*

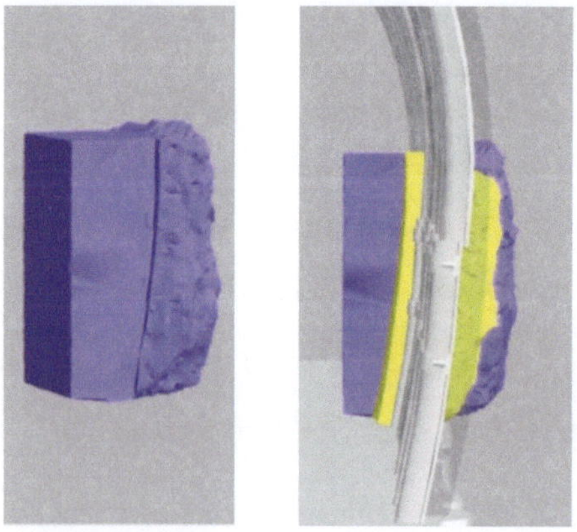

***Figure 3:** 3D-Scan of the BBP#1601 and #1602 concrete elements (T_{CV} formulation – from the gallery to the geological medium): thick wall **(left)** - thin wall vs thick wall **(right)**.*

<u>*Sensors*</u>

To monitor the physical evolution (especially temperature and shrinkage), a significant number of sensors have been placed in each element to measure the temperature evolution with time, the strain in relation to the endogenous shrinkage and the drying (Figure 4 and Figure 5).

Andra drives qualification tests on embedded sensors to be able to provide long term monitoring in its future disposal structures. For combined temperature and strain measurements, the most advanced qualified sensor is the vibrating wire extensometer (abbreviated as VWS for vibrating wire sensor), which benefits from a significant feedback (used for decades for the physical monitoring of several dams in a lot of countries). Temperature measurements are provided by a thermistor, embedded in the VWS casing.

Nineteen VWS with integrated thermometers were embedded into the concretes, to measure temperature and strain in both thick walls. In addition, two other VWS were embedded in two reference samples, cast at the same time as the walls, in semi adiabatic containers. They are used as reference to calibrate the walls' sensors.

For moisture measurements (H), sensing technologies are not as mature as sensors for Thermal and Mechanical (TM) parameters. Four TDR (Time Domain Reflectometry) sensors were selected and placed at different positions (various heights and depths) in the structures, to assess the water saturation gradients in the concretes. Samples were also taken and placed in humidity-controlled chambers. They will provide a calibration: the empirical determination of several concrete saturation levels versus raw measurement, namely the time of flight of an electromagnetic wave that propagates in the TDR sensing line.

More details on sensing technologies and qualification procedure are available from the MODERN project (MODERN, 2013).

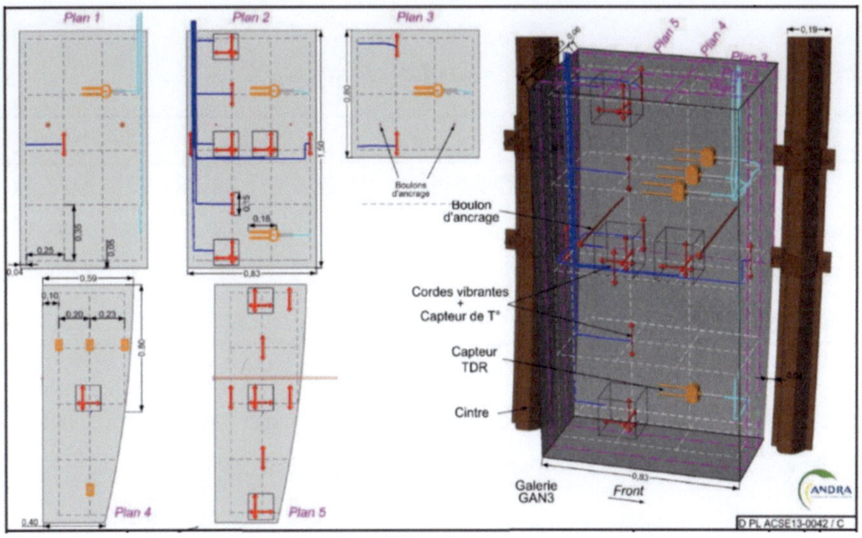

Figure 4: *Thick wall geometry and monitoring device implementation.*

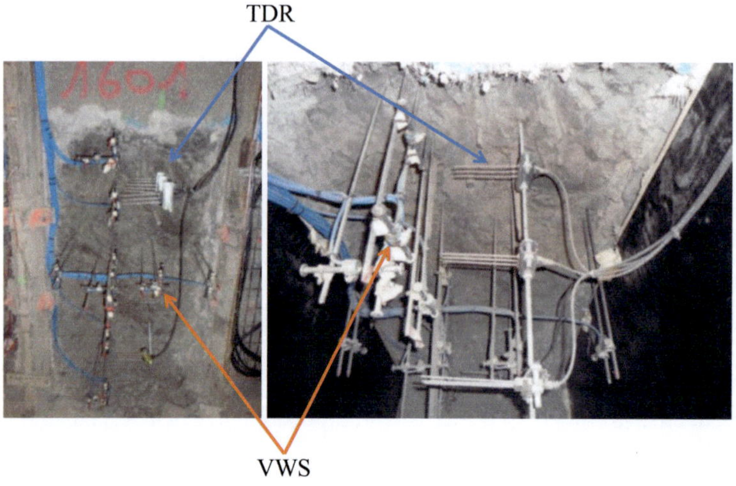

Figure 5: *Sensors location in the thick wall (thick wall #1601 - T_{CV} formulation, face view **(left)** and top view **(right)**).*

Acquired data

The four walls have been casted with industrial devices to be representative of an industrial production. Additional samples have been prepared to measure the mechanical properties of each concrete (Table 3). At present all the sensors are operating and provide information on the thermal, mechanical and hydraulic evolution of the two concretes.

Two days after casting, the maximum temperature was reached. The $T_{max.}$ was in the range 36.5/37.5°C for both concretes, at the central point of each element (maximum temperature increase ~ 16.5 ± 0.5°C in both cases). This confirms the "low hydration heat" property of these concretes.

Table 3: *Mechanical properties of the two low hydration heat/low pH concretes.*

Mechanical properties	**T_{CV}**		**T_L**	
	28 days	90 days	28 days	90 days
Compressive strength (MPa)	64 ± 4	71 ± 5	70 ± 1	80 ± 2
Tensile strength (MPa)	3.3 ± 0.5	3.7 ± 0.5	3.8 ± 0.5	3.6 ± 0.5
Young's Modulus (GPa)	41 ± 1	42 ± 2	39.5 ± 1.5	40.5 ± 1.5

From a kinetic as well as a thermal point of view, these two materials have a similar behaviour. A week has been necessary for the concrete to cool and to reach a temperature below 25°C.

A temperature gradient is highlighted by the different sensors from the bottom to the top of the wall with a maximum value at the centre of each element (between 5 to 6°C from the centre to the upper or the lower surface, see Figure 6). A smaller difference (1 to 2°C, see Figure 7) is measured between the back and the front side.

The highest thermal gradient in both cases is in the volume from a surface (upper or lower) to a depth of nearly 30/40 cm in the concrete in the vertical axe. At the maximum temperature reached one day after pouring, the gradient from the top to the centre is around 11.5°C/m and 20°C/m from the bottom to the centre of each wall. In the thickness of each wall the temperature gradient is between 7.5 to 9.0°C/m. These measurements point out the thermal isolation generated by the host rock interface, compared to the atmospheric interface.

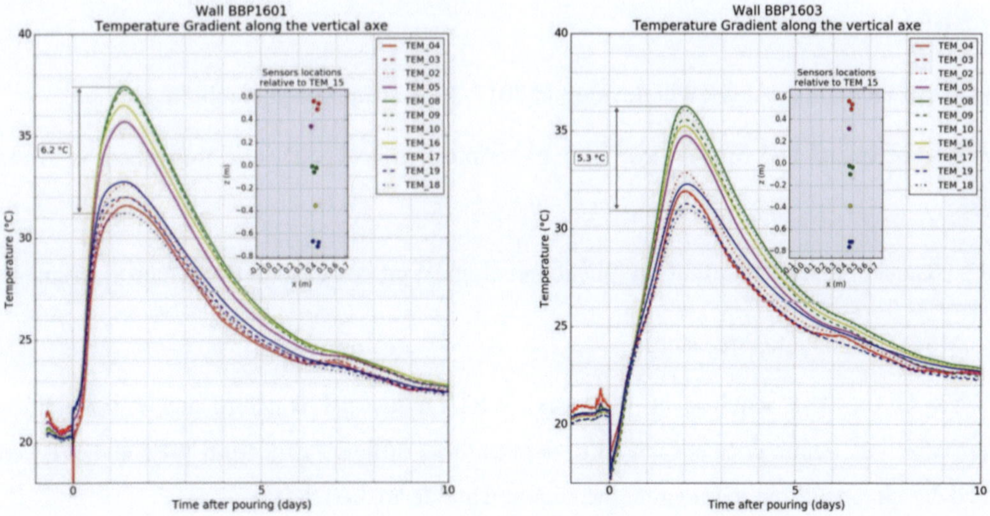

Figure 6: *Temperature evolution of the thick walls T_{CV}* ***(left****, wall BBP #1601) and T_L* ***(right****, wall BBP #1603).*

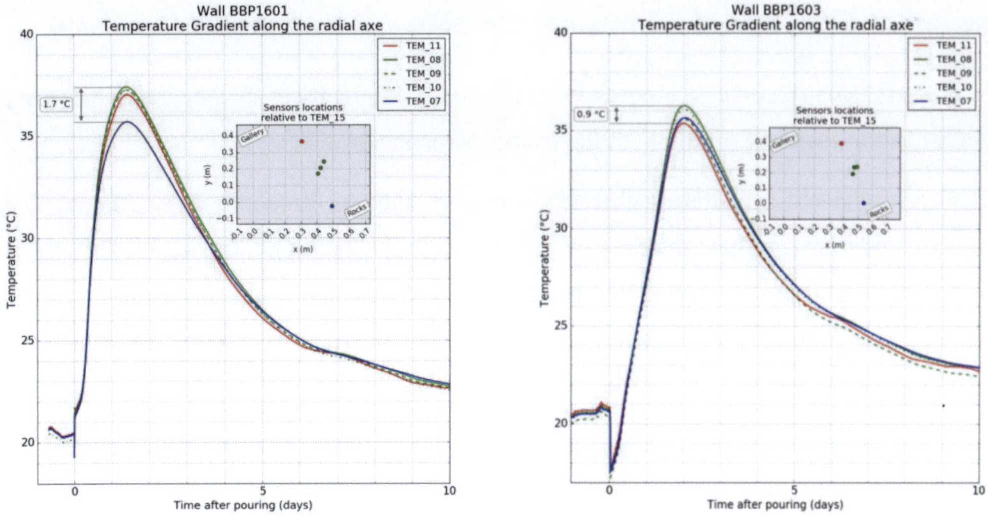

Figure 7: *Temperature evolution of the thick walls T_{CV}* ***(left****, wall BBP #1601) and T_L* ***(right****, wall BBP #1603).*

Note: the reference position sensor is the TEM-15, the VWS located at mid-height on the right side, in the thickness ('plan 2' see Figure 3) of the walls.

Perspectives

The following actions are yet to come:

- Ongoing data acquisition and assessment of the THM behaviour;
- Sampling: the first drilling to analyse the atmospheric interface (atmospheric carbonation) and the host rock interface (chemical reactivity at the host rock interface) will be done during the second part of the year 2016.

In the context of modelling, Andra will develop in 2017-2018 an approach combining:

- A change of spatial and time scales: from experiment to host rock, and from months to thousands of years;
- And a change in representation: from phenomenological processes to performance assessment (PA).

This approach is divided into three steps:

- Analysis of the data acquired in WP1 and WP2 on low-pH concrete and radionuclide behaviour characterization. Andra does not intend to deduce these different data from WP1 and WP2 experiments (model development) but to integrate the acquired and formatted data;
- Integration of the up-scaled data/model in H(M)CR modelling of intermediate long-lived waste cell and drift seal evolution (reactive-transport approach);
- Large scale modelling through a performance assessment approach (simplified chemistry).

The modelling strategy is based on a progressive increase in complexity of the integrated processes. First, the models will consider the chemical reactivity of the materials (concrete and clay).

This reactivity is rendered by different chemical processes such as dissolution/precipitation of minerals at thermodynamic equilibrium or through kinetics reactivity, ionic exchange especially for clay or surface complexation for radionuclides. The system will be considered at water saturated conditions and solute transport will be limited diffusive processes. According to the evolution of the numerical tools, the transient resaturation period will be modelled, as boundary or environmental conditions.

Main expected results are mineral evolution of the different materials, especially the low-pH concrete minerals (C-S-H, C-A-S-H, magnesium-bearing phases…). Other kinds of data will be produced such as the distribution of cations in the ionic exchanger of the swelling clay. Specific focus will be made on radionuclide speciation in cementitious environment linked with redox conditions.

As a complement to the acquired data, the reactivity between the different chemical species will be modelled with the thermodynamic database ThermoChimie v9 (Duro et al., 2012).

These modelling actions will start on the basis of WP1 and WP2 data and have not started yet.

Acknowledgement

The research leading to these results has received funding from the European Union's European Atomic Energy Community's (Euratom) Horizon 2020 Programme (NFRP-2014/2015) under grant agreement, 662147 – Cebama.

References

MODERN. 2013. “State of Art Report on Monitoring Technology” MODERN European Project n° 232598 D-n° 2.2.2.

Duro, L., Grivé, M., Giffaut, E. (2012). ThermoChimie, the ANDRA Thermodynamic Database. MRS Proceedings, 1475.

Core infiltration coupled with X-ray CT: a state-of-the-art technique for comprehensive characterisation of coupled flow and reaction in compound material systems

Urs Mäder[1*], Andreas Jenni[1]

[1] University of Bern, Institute of Geological Sciences (CH)

* Corresponding author: urs.maeder@geo.unibe.ch

Abstract

Experimental devices for rock cores held under confining pressure and with independent control on hydraulic gradient are an elegant means of probing reactive transport properties of compound systems (rocks, barrier components) at a relevant scale (µm to dm) and chemical complexity. A small component of induced advective flux allows continuous sampling of fluid (chemical evolution) and hydraulic properties, and when combined with repeat X-ray computed tomography (CT) provides a comprehensive 4D system evolution. Further, intended combination with positron emission tomography (PET) techniques also allows imaging of the mobile phase. The focus of this contribution is on equipment design and technical details, with capabilities of this method illustrated by two examples.

Introduction

A key issue of addressing reaction progress in slowly reacting complex geological and engineered systems by long-term experiments is the unravelling of system evolution near interfaces, both in terms of solute transport and the underlying progressive alteration of material components. Usually, a compromise has to be made between specimen size (spatial resolution), acceleration of transport (chemical resolution, experimental duration), a focus on solid phase characterisation (e.g., X-ray computed tomography, CT) or on recording the mobile phase (e.g., neutron scattering, positron emission tomography). Most detailed characterisations of the solid phase rely on post-mortem analysis, while the internal evolution of the pore fluid largely remains unresolved. Recent developments in synchrotron techniques also allows highly resolved bulk chemical (µXRF, X-ray fluorescence) or mineralogical information (µXRD, X-ray diffraction) to be acquired in tomographic mode, but with limits on sample size and experimental duration.

Upscaling of results from small-scale and short-term experiments is a major issue in transferring system understanding and quantitative modelling of system evolution to the scale of barrier components or an entire repository and to long time scales of 1 000's to 100 000's of years. It is in this context that intermediate-scale laboratory experiments play an important role in bridging scales, by providing relatively highly-resolved information but in a set-up of sufficient complexity and at relevant boundary conditions. The challenges are manifold with respect to equipment design and analytical techniques, and this is detailed and illustrated below.

Equipment design

Types of equipment: triaxial, column, diffusion

Three types of experimental equipment for mechanical behaviour and reactive transport cover a large breath of relevant boundary conditions and measurement parameters. (1) Triaxial testing equipment is widely used in rock and soil deformation studies whereby a hydrostatic confining pressure (total stress) is imposed on a cylindrical sample, and axial/radial displacement and load are controlled or measured. Such or similar equipment had been combined with X-ray CT (computed tomography) at a relatively early stage, as reviewed and illustrated in a collection of papers edited by Mees et al. (2003). (2) Likewise, simple low-pressure flow-through column experiments with relatively porous rock/soil or granular materials are used to study transport and reactive transport, and the set-up is amenable to combining with CT. (3) Through-diffusion cells are a third type of apparatus used in the study of solute transport by diffusion and retardation, mainly for low-permeability materials such as claystone, compacted bentonite or cement paste/mortar. Combination with CT is usually not desired because the evolution of the stationary phase is most often not a prime target of study. Hybrid equipment derived from the above three types (triaxial, column, diffusion) include notably, rock deformation with separate control of fluid pore pressure or fluid flow, high-pressure flow-through column experiments with low-permeability materials under confining pressure, or diffusion-type cells coupled to load measurement for swelling materials.

Core infiltration apparatus

The robust steel design of triaxial cells for high pressures (rock/soil deformation) was adopted but with a fixed axial length and omitting load measurement (Figure 1). The result is a column reactor that allows the application of a hydraulic confining pressure. The cylindrical sample is isolated from the confining medium (water) by a Teflon membrane (chemical isolation) and a latex rubber sleeve (hydraulic barrier). The core sample is mounted between two titanium adapters and titanium filter discs (Figure 2). The Teflon layer covers the transition to filter and adapter, and the rubber sleeve is additionally sealed with silicone to the adapter to provide a good initial seal. Shrink tubing may be applied to provide extra stability if required (samples with limited cohesion) and if thermal effects are not detrimental to the sample. The core assembly (sample + adapters) is coupled to inserts and mounted in a pressure vessel. A spindle is used to adjust to the length of the core assembly and provides the axial confinement (fixed length). The system is self-sealing once the confining pressure is stepped up. Pressure is quasi-isostatic, with the axial load exerted by the adapters somewhat reduced compared to total stress by a small area where the adapters connect to the inserts.

We built equipment in this fashion to hold 20 - 50 MPa confining pressure. The confining pressure is best generated in a water/argon separation cylinder that also provides effective compensation for changes in laboratory temperature. Such systems can maintain stable pressures for years without control.

PEEK (polyether ether ketone) or stainless steel capillary tubes of 1/16" OD (outer diameter) are fed to the surface of the filter and are sealed and held in place by compression fittings. A fluid of desired composition is injected on one side for driving advective flow, and the outflow is collected on the other side. It is best to use helium for pressurizing a supply tank and drive fluid flow in this manner at constant pressure during long-term experiments with low-permeability media. Helium is the gas of choice because of its low solubility and low pressure-dependence of solubility to avoid de-gassing phenomena. The mass flow rate is determined at each sampling time, and an accurate hydraulic conductivity can be calculated for the sampling period according to Darcy's law. An alternative means to induce flow are high-pressure syringe pumps with control on pressure and injection rate. We employ such systems when working with more permeable media (e.g., fracture flow) at lower pressures.

Sample collection at ambient pressure is easiest done by syringes, and this also provides good sample protection from atmosphere and evaporation, a requirement when using water isotopes as tracers, for example. Alternatively, back-pressure regulators or closed sampling devices that operated at a defined back-pressure can be constructed, that also allow for periodic sample extraction.

Finally, it is relatively straightforward to keep temperature constant if required by a heater/cooler system and a thermostat coil around the pressure vessel. Working at supercritical conditions requires an appropriate re-design of the sealing systems, and sampling at sufficient back-pressure, as well as choosing a non-boiling confining fluid for safety reasons.

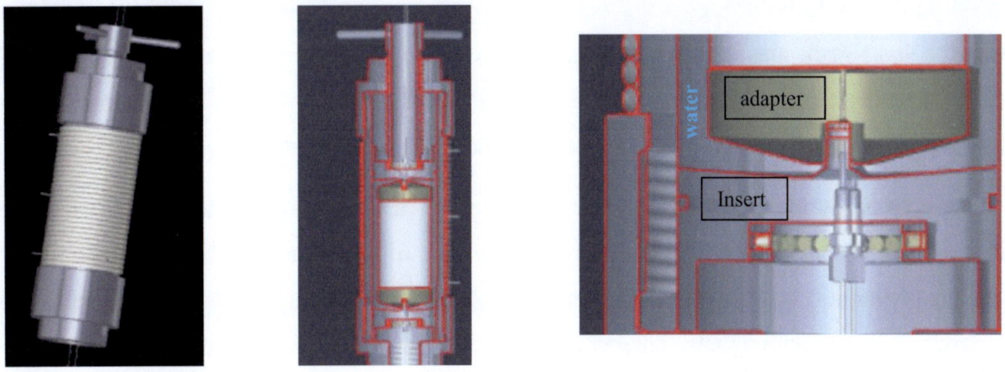

Figure 1: *Core infiltration apparatus with thermostat coil* ***(left)****, cross section* ***(middle)*** *and details* ***(right)****: sample (white), titanium adapter (grey-green metallic), steel insert (label) and fittings to feed a capillary to the filter disc located between sample and adapter.*

Equipment details, materials, in-line measurement, sample analysis

Geochemical experiments require attention to chemical compatibilities of fluid-wetted materials and isolation from the atmosphere. Appropriate choices are PEEK and titanium for tubing and filter materials. Titanium, PEEK and POM (polyoxymethylene) are good chemically inert high-strength construction materials.

Low permeability samples (10^{-12} to 10^{-14} m/s) such as claystone, compacted bentonite and many cementitious materials result in low fluxes (ca. 0.01 - 1 mL/d) even at large hydraulic gradients (> 1000 m/m) and core diameters (30 - 100 mm). The challenges are therefore to minimize dead volume in all equipment parts: flow capillaries, filters and any flow-through measurement cells. We prefer PEEK capillary tubing of 1/16" (1.6 mm) OD and either 0.5 or 0.75 mm ID. Titanium filter discs of up to 50 mm diameter can be manufactured at a thickness of 0.5 mm (Figure 2), but still form a major contribution to the total dead volume. Thinner porous polymer filters commonly may not support the high loads exerted at confining pressures above ca. 3 MPa. An alternative is to use three small and thin titanium filters embedded in a disc of non-porous PTFE (polytetrafluoroethylene) – the titanium frits support the load and function as fluid pathways, and the PTFE reduces dead volume. Thin fibre-glass filter discs are also an alternative used in some rock squeezing equipment, but attention has to be made to contaminants possibly released from binders. Very stiff rocks (granite) do not require filters and fluid flow and fluid distribution can be facilitated by groove patterns on the surfaces of the adapters.

Electric conductivity (EC) can be measured in-line with very small-volume flow-through cells such as used in ion chromatography equipment (Figure 2). Likewise, some small-volume flow-through cells such as used in amperometric detectors can be used for either pH or Eh measurement, using simply a reference electrode and an appropriate measuring electrode. The commonly used flat-bottom electrodes result in a thin film of fluid and thus a small internal volume in such cells. We build our own flow-through cells for small combination pH electrodes (Figure 2). These are fragile sensors and calibration should be done in the built-in state. Electrode measurement except for EC cannot be done continuously, but requires periodic calibration and electrode maintenance. Also, the sample stream is contaminated by the electrode reference electrolyte (e.g., KCl). We have also used small pH flow-through cells under back-pressure, enclosing the entire electrode set-up and exposing it to the same back pressure. An attractive option is to directly couple an ion chromatograph to the outflow system, whereby the injection loop of the IC is being filled, and periodically an injection is performed to either analyse cations or anions via column separation and conductivity detection.

Normal O-ring fitted syringes are appropriate for sampling for most applications. The water loss due to evaporation during sample collection is negligible for a duration of a few months, but in-diffusion of oxygen is an issue and its prevention would require a protected-atmosphere system. It is also possible to build an automated fraction collector with a many-position motor-driven valve to which syringes are attached, with these being sequentially filled.

Figure 2: From left: *electric conductivity cell; pH cell; sample with Ti filter (grey) and PTFE annulus; core assembly with sample (Opalinus Clay) between filters and Ti adapters; sample assembly wrapped in Teflon tape.*

Sample analysis for liquid aliquots needs to be optimized for small samples of only 0.5 - 2 mL. We optimized ion chromatography, titration (alkalinity), pH measurement, carbon analysis (TC/total carbon, DOC/dissolved organic carbon, low-molecular weight organic acids), stable water isotopes and select minor components by ICP-OES (inductively coupled plasma optical emission spectrometry), AAS (atomic absorption spectroscopy) and photometry. Post mortem sample characterisation includes methods for clay analysis (XRD for identity, CEC/cation exchange capacity, cation selectivity), mineralogical analysis (XRD), physical parameters (water content, densities, porosities), and leaching methods for certain non-reactive pore water components. An example of such work is included in Mazurek et al. (2012) for clay-rich rocks.

It is thus possible to extract information on the in-situ pore water and isotopic composition, transport properties of water vs. anions, as well as ion-exchange properties and controls by mineral dissolution / precipitation (carbonates, sulphates) from a single long-term experiment. The key is a well preserved drill core sample that has not been exposed to atmosphere or to drying.

Coupling to X-ray tomography

Coupling to X-ray CT aims primarily to better examine reactive systems where notable changes in mineralogy and porosity may be expected, and this can be spatially resolved as a function of time/reaction progress. Performing X-ray CT during a running advective transport experiment either requires connections of fluid/pressure lines to the apparatus at the CT location, or a self-sustained unit integrated in the apparatus that maintains confining pressure, infiltration pressure and sampling. While the former option is feasible for medical CT equipment (no rotation of the set-up) or if stage rotation is not continuous, the latter is more versatile but requires some kind of docking station where the experiment is sustained in between CT scans.

Literature contains a large range and size of different implementations for combining experimental equipment with X-ray tomography (e.g., Mees et al., 2003; Fusseis et al., 2014). Materials that are reasonably transparent to X-ray and stable under radiation include glass, titanium and aluminium (if sufficient X-ray energy is available). More transparent are polymer plastics like PEEK, POM, PMMA (poly methyl methacrylate), and a range of carbon-fibre reinforced plastics (CFK/CFRP). High-strength polymers that are adequate for high pressure equipment include CFK, PEEK (also as glass-filled or carbon-filled variety), fibre glass compounds, and also POM. A supreme material of choice is CFK combining excellent tensile strength and optimal X-ray transparency. It does require the use of adhesives to attach machinable parts such as flanges to a carbon-fibre cylinder. Suitable high-strength machinable polymer materials include epoxy resin glass laminates and fabric laminates.

We built a first-generation of core infiltration equipment of carbon-fibre tubes designed for up to 100 bar confining pressure and sample diameters of up to 50 mm (Dolder et al., 2014, Figure 3). The apparatus is operated on a docking station and can be detached and transported on a mobile rack and placed on a medical CT. Small liquid/gas tanks sustain the confining pressure and the infiltration pressure for several hours. Also, the sampling syringe remains attached to the outflow capillary at all times. We opted for a metal-free design of the entire pressure vessel. The outer shell was made of CFK with fabric orientations optimized to withstand both hoop stress and axial tension, and an inner sleeve of PE for water tightness. Threaded flanges made from epoxy resin fabric laminate were bonded to the CFK cylinder. The internal parts were machined from POM, similar in design as shown in Figure 1. There is relatively little support from industry for high-pressure engineering with polymers, and few accepted standard solutions for sealing systems.

A second-generation of core infiltration equipment is presently being built and tested to fit into an industrial tomograph (Bruker SkyScan 2211 Multiscale Nanotomograph) with a much higher spatial resolution while maintaining a convenient sample diameter of 50 mm (Figure 3). This development is part of our CEBAMA project. The equipment was shortened, and small pressure containers (confining pressure, infiltration pressure, sampling) are integrated in a compact unit placed above the sample. A test with a partially saturated compacted sand/bentonite core (50 x 50 mm) showed that it is possible to image the entire sample at 35 μm voxel resolution, limited by how close the X-ray source can be moved to the sample (separated by the pressure vessel and confining water). Distinctly higher resolution is possible by imaging a sub-volume in focus mode with a different detector without changing the sample/experiment position.

A small-scale version of a similar apparatus was designed in Bern for synchrotron based tomography of rock deformation experiments by Fusseis et al. (2014) (Figure 3), including options for heating. A small pressure cell (100 bar) was fabricated from an optically pure thick-walled glass cylinder and end-pieces machined from carbon-filled PEEK were tensioned by thin aluminium-alloy rods (blue-print provided in reference). The applications of this cell are directed towards studying rock deformation at elevated temperature and at controlled fluid conditions with highest-possible spatial resolution.

A conclusion from all these design and construction efforts is that versatile equipment appropriate for different scales can be built for withstanding relevant physical and chemical boundary conditions, but attention to many small details is required. In-line electrode measurements are tedious and signal conditioning for small-volume cells and miniature electrodes remain a challenge. Coupling to CT is in principle possible for most configurations, but is machine-specific. Integration of long-term experiments with synchrotron CT is presently hampered by the procedure for obtaining beam time (single sessions, non-predictable schedule) for most facilities.

Processing of CT data is not treated here, but is an active field of research. Modern equipment with detectors with a linear resolution of 2 000 – 3 000 pixels generate a large amount of data, and data treatment has become a challenge in itself, especially when processing time series derived from repeat runs with associated image/volume registration issues. While pore-scale resolution is often referred to as motivation for high-resolution CT work, one should keep in mind that a large portion of porosity and its connectivity, and therefore also pore-scale processes still remain below detection in clay and cementitious materials.

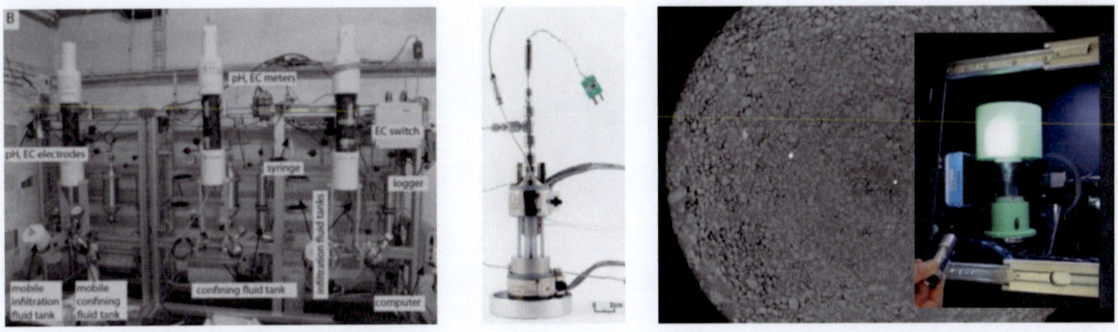

Figure 3: Left: *three core infiltration devices made of CFK and polymer materials for use in a medical tomograph.* ***Middle:*** *miniature device developed for synchrotron CT (from Fusseis et al., 2014).* ***Right:*** *CT section of compacted sand/bentonite (50 mm DM) scanned in a Bruker SkyScan211 at 35 µm voxel resolution. Sand is light, saturated bentonite grey and gas-filled porosity dark, white specs are pyrite and Fe-oxides. Insert: set-up, with short CFK tube between green equipment parts, and flat-panel detector on left (blue).*

Sample applications

A first application of long-term reactive transport in steel/titanium/PEEK equipment with Opalinus Clay and a high-pH solution (OPC pore fluid) was carried out in our laboratory during 1998 - 2000 (Adler, 2001). This was followed by extraction of pore water by advective displacement from Opalinus Clay, and determination of transport properties for water compared to anions (Mäder et al., 2004). Since then, a number of experiments with different clay stones, compacted bentonite and fractured granitic rock were carried out, with two examples detailed below. Higher-temperature experiments (90 and 140°C) were carried out with altered volcanic rock and compacted bentonite. Coupling to X-ray CT was initiated in 2010/2011.

Displacement of pore water from claystone and transport properties

Both the initial pore water composition and transport properties can be derived from a single core infiltration experiment. First, in-situ pore water is displaced and sampled, and subsequently the breakthrough of tracers injected with the artificial pore water is monitored to derive transport properties. Figure 4 (left) illustrates select data on breakthrough of water (as deuterium) and bromide to illustrate the anion exclusion effect that is pronounced in claystone (here Opalinus Clay, Mäder et al., 2004). Even small differences in ion-specific transport properties between monovalent anions can be resolved, as illustrated with chloride and bromide in an experiment with clayey rock from the Effingen Member (Figure 4, right, in Mazurek et al., 2012). Sulphate behaves not as a conservative component and is affected by dissolution/precipitation of a sulphate phase. Details on cation behaviour and some other dissolved species are described in the respective report, where data are also included and synthesized from many different experimental and analytical methods.

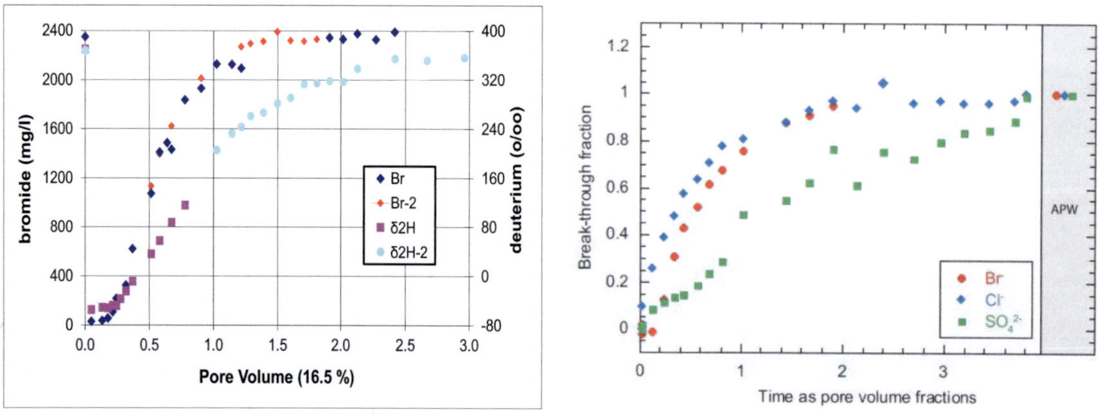

Figure 4: Left: *breakthrough of water vs. bromide in a core infiltration experiment with Opalinus Clay from Mont Terri (Mäder et al., 2004), illustrating the anion-exclusion effect. Composition of APW is plotted at time zero.* ***Right:*** *small difference in anion transport between bromide and chloride, in comparison to sulphate that is controlled by reaction with a sulphate phase. Pore volume fraction is based on total water content.*

This anion-exclusion behaviour is direct evidence for a restricted anion-specific transport porosity, and modelling the experiment as a chemically and thermodynamically consistent multi-component transport process required the implementation of anion-exclusion also into numerical reactive transport models, for example using a dual-porosity approach with an electrostatic treatment of negatively charged clay surfaces (e.g., Alt-Epping et al., 2015; Tournassat and Steefel, 2015).

Interface of compacted bentonite infiltrated by high-pH fluids

A series of experiments were performed with OPC-type high pH solution and a pore fluid typical for a low-alkali cementitious product, infiltrating compacted bentonite and sand/bentonite for extended periods of time. These core infiltration experiments were all carried out in X-ray transparent CFK equipment (Figure 3, left) and scanned periodically on a medical tomograph. A comprehensive data set of chemical evolution of the outflow, mineralogical/porosity changes imaged during the experiment and analysed post mortem, as well as volumetric response of core volume to changes in ionic strength quantified with CT allowed for a detailed discussion of the complex coupled processes involved (Dolder et al., 2014, 2016).

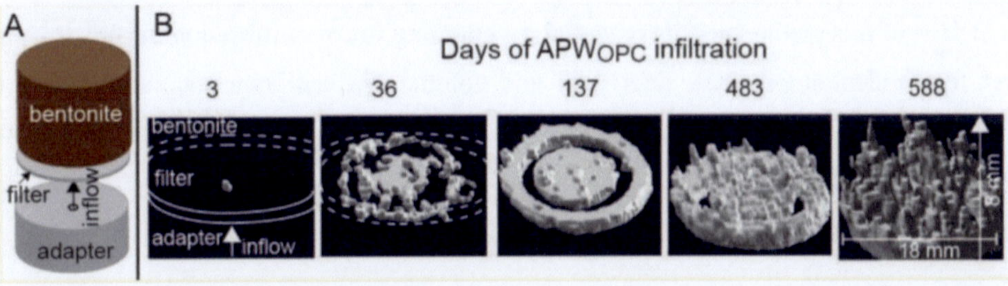

Figure 5: *Time sequence of CT images taken with a medical tomograph over nearly 600 days of the interface region between compacted MX-80 bentonite and infiltrating high-pH fluid. A growing ring-shaped complex region of porosity plugging and mineral alteration is segmented from bentonite portions that did not change significantly in X-ray absorbance. Modified from Dolder et al. (2014).*

A conclusion was (Dolder et al., 2016) that infiltration of a high-pH pore fluid after 4.5 months of percolation time led to a decrease in hydraulic conductivity from $1.1 \cdot 10^{-13}$ to $4.2 \cdot 10^{-15}$ ms^{-1}. The former advective dominated flow changed to diffusion-dominated flow with column Péclet numbers close to 1. This decrease resulted from a combination of mineral precipitation in the inlet filter and in the adjacent bentonite, as well as from an increase in ionic strength of the pore water mainly by extensive gypsum dissolution. The high-pH infiltration led to a core shrinkage of approximately 0.9 vol%, detected by CT measurements, induced by an increase in ionic strength with subsequent compaction and gypsum dissolution. The experiment attested an effective buffering capacity for bentonite and a progressing coupled hydraulic–chemical sealing process. Also, the physical integrity of the interface region was preserved in this set-up with a total-pressure boundary condition on the core sample. Constant-confining-pressure constraints may be more realistic than constant-volume for a repository environment due to the ability of the surrounding and unaltered bentonite to compensate volumetric changes at the interface.

Application to CEBAMA

Samples with cement/claystone interfaces

Existing samples from interfaces between mortar and Opalinus Clay, drilled in 2015, are foreseen for core infiltration experiments carried out under the CEBAMA work plan. These samples derive from the Mont Terri CI project, and similar interfaces from an earlier campaign had been characterized previously (Jenni et al., 2014). The interfaces of current interest are 3 - 4 years old and had been recovered in a physically and chemically undisturbed state. The plan is to use such samples for core infiltration experiments whereby the hydraulic and chemical effect of the interface can be probed.

Combination with PET (positron emission tomography)

Combination of the core infiltration method with positron emission tomography is a means to also image the transport of the mobile phase that normally cannot be observed with X-ray CT and normal pore water compositions. Details of this method and its potential for coupling to core infiltration experiments are detailed in the S+T report by Kuhlenkampf et al. (2016) in this volume. Recent improvements in imaging and data processing offer essential complementary information to methods that focus more on the stationary phase.

Relevance to deep disposal of radioactive waste

Any chemical interactions that involve materials at the cement-claystone/bentonite interfaces, and which may affect the chemical environment and may alter any of the safety-relevant properties of cement/concrete and clay-based materials must be assessed in the context of performance assessment. For a repository in Switzerland, Bradbury et al. (2014) showed that the interfaces between the concrete liner and both the Opalinus Clay and bentonite have a strong tendency to clog after a few thousand years. However, it remains difficult to predict long-term behaviour and materials properties from short-term experimental data. The main objective of the work of this project within CEBAMA is to assess the chemico-physical evolution of the Opalinus Clay adjacent to cementitious materials and to understand the impact of this evolution on safety-relevant properties of rock, including its radionuclide transport and retention properties. Over the last 10 years, experimental and modelling work (e.g., within the CI experiment at Mont Terri URL, Mäder et al., 2017) has shown that interactions at cement-clay interfaces start almost immediately upon contact, but progress very slowly. For example, experiments involving 5-year old samples show only up to few mm of reacted materials (Jenni et al., 2014). This project plans to use such aged interface samples to also more directly asses the physical and hydraulic properties via experiments using techniques as illustrated in this S+T contribution.

Acknowledgement

The research leading to these results has received funding from the European Union's Horizon 2020 Research and Training Programme of the European Atomic Energy Community (EURATOM) (H2020-NFRP-2014/2015) under grant agreement n° 662147 (CEBAMA). Anneleen Foubert and Christoph Neururer supported the X-ray CT work on the Bruker Skyscan 2211 instrument at the University of Fribourg (Swiss National Science Foundation R'EQUIP Grant n° 150731).

References

Adler, M., Mäder, U.K., Waber, H.N. (2001). Core infiltration experiment investigating high-pH alteration of low-permeability argillaceous rock at 30°C. Proceedings WRI-10 (10th International Symp. Water Rock Interaction, Villasimius, Italy). Balkema, 1299-1302.

Adler, M. (2001). Interaction of clay stone and hyperalkaline solutions at 30 °C: a combined experimental and modeling study. Dissertation, University of Bern, Switzerland.

Alt-Epping, P., Tournassat, C., Rasouli, P., Steefel, C.I., Mayer, K.U., Jenni, A., Mäder, U., Sengor, S.S., Fernández, R. (2014). Benchmark reactive transport simulations of a column experiment in compacted bentonite with multispecies diffusion and explicit treatment of electrostatic effects. Computational Geosciences, 1-16.

Bradbury, M.H., Berner, U., Curti, E., Hummel, W., Kosakowski, G., Thoene, T. (2014). The long term geochemical evolution of the nearfield of the HLW repository. Nagra Technical Report, NTB 12-01.

Dolder, F., Mäder, U., Jenni, A., Schwendener, N. (2014). Experimental characterization of cement–bentonite interaction using core infiltration techniques and 4D computed tomography. Physics and Chemistry of the Earth, Parts A/B/C, 70-71, 104-113.

Dolder, F., Mäder, U., Jenni, A., Münch, B. (2016). Alteration of MX-80 bentonite backfill material by high-pH cementitious fluids under lithostatic conditions – an experimental approach using core infiltration techniques. Geological Society, London, Special Publications, 443.

Fusseis, F., Xiao, X., Schrank, C., De Carlo, F. (2014). A brief guide to synchrotron radiation-based microtomography in (structural) geology and rock mechanics. Journal of Structural Geology, 56, 1-16.

Fusseis, F., Steeb, H., Xiao, X., Zhu, W.L., Butler, I.B., Elphick, S., Mäder, U. (2014). A low-cost X-ray-transparent experimental cell for synchrotron-based X-ray microtomography studies under geological reservoir conditions. Journal of Synchrotron Radiation, 21, 251-253.

Jenni, A., Mäder, U., Lerouge, C., Gaboreau, S., Schwyn, B. (2014). In situ interaction between different concretes and Opalinus clay. Physics and Chemistry of the Earth, Parts A/B/C, 70-71, 71-83.

Kulenkampff, J., Zakhnini, A., Gründig, M., Lippmann-Pipke, J. (2016). Quantitative experimental monitoring of molecular diffusion in clay with positron emission tomography. Solid Earth, 7, 1207-1215.

Kulenkampff, J., Mäder, U., Gründig, M., Eichelbaum, S., Lippmann-Pipke, J. (2016). PET/CT during degradation processes at the cement-clay interface and derivation of process parameters. S+T Report, 1st Annual Workshop Proceedings, CEBAMA Project.

Mäder, U.K., Waber, H.N., Gautschi, A. (2004). New method for porewater extraction from claystone and determination of transport properties with results for Opalinus Clay (Switzerland). In: R.B. Wanty & R.R. Seal II (eds), Proceedings of the 11th International Symposium on Water-Rock Interaction, WRI-11, 27 June – 2 July 2004, Saratoga Springs (N.Y.), Balkema, 445-448.

Mäder, U., Jenni, A., Lerouge, C., Gaboreau, S., Miyoshi, S., Kimura, Y., Cloet, V., Fukaya, M., Claret, F., Otake, T., Shibata, M., Lothenbach, B. (2017). 5-year chemico-physical evolution of concrete-claystone interfaces. Swiss Journal of Geosciences, special issue 110/1 (accepted).

Mazurek, M., Waber, H.N., Mäder, U.K., Gimmi, T., De Haller, A., Koroleva, M. (2012). Geochemical Synthesis for the Effingen Member in Boreholes at Oftringen, Gösgen and Küttigen. Nagra Technical Report, NTB 12-07.

Mees, F., Swennen, R., Van Geet, M., Jacobs, P. (2003). Applications of X-Ray Computed Tomography in the Geosciences. Geological Society, London, Special Publications, 215.

Tournassat, C. and Steefel, C.I. (2015). Ionic transport in nano-porous clays with consideration of electrostatic effects. Reviews in Mineralogy and Geochemistry, 80, 287-329.

Geochemical evolution of cementitious materials in contact with a clayey rock at 70°C

Alexandre Dauzeres[1*], Philippines Lalan[1,2], Laurent De Windt[2], Valéry Detilleux[3], Daniel Bartier[4], Isabelle Techer[5]

[1] IRSN, Institute for Radiological Protection and Nuclear Safety (FR)
[2] MINES ParisTech (FR)
[3] BELV (BE)
[4] CNRS UMR 7566, Université de Lorraine (FR)
[5] EA 7352 CHROME, Université de Nîmes (FR)

* Corresponding author: alexandre.dauzeres@irsn.fr

Abstract

Radioactive wastes in future clayey underground disposal sites will induce a temperature increase at the interface between the cementitious materials and the host rock. To understand the evolution of cementitious materials (Portland and low-pH cements) in this environment, an in situ specific device named CEMTEX (CEMent Temperature EXperiment) was developed in the Underground Research Laboratory in Tournemire (France). OPC cement paste was put into contact with clayey rock under water-saturated conditions at 70°C. As part of the CEBAMA project, two different parts are suggested.

Firstly a set of 3 experiments was started with OPC paste before the design of the CEBAMA project.

Secondly, inside the CEBAMA project, a new set of three experiments were launched with new formulation of low-pH cementitious materials (Mix of CEM III/silica fume and filler), based on the formulation developped by Andra in the FSS experiment by Poyet et al. (2014). The pore water chemistry was measured in laboratory on reference samples and the three tests were started in March 2016.

In 2016 and 2017, the post mortem characterizations will be focused on the OPC sample after one year of interaction with the clayey rock. These characterizations will be dedicated to the understanding of the solid chemistry and microstructure evolutions.

Introduction

In France, the decision has been taken to design a radioactive waste disposal facility in a natural clayey environment with favourable radionuclide containment properties. This solution is also being considered in Belgium and Switzerland by the waste management organisations. However, for stability reasons during the excavation work and for facilitating the installation of radwastes, galleries excavated in clayey rocks must generally be mechanically supported by concrete components. The contact between the clayey rock and the concrete inevitably leads to an alkaline plume spreading from the concrete toward the host rock, while a multi ionic attack occurs from the clayey pore water against the cementitious material side.

Two types of cementitious binders are considered today for the design of deep disposal facilities: an ordinary Portland cement binder and a low-pH binder. The latter was formulated to induce a less aggressive alkaline plume mainly due to a lower pore water pH. Over the last decades, interest has grown about the geochemical behaviour of cementitious materials in a clayey environment. Studies have focused mainly on the physico-chemical evolution of clay under alkaline conditions and not on the material interface. Only a handful of studies have discussed the interactions at the interface between cement binder and claystone. Such interfaces have been created at ambient temperature by pouring cement mixes into boreholes (Gaboreau et al., 2011; Jenni et al., 2014; Bartier et al., 2013) or by putting disks of material into contact in transport cells (Dauzeres et al., 2010). In the current design of the French radioactive waste deep disposal facility (Andra, 2005), the thermal transient, due to the presence of heat-emitting waste, is expected to entail a temperature increase of up to 70°C in the concrete plugs located in the high-level waste cells and potentially also in the intermediate long-lived waste cells. Based on the existing literature, only a few laboratory studies have focused on the impact of such a temperature level on the evolution of the interface between natural rock and anthropic material, and even fewer in situ studies have been carried out.

The CEMTEX experiment (CEMent Temperature EXperiment) aims to fill the lack of knowledge about the geochemical and micro-structural evolutions of these interfaces under representative thermo-hydraulic conditions (i.e., 70°C and full water saturation), which should provide input for future studies on the durability of cementitious materials. Within this framework, three in situ experiments have been set up in IRSN's underground research laboratory in Tournemire (France) in 2012, before the starting of the CEBAMA project and three other with low-pH cement paste were started inside the CEBAMA project in 2016. This paper presents the developed heating device, the cementitious materials, the protocols, the temperature regulation results and the planning of the post-mortem characterizations.

Experimental device

The heating system

A study prerequisite was to achieve a temperature of 70°C for a long time around an interface between the clayey rock and the cementitious material. A numerical scoping simulation was carried out to design the heating device (Lalan et al., 2016). The considered option (called the "coil heating device") was a cylinder in contact with the borehole wall and a rod located in its centre for simulating a heating coil with a cement paste poured in between.

The coil heating device model led to temperatures of approximately 70°C at the interface (close to the coil) and about 60°C in the centre. Based on this result, the coil heating device design seemed to be a good option to prescribe temperature at the interface between both materials.

The selected heating design (coil device) (Figure 1) was optimised to resist the in situ conditions and duration (up to 5 years) of the test: the inner and outer sides of the coil tubing were, respectively, in contact with water at around 70°C and a hyperalkaline environment. The choice of the coil material was therefore a critical issue. Aluminium and copper have high thermal conductivity, but they are very sensitive to hyperalkaline pH, whereas stainless steel resists to alkaline medium but is a poor thermal conductor. The compromise was to coat the copper coil with nickel, using electrochemical plating. A 10 µm thick nickel plating was deposited on the coil surface. Each spin of the coil was butt-jointed to its neighbours to form a pseudo cylinder of 18 cm. After installation, the coil was connected to an electric boiler providing hot water.

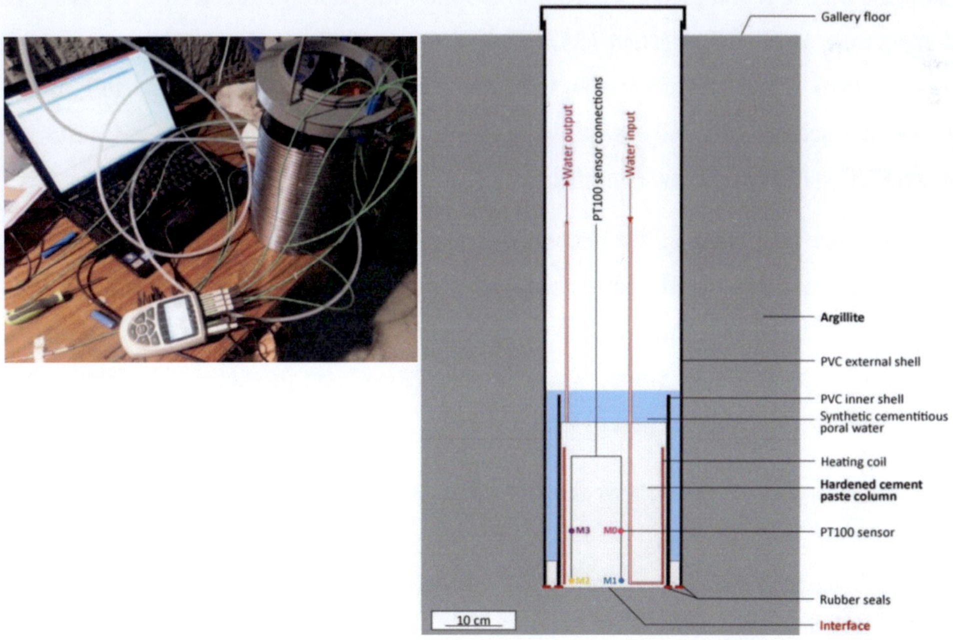

Figure 1: *View of the heating coil device connected to the temperature monitoring system* ***(left)*** *and outline of the CEMTEX in situ device* ***(right)***.

The protocol

Six downward vertical boreholes of 1 m depth and 25 cm of diameter were drilled into the argillite in the Tournemire tunnel (3 in 2012 and 3 in 2016). This depth was chosen to ensure that the devices were located out of the excavation disturbed zone (EDZ) created while the gallery was dug. The bottom of each borehole was polished with a specific tool and all the dust was carefully vacuum cleaned to obtain an as perfect as possible contact surface between the rock and the cementitious material. Three cups filled with water and humidity probes were placed into the borehole, which was then tightly closed to resaturate the rock by vapour phase. The rock was considered saturated when the measured relative humidity within the borehole reached a value of 99.5%. In each borehole, the relative humidity reached 97.5% and 99.5% after 7 and 45 days, respectively. To force a one-dimensional mass transport (vertical) across the interface and to protect the borehole sides from hyperalkaline solution, a PVC tube (25 cm outer diameter, 0.5 cm thick, 125 cm height) was placed within the borehole. A rubber seal was placed at the bottom of the PVC tubing to seal the bottom of the tube from the borehole wall.

At the same time, heater devices were built outside the boreholes. Four temperature sensors (PT 100) were attached to a single support in order to monitor the temperature during the experiment. Two sensors were close to the centre of the coil, one at 1 cm height from the future interface (M1) and the other at 20 cm (M0). The two other sensors were placed just inside the coil, again at 1 cm (M2) and 20 cm (M3). The coil top and the temperature sensor support were fixed to a PVC shell of 20 cm diameter. A rubber seal provided sealing between the argillite borehole bottom and this latter PVC shell. After the argillite saturation, the entire device was placed into the borehole and then connected to a boiler via 12 mm diameter PFA pipes (perfluoroalkoxy) and to the temperature acquisition system (ALMEMO). Once all these steps had been achieved, the cement paste was poured onto the heater device and the temperature sensors to fill the full inner PVC volume constituting a 30 cm height OPC and low-pH paste plug. A small amount of cement paste was also poured between the two PVC shells over a few centimetres.

One month after the beginning of the cement hydration, heating was started while the device was being maintained under the equilibrium solution at high pH with the considered cementitious material in order to ensure water saturation throughout the test. The water reservoir is trapped between the cement paste and a layer of Resin to limit the evaporation of water. The system is closed and protected. The Figure 2 shows a set of pictures of the 3 new tests launched as part of the CEBAMA project.

***Figure 2:** Pictures of the 3 CEBAMA tests setting up (low-pH cement paste).*

The materials

Ordinary Portland Cement paste

An Ordinary Portland Cement, Sulphate Resisting Portland Cement (OPC-SRPC) from Val d'Azergues (France, Lafarge®) was used. Lafarge provided the cement's composition. The cement paste had a water/cement ratio of 0.42. Cement paste samples were poured from the mix used for the experiment in order to characterise the initial sound material. XRD analysis of a cement paste cured for 5 months showed presence of cement hydration products, portlandite, C-S-H, Si-katoite (C_3ASH_4) and ettringite, as well as calcite (cement filler) and anhydrous cement phases. Katoite is part of a solid series between two endmembers, grossular ($Ca_3Al_2Si_3O_{12}$) and synthetic phase ($Ca_3Al_2(OH)_{12}$), also known as hydrogarnet. The connected total porosity measured on hardened OPC paste by water method was about 35% (± 1) and by autoradiography method after impregnation with ^{14}C-Polymethylmethacrylate about 30% (± 3). The difference between these two values can be easily explained: the MMA molecule cannot penetrate into the C-S-H interlayer space, entailing an underestimate of the total porosity of a cementitious material.

The low-pH cement paste

The low-pH binder formulation chosen for the 3 new tests is detailed in Poyet et al. (2014) (DOPAS European project – FSS experiment). It was chosen as formulation with the lower pH in the pore solution after 28 days of hydration (based on this study). It is a mix of CEM III Rombas (Calcia) at 20% wt, Silica Fume (Condensil) at 20% wt and 60% of calcareous filler (Carmeuse). The Super Plasticizer Glenium 537 is added (2.3% in wt). The equivalent w/b ratio (reactivity index is equal to 1 for the cement and the silica fume and 0.2 for the fillers) used is equal to 0.60. The characterization of the pore solution and the hydrates composition is in progress.

The temperature evolution

The Figure 3 illustrates the good regulation of the temperature during the first year of interaction of the third test on the OPC cement paste (duration: 5 years) and also the exothermic peak during the hardening of the cement paste at the early age. The tests is still in progress and will be dismantled for characterizations in February 2017. Whatever the location of the probe in the cement plug, the temperature is systematically in a range between 65°C and 70°C, validating the heating device.

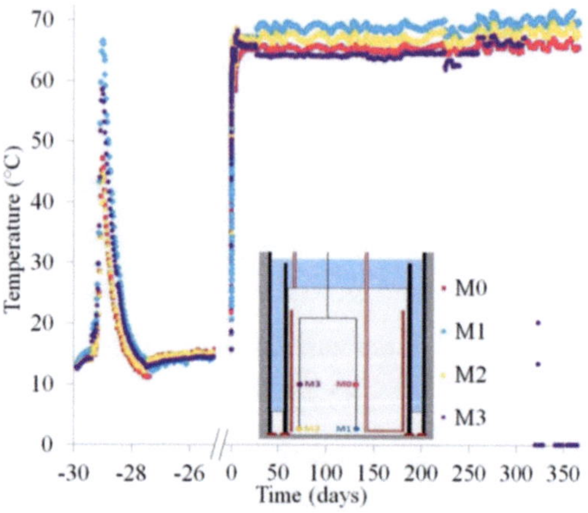

Figure 3: *Continuous thermal monitoring during the cement hydration phase and the experiment duration.*

Conclusions and Future work

Four of the six CEMTEX experiments (2 for OPC and 2 for low-pH after one and two years of interaction) are finished and 2 are still in progress (5 years of interaction). The one and two years experiments for the OPC are already finished and samples are ready for characterizations. The in situ thermal regulation on these two tests was a success. The next steps of the present study will be to characterize the low-pH sound material, the evolution of the solid chemistry and the microstructure in the cementitious materials (firstly in OPC and secondly in low-pH).

Acknowledgement

The research leading to these results has received funding from the European Union's European Atomic Energy Community's (Euratom) Horizon 2020 Programme (NFRP-2014/2015) under grant agreement, 662147 – Cebama.

References

Andra (2005). Dossier 2005 Argile - Evaluation de la faisabilité du stockage radioactif en formation argileuse. Site de Meuse/Haute-Marne, Andra.

Bartier, D., Techer, I., Dauzères, A., Boulvais, P., Blanc-Valleron, M.-M., Cabrera, J. (2013). In situ investigations and reactive transport modelling of cement paste/argillite interactions in a saturated context and outside an excavated disturbed zone. Applied Geochemistry, 31, 94-108.

Dauzères, A., Le Bescop, P., Sardini, P., Cau-Dit-Coumes, C. (2010). Physico-chemical investigation of clayey/cement-based materials interaction in the context of geological waste disposal: Experimental approach and results. Cement and Concrete Research, 40, 1327-1340.

Gaboreau, S., Prêt, D., Tinseau, E., Claret, F., Pellegrini, D., Stammose, D. (2011). 15 years of in situ cement–argillite interaction from Tournemire URL: Characterisation of the multi-scale spatial heterogeneities of pore space evolution. Applied Geochemistry, 26, 2159-2171.

Jenni, A., Mäder, U., Lerouge, C., Gaboreau, S., Schwyn, B. (2014) In situ interaction between different concretes and Opalinus Clay. Physics and Chemistry of the Earth, Parts A/B/C, 0-71, 71-83.

Lalan, P., Dauzères, A., De Windt, L., Bartier, D., Sammaljärvi, J., Barnichon, J.-D., Techer, I. Detilleux, V. (2016). Impact of a 70°C temperature on an Ordinary Portland Cement paste / claystone interface: an in situ experiment. Cement and Concrete Research, 83, 164-178.

Poyet, S., Le Bescop, P., Touze, G., Moth, J. (2014). Formulating a low-alkalinity and self-consolidating concrete for the DOPASS-FSS experiment. NUWCEM 2014, Avignon.

Solubility, hydrolysis and sorption of beryllium in cementitious systems

Xavier Gaona[1*], Melanie Böttle[1], Thomas Rabung[1], Marcus Altmaier

[1] KIT-INE, Institute for Nuclear Waste Disposal, Karlsruhe Institute of Technology (DE)

* Corresponding author: xavier.gaona@kit.edu

Abstract

Beryllium is a highly chemotoxic element expected in certain wasteforms to be disposed of in repositories for radioactive waste. The amphoteric behaviour of Be(II) is widely accepted in the literature, although the number of experimental studies focussing on the alkaline to hyperalkaline pH range is very scarce. The formation of ternary Na–Be(II)–OH and Ca–Be(II)–OH solid phases under hyperalkaline pH conditions has been described in the literature. These solid phases may eventually control Be(II) solubility in cementitious environments, but no thermodynamic data are available so far for these systems. In spite of the lack of dedicated studies assessing the uptake of Be(II) by cementitious materials, a weak sorption is conservatively predicted based on the formation of negatively charged species in the aqueous phase.

This contribution summarizes the state of the art on the solubility, hydrolysis and sorption of Be(II) in the alkaline to hyperalkaline pH conditions relevant in cementitious systems, and provides an overview of the preliminary results obtained by KIT-INE for the solubility of BeO(cr) under weakly acidic to hyperalkaline pH conditions.

Literature available on the solubility, hydrolysis and sorption of Be(II)

Solubility and hydrolysis of Be(II) in alkaline to hyperalkaline pH conditions

This section summarizes those experimental studies investigating the solubility and hydrolysis of Be(II) under alkaline to hyperalkaline pH conditions. A larger number of publications focus on the chemistry of Be(II) under acidic to near-neutral pH conditions, but these conditions are out of the scope of CEBAMA project and consequently are not discussed here.

Gilbert and Garrett (1956) conducted a comprehensive solubility study with α-Be(OH)$_2$(cr) in weakly acidic ($4.8 \leq pH \leq 5.3$) and hyperalkaline pH conditions ($0.02\ m \leq [NaOH] \leq 0.71\ m$). In the case of alkaline samples, the authors worked with nitrogen-filled flasks to avoid carbonate contamination. Samples for solubility studies were equilibrated for one week. Phase separation was achieved by sedimentation for at least seven days. Experimental data collected in NaOH solutions were interpreted by the authors as the formation of $Be(OH)_3^-$ and $Be(OH)_4^{2-}$ according to the equilibrium reactions (1) and (2). Hydrolysis constants for these species recalculated in Baes and Mesmer (1976) and Bruno (1987) from experimental solubility data in Gilbert and Garrett (1956) and using estimated corrections for activity coefficients are summarized in Table 1.

$$\alpha\text{-}Be(OH)_2(cr) + OH^- \Leftrightarrow Be(OH)_3^- \qquad (1)$$

$$\alpha\text{-}Be(OH)_2(cr) + 2\ OH^- \Leftrightarrow Be(OH)_4^{2-} \qquad (2)$$

Green and Alexander (1963, 1965) performed solvent extraction experiments with $^7Be(II)$ at $5 \leq pH \leq 13$ using *N*-n-butylsalicylideneimine dissolved in toluene as the extractant system. Concentrations of $^7Be(II)$ in the organic and aqueous phase were quantified by γ-spectroscopy. The authors interpreted their extraction data according with the chemical reactions (3) and (4), although acknowledging that the decrease in the distribution coefficients observed at pH > 9 (assigned to the formation of $Be(OH)_3^-$) could be attributed also to the decomposition of *N*-n-butylsalicylideneimine. The stability constants reported in Green and Alexander (1965) for the chemical reactions (3) and (4) are summarized in Table 1.

$$Be(H_2O)_4^{2+} \Leftrightarrow Be(H_2O)_2(OH)_2(aq) + 2\ H^+ \qquad (3)$$

$$Be(H_2O)_4^{2+} \Leftrightarrow Be(H_2O)(OH)_3^- + 3\ H^+ \qquad (4)$$

Soboleva et al. (1977) investigated the solubility of α-BeO(cr) in acidic to hyperalkaline pH conditions at T = 150, 200 and 250°C. The authors explained their experimental observations with the equilibrium reactions (5) to (7). The experimentally determined $\log_{10} K_{s,(1,x)}$ (with x = 1 - 3, T = 150, 200 and 250°C) and the accordingly extrapolated constants at T = 25°C are summarized in Table 1. Chemical and thermodynamic models considered by the authors at T = 25°C include also the species Be^{2+}, Be_2OH^{3+} and $Be_3(OH)_3^{3+}$ previously reported in the literature. Soboleva and co-workers estimated also the stability of the Be(II) solid phases at T = 25°C as α-BeO(cr) > β-$Be(OH)_2$(cr) > α-$Be(OH)_2$(cr).

$$\alpha\text{-}BeO(cr) + H^+ \Leftrightarrow BeOH^+ \qquad (5)$$

$$\alpha\text{-}BeO(cr) + H_2O(l) \Leftrightarrow Be(OH)_2(aq) \qquad (6)$$

$$\alpha\text{-}BeO(cr) + H_2O(l) + OH^- \Leftrightarrow Be(OH)_3^- \qquad (7)$$

Bruno (1987) and Bruno et al. (1987a, 1987b) performed the most comprehensive investigation available to date on the system Be(II)–H_2O–CO_2(g) under acidic to weakly alkaline pH conditions. The authors used a combination of e.m.f. (electromotive force) measurements and solubility experiments in the range $2.0 \leq -\log_{10} [H^+] \leq 8.5$ in the absence and presence of CO_2(g). Besides the determination of the stoichiometry and stability of several ternary Be(II)–OH–CO_3 aqueous species, Bruno and co-workers quantified the solubility product of the crystalline phase β-$Be(OH)_2(cr)$ and the stability constants for the hydrolysis species $Be_x(OH)_y^{2x-y}$ with (x,y) = (1,1), (1,2), (2,1), (3,3), (5,6) and (6,8). The authors re-interpreted also the solubility data previously reported in Gilbert and Garrett (1956), and accordingly provided updated values for $\log_{10} K°_{s,0}\{\alpha\text{-}Be(OH)_2(cr)\}$ and hydrolysis constants of $Be(OH)_3^-$ and $Be(OH)_4^{2-}$ (see Table 1).

Table 1: *Summary of thermodynamic data reported in the literature for the solubility and hydrolysis of Be(II). Only aqueous species forming in alkaline to hyperalkaline pH conditions are reported.*

Reference	Method	Medium	T (°C)	$\log_{10}$ K	$\log_{10}$ K°
α-$Be(OH)_2(cr) + 2 H^+ \Leftrightarrow Be^{2+} + 2 H_2O(l)$					
Gilbert and Garrett (1956)	solubility	$HCl / HClO_4$	25		(6.86 ± 0.05) (6.69 ± 0.02)[a] (6.87 ± 0.05)[b]
β-$Be(OH)_2(cr) + 2 H^+ \Leftrightarrow Be^{2+} + 2 H_2O(l)$					
Bruno et al. (1987)	solubility	3.0 M $NaClO_4$	25	(6.18 ± 0.03)	(5.9 ± 0.1)
α -$BeO(cr) + 2 H_2O(l) \Leftrightarrow Be(OH)_2(aq)$					
Soboleva et al. (1977)	estimated	$I = 0$	25		– 7.13[c]
α -$BeO(cr) + 3 H_2O(l) \Leftrightarrow Be(OH)_3^- + 3 H^+$					
Soboleva et al. (1977)	estimated	$I = 0$	25		– 17.60[c]
$Be^{2+} + 2 H_2O(l) \Leftrightarrow Be(OH)_2(aq)$					
Kakihana and Sillén (1956)	e.m.f.	3.0 M $NaClO_4$	25	– 10.9	
Green and Alexander (1965)	sol. ext.	"self medium"	25	– (13.65 ± 0.04)	– (13.65 ± 0.05)[a]
Kakihana and Maeda (1970)	e.m.f.	3.0 M $NaClO_4$	25	– 11.16	
Brown et al. (1983)	e.m.f.	0.1 M KNO_3	25	– 11.32	
Bruno (1987)	e.m.f.	3.0 M $NaClO_4$	25	– 11.07	– (11.00 ± 0.05)
Chinea et al. (1997)	e.m.f.	0.5 M $NaClO_4$	25	– (11.68 ± 0.06)	
$Be^{2+} + 3 H_2O(l) \Leftrightarrow Be(OH)_3^- + 3 H^+$					
Gilbert and Garrett (1956)	solubility	NaOH	25		– (23.26 ± 0.04)[a] – (23.46 ± 0.05)[b]
Green and Alexander (1965)	sol. ext.	NaOH	25	– (24.11 ± 0.03)	
$Be^{2+} + 4 H_2O(l) \Leftrightarrow Be(OH)_4^{2-} + 4 H^+$					
Gilbert and Garrett (1956)	solubility	NaOH	25		– (37.4 ± 0.2)[a] – (37.59 ± 0.05)[b]

a. recalculated in Baes and Mesmer (1976)

b. recalculated in Bruno (1987)

c. extrapolated in Soboleva et al. (1977) from experimental results at T = 150, 200 and 250°C.

A number of experimental studies in the 60's investigated the formation of ternary M–Be(II)–OH solid phases (with M = Na^+, Ca^{2+}, Sr^{2+} and Ba^{2+}) under hyperalkaline pH conditions (Everest et al., 1962; Scholder et al., 1968, among others). The structural definition of these solid phases was based on the predominant role of the anion $Be(OH)_4^{2-}$ (i.e., $Na_2[Be(OH)_4](cr)$, $Ca[Be(OH)_4](cr)$, etc.), although no proof of concept other than quantitative chemical analysis was provided in these publications. Later studies have demonstrated that the structure of these solid phases is more complex than originally proposed (Schmidbaur et al., 1998; Schmidt et al., 1998; Schmidbaur 2001), and contain polyatomic moieties of Be(II) as those forming under acidic pH conditions. Hence, the compound $Na_2[Be_4(OH)_{10}]\cdot 5H_2O(cr)$ was reported to form in concentrated NaOH solutions (Schmidbaur et al., 1998), whereas the solid phase $Ca_2[Be_2(OH)_7][H_3O_2]\cdot 2H_2O(cr)$ was observed in alkaline $CaCl_2$ solutions (Schmidt et al., 1998). Note that the latter structure contains the hydrated OH^- ion, $H_3O_2^-$, which was previously reported in the literature (Ruf et al., 1996; Tuckerman et al., 1997, among others). The role of these polyatomic Be(II) moieties in the aqueous phase and the corresponding equilibria with the monomeric $Be(OH)_4^{2-}$ species and solid phases remain so far unknown:

$$2\ Be(OH)_4^{2-} \Leftrightarrow [Be_2(OH)_7]^{3-} + OH^- \qquad (8)$$

$$4\ Be(OH)_4^{2-} \Leftrightarrow [Be_4(OH)_{10}]^{2-} + 6\ OH^- \qquad (9)$$

High level nuclear waste resulting from plutonium production is stored in more than 170 tanks in the Hanford Site, WA (USA). Most of these wastes are characterized by high pH (12 - 13.5) and the presence of concentrated salts ($NaNO_3$, $NaNO_2$, etc.). As a result of its use as a component of the Zircaloy 2 fuel cladding processed in the PUREX plant and as a feed constituent in the plutonium finishing plant (PFP), beryllium is found in both solid and liquid phases of the Hanford waste (Reynolds, 2013). Although with a very heterogeneous distribution, Be(II) concentrations of up to 180 ppm and 3 ppm ($\approx 3.3\cdot 10^{-4}$ M) are found in solid and liquid wastes, respectively. Provided the high concentration of carbonate and fluoride in the alkaline wastes, Reynolds speculated on the possible predominance of the species $Be(OH)_2CO_3^{2-}$ and $BeF_y(OH)_m^{2-y-m}$ in the aqueous phase of the Hanford tanks.

^{9}Be nuclear magnetic resonance (NMR) and density functional theory-based (DFT) calculations have been also used in the literature to assess the aqueous speciation of Be(II) (Chinea et al., 1997; Alderighi et al., 1998; Rozmanov et al., 2004; among others). So far most of these studies have focused on the acidic to near-neutral pH range where cationic hydrolysis species prevail, but may also prove to be helpful in the assessment of Be(II) aqueous speciation under alkaline to hyperalkaline pH conditions where anionic hydrolysis species are known to form.

Sorption of Be in cementitious systems

No experimental studies investigating the uptake of Be(II) by cement and cementitious materials are available in the literature. Due to the high charge-to-size ratio of Be^{2+} ($z/d = 1.21$, with $d = r_{Be^{2+}} + r_{OH^-}$) caused by its very small size ($r_{Be^{2+}} = 0.27$ Å), Wieland and Van Loon (2003) speculated that moderate sorption could be expected in hardened cement paste (HCP). In spite of this, the authors conservatively proposed a $R_d = 0$ accounting for the predominance of negatively charged hydrolysis species $Be(OH)_3^-$ and $Be(OH)_4^{2-}$ in the pore water conditions expected in cement systems (Wieland and Van Loon, 2003; Wieland, 2014). A similar approach was proposed in the recent review work by Ochs and co-workers (Ochs et al., 2016).

A) Solubility and hydrolysis of Be(II) under weakly acidic to hyperalkaline pH conditions: preliminary experimental results of KIT–INE within the CEBAMA project

Experimental

NaCl (p.a.), NaOH (Titrisol®), and HCl (Titrisol) were purchased from Merck. Carbonate impurities in 1 M NaOH prepared from NaOH Titrisol were quantified as $(3.1 \pm 0.2)\cdot 10^{-5}$ M. All solutions were prepared with ultrapure water purified with a Milli-Q-academic (Millipore) apparatus and purged with Ar before use. All solutions and samples were prepared, stored, and handled inside an inert gas (Ar) glovebox at $T = 22 \pm 2$°C.

A combination glass pH electrode (type ROSS, Orion), freshly calibrated against dilute standard pH buffers (pH 1 - 13, Merck), was used to determine the molal H^+ concentration, m_{H^+} (with $pH_m = -\log_{10} m_{H^+}$). In salt solutions of ionic strength $I \geq 0.1$ mol/kg, the measured pH value (pH_{exp}) is an operational apparent value related to m_{H^+} by $pH_m = pH_{exp} + A_m$, where A_m is given as a function of the NaCl concentration in molal units (Altmaier et al., 2003). In NaCl–NaOH solutions with $[OH^-] > 0.03$ M, the H^+ concentration was calculated from the given $[OH^-]$ and the conditional ion product of water.

Solubility experiments with commercial BeO(cr) (Alfa Aesar, 99.95%) and synthetic $Be(OH)_2$(cr) are planned at KIT–INE within the framework of the CEBAMA project. At the present stage of the project, preliminary experiments have been conducted with commercial BeO(cr). The strategy for the synthesis of $Be(OH)_2$(cr) includes as a first step the dissolution of a given amount of BeO(cr) (*ca.,* 500 mg) in HCl, and the subsequent precipitation of $Be(OH)_2$(cr) under hyperalkaline pH conditions. The dissolution of BeO(cr) in HCl solutions (up to 8 M) is affected by very slow kinetics, and thus no $Be(OH)_2$(cr) is available yet for the solubility experiments with this material.

Preliminary solubility experiments with commercial BeO(cr) were performed from undersaturation conditions. Five independent batch samples were prepared in 0.1 M NaCl–NaOH solutions at $5 \leq pH_m \leq 12.8$. The total volume in each sample was either 10 mL ($pH_m = 5$) or 20 mL ($pH_m \geq 7$), to which approximately 5 mg of BeO(cr) were added. Although the focus of our studies is on the alkaline to hyperalkaline pH-range, experiments conducted under acidic conditions are required for the quantification of $\log_{10} {}^*K^\circ_{s,0}$\{BeO(cr)\}. Total concentration of Be(II) in the aqueous phase was quantified by inductively coupled plasma mass spectrometry (ICP–MS, X-Series II, Thermo Scientific) after ultrafiltration with 10 kD filters (~ 1.5 nm, Pall Life Sciences). Concentration of Be(II) in the clear supernatant (without phase separation) of selected solubility samples was also quantified by ICP–MS with the aim of assessing the possible presence of colloidal Be(II) species. The characterization by X-ray diffraction (XRD), scanning electron microscopy coupled with energy-dispersive X-ray spectroscopy (SEM–EDS), X-ray photoelectron spectroscopy (XPS) and quantitative chemical analysis of solid phases in selected solubility experiments is planned after attaining thermodynamic equilibrium (assumed after obtaining constant pH_m and $m_{Be(II)}$ measurements).

Results and discussion

Figure 1 shows the experimentally determined solubility data of BeO(cr) in 0.1 M NaCl after a relatively short contact time of $t \leq 18$ days. The figure also includes a number of thermodynamically calculated solubility curves for α-BeO(cr), β-$Be(OH)_2$(cr) and α-$Be(OH)_2$(cr), extrapolated to I = 0.1 M NaCl using the Davies equation for calculating activity coefficients:

- Green line: solubility curve of α-BeO(cr), as calculated with solubility product and hydrolysis constants reported in Soboleva et al. (1977). Note that these authors performed their solubility experiments at T = 150, 200 and 250°C, and thus a large uncertainty (not provided in the original publication) is expected for the data extrapolated to T = 25°C.

- Solid blue and red lines: solubility curves of β-$Be(OH)_2$(cr) and α-$Be(OH)_2$(cr), as calculated with solubility products and hydrolysis constants reported in Bruno (1987) and Bruno et al. (1987a, 1987b) (partially based on the reinterpretation of solubility data in Gilbert and Garrett (1956)).

- Dashed blue and red lines: same as solid blue and red lines, except for the hydrolysis constant of $Be(OH)_2$(aq), which is taken from Green and Alexander (1965) instead of Bruno et al. (1987a) (see Table 1).

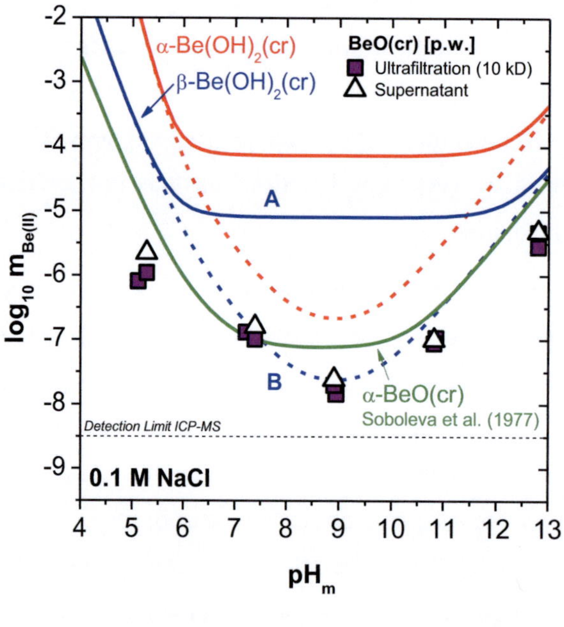

Figure 1: *Solubility of Be(II) in 0.1 M NaCl. Symbols: experimental data obtained in the present work. Solid/dashed lines: thermodynamically calculated solubility of α-BeO(cr), β-$Be(OH)_2$(cr) and α-$Be(OH)_2$(cr) (see text).*

Experimental solubility data collected in the present work confirm the amphoteric character of Be(II), with the formation of cationic and anionic hydrolysis species below and above $pH_m \approx 9$, respectively. No relevant differences are observed between $m_{Be(II)}$ quantified in the clear supernatant and after 10 kD ultrafiltration, thus excluding any significant contribution of dissolved colloidal beryllium species to the total solubility.

Experimental data obtained at $pH_m \geq 10$ follow a well-defined slope of +1, which supports the predominance of the aqueous species $Be(OH)_3^-$ in equilibrium with BeO(cr). The minimum of the solubility curve is observed at $pH_m \approx 9$, and points towards a relatively small stability field of the charge neutral $Be(OH)_2(aq)$ species in solution. Note that the hydrolysis constant for this species is treated controversially in the literature, with two very discrepant groups of values reported: $\log_{10} \beta^\circ_{(1,2)} \approx -11$ (Kakihana and Sillen, 1956; Schwarzenbach and Wenger, 1969; Kakihana and Maeda, 1970; Brown et al., 1983; Bruno, 1987; Chinea et al., 1997) and $\log_{10} \beta^\circ_{(1,2)} = -13.65$ (Green and Alexander, 1965; selected also in Baes and Mesmer, 1976). The two groups of constants were obtained by potentiometry and solvent extraction methods, respectively. Preliminary solubility data collected in the present work seem to indicate that $\log_{10} \beta^\circ_{(1,2)}$ obtained by potentiometric methods is overestimated. Additional experimental studies are on-going at KIT-INE to clarify and eventually solve this issue.

Concentrations of Be(II) measured at $pH_m \approx 5$ fall clearly below any thermodynamic prediction of the solubility of Be(II) based on literature. Accounting for the relatively short equilibration time considered so far ($t \leq 18$ days) and the kinetically hindered dissolution of BeO(cr) observed at very acidic conditions (see experimental description above), we hypothesize that significantly longer equilibration time is likely needed in this pH region to reach equilibrium.

Outlook

Beyond the preliminary solubility results summarized above, the following steps are foreseen at KIT–INE within this experimental study on Be(II) solubility:

- Solubility experiments with BeO(cr) and $Be(OH)_2(cr)$ in dilute to concentrated NaCl, KCl and $CaCl_2$ solutions ($I \leq 3.0$ M). (Use of high ionic strength is required to derive comprehensive thermodynamic treatment including ion-interaction processes).

- Extensive solid phase characterization of the solubility-controlling Be(II) solid phases: XRD, SEM–EDS, XPS and quantitative chemical analysis.

- Investigation of the ternary solid phases Na–Be(II)–OH and Ca–Be(II)–OH eventually forming under hyperalkaline pH conditions.

- Attempt to analyse aqueous beryllium speciation under hyperalkaline pH conditions by ^{9}Be NMR.

- Development of comprehensive chemical, thermodynamic and SIT activity models based on the newly generated solubility data and available studies in the literature.

- The solubility of Be(II) in $NaCl–NaHCO_3–Na_2CO_3$ solutions will be investigated in a later stage of the project.

B) Uptake of Be(II) by cement and cementitious materials: experiments planned at KIT–INE within the CEBAMA project

The uptake of Be(II) by ordinary Portland cement (OPC, fresh) and C–S–H phases with different Ca:Si ratio will be investigated in a later stage of the project. In a preliminary step, the possible use of ^{7}Be will be assessed ($t_{1/2}$ = 53.12 days). Sorption isotherms with increasing [Be] will be quantified with these materials. In the case of C–S–H phases, different pH conditions ($10.0 \leq pH \leq 13.3$) will be investigated as imposed by the corresponding Ca:Si ratio. The preparation of these solid phases will be conducted in close co-operation with BRGM, PSI–LES and EMPA. The main goal of this part of the study is to obtain an appropriate quantitative description of Be(II) uptake by cementitious materials within different degradation stages of cement.

Acknowledgements

The contribution of F. Geyer (KIT–INE) to ICP–MS measurements is highly appreciated.

The research leading to these results has received funding from the European Union's European Atomic Energy Community's (Euratom) Horizon 2020 Programme (NFRP-2014/2015) under grant agreement, 662147 – Cebama.

References

Alderighi, L., Bianchi, A., Mederos, A., Midollini, S., Rodriguez, A., Vacca, A. (1998). Thermodynamic and multinuclear NMR study of beryllium(II) hydrolysis and beryllium(II) complex formation with oxalate, malonate, and succinate anions in aqueous solution. European Journal of Inorganic Chemistry, 1209-1215.

Altmaier, M., Metz, V., Neck, V., Müller, R., Fanghänel, T. (2003). Solid-liquid equilibria of $Mg(OH)_2$(cr) and $Mg_2(OH)_3Cl\cdot4H_2O$(cr) in the system Mg–Na–H–OH–Cl–H_2O at 25°C. Geochimica et Cosmochimica Acta, 67, 3595-3601.

Baes, C.F. and Mesmer, R.E. (1976). The hydrolysis of cations. Wiley.

Brown, P.L., Ellis, J., Sylva, R.N. (1983). The hydrolysis of metal ions. Part 7. Beryllium(II). Journal of the Chemical Society, Dalton Transactions, 2001-2004.

Bruno, J. (1987) Beryllium(II) hydrolysis in 3.0 mol dm^{-3} perchlorate. Journal of the Chemical Society, Dalton Transactions, 2431-2437.

Bruno, J., Grenthe, I., Sandström, M., Ferri, D. (1987a). Studies of metal carbonate equilibria. Part 15. The beryllium(II)–water–carbon dioxide(g) system in acidic 3.0 mol dm^{-3} perchlorate media. Journal of the Chemical Society, Dalton Transactions, 2439-2444.

Bruno, J., Grenthe, I., Munoz, M. (1987b). Studies of metal carbonate equilibria. Part 16. The beryllium(II)–water–carbon dioxide(g) system in neutral-to-alkaline 3.0 mol dm^{-3} perchlorate media at 25°C. Journal of the Chemical Society, Dalton Transactions, 2445-2449.

Chinea, E., Dominguez, S., Mederos, A., Brito, F., Sanchez, A., Ienco, A., Vacca, A. (1997). Hydrolysis of beryllium(II) in DMSO:H_2O. Main Group Metal Chemistry, 20, 11-17.

Everest, D.A., Mercer, R.A., Miller, R.P., Milward, G.L. (1962). The chemical nature of sodium beryllate solutions. Journal of Inorganic and Nuclear Chemistry, 24, 525-534.

Green, R.W. and Alexander, P.W. (1963). Hydrolysis of bis-(acetylacetonato)-beryllium(II). Journal of Physical Chemistry, 67, 905-907.

Green, R.W. and Alexander, P.W. (1965). Schiff base equilibria. II. Beryllium complexes of N-n-butylsalicylideneimine and the hydrolysis of the Be^{2+} ion. Australian Journal of Chemistry, 18, 651-658.

Gilbert, R.A. and Garrett, A.B. (1956). The equilibria of the metastable crystalline form of beryllium hydroxide. $Be(OH)_2$ in hydrochloric acid, perchloric acid and sodium hydroxide solutions at 25°. Journal of the American Chemical Society, 78, 5501-5505.

Kakihana, H. and Maeda, M. (1970). The hydrolysis of the beryllium ion in heavy water. Bulleting of the chemical society of Japan, 43, 109-113.

Kakihana, H. and Sillén, L.G. (1956). Studies on the hydrolysis of metal ions. XVI. The hydrolysis of the beryllium ion, Be^{2+}. Acta Chemica Scandinavica, 10, 985-1005.

Ochs, M., Mallants, D., Wang, L. (2016). Radionuclide and Metal Sorption on Cement and Concrete. Springer International Publishing.

Reynolds, J.G. (2013). The concentration and distribution of beryllium in Hanford high-level waste. Proceedings of the IHLRWMC, Albuquerque, NM.

Rozmanov, D.A., Sizova, O., Burkov, K.A. (2004). Ab initio studies of beryllium aquahydroxocomplexes. Journal of Molecular Structure: THEOCHEM, 712(1-3), 123-130.

Ruf, M., Weis, K., Vahrenkamp, H. (1996). Zn–O_2H_3–Zn: a coordination mode of the hydrolytic zinc–aqua function and a possible structural motif for oligozinc enzymes. Journal of the American Chemical Society, 118, 9288-9294.

Schmidbaur, H. (2001). Recent contributions to the aqueous coordination chemistry of beryllium. Coordination Chemistry Reviews, 215, 223-242.

Schmidbaur, H., Schmidt, M., Schier, A., Riede, J., Tamm, T., Pyykkö, P. (1998). Identification and structural characterization of the predominant species present in alkaline hydroxyberyllate solutions. Journal of the American Chemical Society, 120, 2967-2968.

Schmidt, M., Schier, A., Riede, J., Schmidbaur, H. (1998). The novel binuclear hydroxyberyllate species $[Be_2(OH)_7]^{3-}$ and the hydroxide hydrate anion $[H_3O_2]^-$ as components of beryllate equilibria. Inorganic Chemistry, 37, 3452-3453.

Scholder, R., Hund, H., Schwarz, H. (1968). Über Hydroxoberyllate des Natriums und der Erdalkalimetalle. Zeitschrift für anorganische und allgemeine Chemie, 361, 284-294 (in German).

Schwarzenbach, G. and Wenger, H. (1969). Die Deprotonierung von Metall-Aquoionen I.: $Be{\cdot}aq^{2+}$ Solvatations-Isomerie. Helvetica Chimica Acta, 52, 644-665 (in German).

Soboleva, G.I., Tugarinov, I.A., Kalinina, V.F., Khodakovskiy, I.L. (1977). Investigation of equilibria in the system $BeO–NaOH–HNO_3–H_2O$ in the temperature interval 25° - 250°C. Geokhimiya, 7, 1013-1024 (in Russian).

Tuckerman, M.E., Marx, D., Klein, M.L., Parrinello, M. (1997). On the quantum nature of the shared proton in hydrogen bonds. Science, 275, 817-820.

Wieland, E. and Van Loon, L.R. (2003). Cementitious near-field sorption data base for performance assessment of an ILW repository in opalinus clay. PSI report 03-06.

Wieland, E. (2014) Sorption Data Base for the Cementitious Near Field of L/ILW and ILW Repositories for Provisional Safety Analyses for SGT-E2. NAGRA Technical report, NTB 14-08.

Synthesis and characterization of C-S-H gels for molybdenum sorption studies in cementitious media

Mireia Grivé[1*], Marta López[1], Javier Olmeda[1]

[1] Amphos 21 Consulting S.L. (ES)

* Corresponding author: mireia.grive@amphos21.com

Abstract

Calcium Silicate Hydrates (C-S-H) gels constitute an example of pure cementitious phases and thus an interesting matrix to evaluate their ability to retain radionuclides in cement-based materials. In this work, single-phase homogeneous C-S-H phases have been prepared and analysed at Ca:Si molar ratios of 1.2 and 0.8 through various techniques. Morphological characterization of C-S-H solid samples to be further used in molybdate retention experiments was conducted. Prior to experimental work with Mo, calculations were performed in order to determine the predominant Mo speciation and its corresponding solubility controlling phase. XRD spectra revealed the presence of tobermorite-like structures in all the C-S-H samples analysed. BET-N_2 isotherms indicated that a decrease of either C/S ratio or S/L ratio results in an increase in surface area. SEM-SEI/EDX characterization was also performed observing a heterogeneous surface composition with variable C/S ratios and significant differences in terms of surface morphology between samples.

Introduction

The long-term chemical evolution of a geological repository near-field will be determined by the chemical processes occurring due to metal corrosion and concrete degradation, as well as interactions with groundwater flowing through the facility. Concrete is included in the repository design not only to stabilize the repository structures and to create hydrodynamic barriers but also to condition the near-field pore water to a high pH and to provide abundant microstructural surfaces for the adsorption of radioelements. Under cementitious conditions ($pH > 9$) solubility of many radioelements is lower than under near neutral pH conditions, which ensures their immobilization. Nevertheless, pH is not the only crucial parameter for radionuclide retention, redox potential is also an important parameter affecting the chemical speciation of some redox-sensitive radionuclides and thus the mechanism which determine their immobilization process.

Among the general purposes of the CEBAMA project, WP2 is focused on the study of radionuclide retention processes in high pH concrete environments on relevant hydrated cement phases and alteration products. More specifically, Amphos 21 will study Mo retention with the aim to gain insight on the acquisition of data and sound understanding of the retention mechanisms of Mo in cement/concrete.

Molybdenum is an important fission product present in nuclear fuel as impurity, which constitutes about 0.4% of spent fuel (Lindgren et al., 2007; Osborne et al., 1984). Moreover, molybdenum-93 is an activation product from the steel with a half-life of 4 000 years. Therefore, from a radionuclide transport perspective, the activated components (such as Mo-93) are more important than those from fuel. The main parameters affecting the mobility of this element are the pH and the Eh of the system, although the calcium concentrations present in contacting waters may play a significant role, as well. The aqueous speciation of Mo is dominated by the thermodynamically highly stable molybdate (MoO_4^{2-}).

Under cementitious conditions both, pure molybdenum phases and sorption/ionic exchange processes could be the ones affecting the mobility of this radionuclide. The most relevant processes expected to control its mobility are sorption and/or its incorporation in the large amounts of dissolving/precipitating cement phases (e.g., co-precipitation and ion-exchange processes with AFm and AFt phases, or surface adsorption by C-S-H gels). Nevertheless, the pure solid phase $CaMoO_4(s)$ could play a role as controlling phase for Mo immobilization.

The term C-S-H phase comprises a group of more than 30 identified phases that are poorly ordered (Taylor, 2002). Due to the poorly crystalline structure of C-S-H gels, no real crystal substitution reactions can occur. However, its irregular stacking of the layers, each 10 - 100 nm, creates a large specific surface area available for sorption (Glasser, 1993). At C/S ratios > 1.2 these C-S-H gels are positively charged, and thus might have the potential to adsorb oxyanions like (MoO_4^{2-}).

The further experimental work planned in the CEBAMA project will be of interest to fully understand the behavior of Mo under cementitious environments. The most important aim of this work is to reduce the uncertainty concerning molybdenum due to the scarcity of thermodynamic data available in the literature as well as decipher the mechanisms of Mo retention onto cement phases.

Pre-experimental calculations

Solubility and speciation of Mo under alkaline/cementitious conditions

Scoping calculations performed with Spana v3 using ThermoChimie (TC) database v.9b0 (www.thermochimie-tdb.com) show that the main parameters affecting molybdenum solubility are the pH, calcium concentration present in contacting waters and, to a lower extent the system redox potential.

Under the conditions imposed by cementitious media (strong alkaline pH), molybdenum is predominant in it VI oxidation state, i.e., molybdate ion (MoO_4^{2-}).

Although molybdenum is a redox-sensitive element, the Eh of the system does not influence Mo concentrations in solution under cement conditions, only the conditions imposed by a system ruled by steel corrosion (very low Eh) could be able to reduce molybdenum from Mo (+VI) to Mo (+IV) (Figure 1).

Looking at the predominance diagram, at 10^{-2} M of dissolved calcium, Powellite ($CaMoO_4(s)$) could govern the solubility of this element covering a wide range of pH and Eh, suggesting the great stability of this phase at cementitious-related conditions (Figure 1). Powellite is a potential solubility-controlling phase with a solubility of ~ 10^{-4} M, being the Mo aqueous speciation completely dominated by MoO_4^{2-} (Ochs et al., 2016).

Under an eventual de-calcification (leaching) of cement, molybdenum concentration in solution could be still controlled by Powellite in all degradation states; however, as the solubility of this phase is strongly associated to calcium availability in solution, it would involve Powellite solubility variations up to 2 orders of magnitude and the possibility of adsorption/coprecipitation processes occurring in that systems increase.

A simple ternary system composed by Ca, Mo and H_2O has been considered within the calculations; so in a more complex system sufficient sulphide content might be present (e.g., from SCMs) so as to form other compounds such as MoS_2.

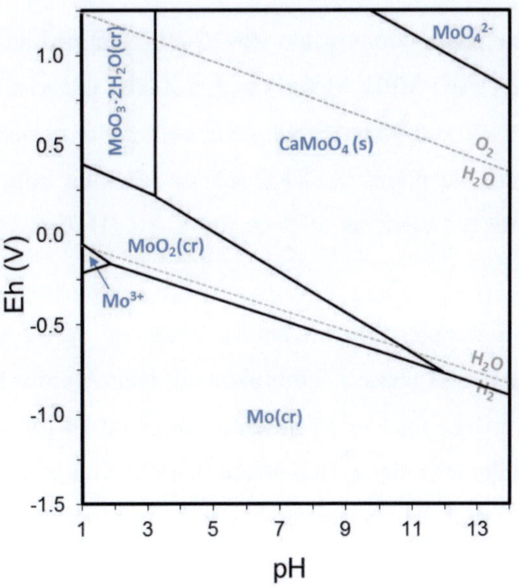

Figure 1: *Predominance Eh/pH diagram of the system Mo-Ca-H_2O at 25°C (Phase-controlled environment) [Ca(II)]=10^{-2} M; [Mo(VI)]=10^{-3} M. (Diagrams calculated in the present work with Spana v3 using ThermoChimie database v.9b0 (https://www.thermochimie-tdb.com)).*

Synthesis and characterization of C-S-H gels

Synthesis procedure

The methodology followed to synthesize C-S-H gels was the one proposed by Pointeau (2000). Two C-S-H phases were synthesized in the laboratory aiming to obtain C-S-H gels with C/S molar ratios of 1.2 and 0.8. The procedure to obtain them was preparing suspensions with two different S/L ratios: 20 and 50 g/L. The C/S ratio remained constant before and after the CSH formation.

The whole synthesis process was conducted in a nitrogen-saturated glove-box with constant temperature (20°C). Reagent grade CaO and silica fume were used at different proportions so as to obtain C-S-H gel suspensions of 15 mL volume with targeted C/S molar ratios. C-S-H gel suspensions were equilibrated during 3 months to constant pH. After this ageing, the solid were filtered and further dried to proceed with the characterization.

Analysis and characterization

<u>XRD</u>

Crystalline and semi-crystalline phase composition was also determined on C-S-H gels by means of X-Ray diffractometer (PANalytical X'Pert PRO MPD Alpha1) with a X-Ray generator formed by a Copper anode and a Wolfram cathode working at 40 kV and 40 mA. Sample scanning was performed on disordered powder with the sample spinning at 2 rps. Figure 2 contains the XRD spectra resulting from the analysis of the four types of C-S-H gels synthesised in the present work as well as the C-S-H(I) X-ray data reported by Taylor (1997) including hkl reflections.

The XRD patterns of the C-S-H samples follow a similar trend with peaks which correspond to tobermorite-like structures. The structure of a C-S-H phase is amorphous in nature, but it keeps some order within a short range. Taylor (1986) proposed a model for C-S-H structure based on 14-nm tobermorite (5-link chain) layers with jennite (2-link chain) placed the inter-layer. In a recent work by Grangeon et al. (2013), it has been proven that the C-S-H structure remains based on tobermorite-like structure even at C/S ratios close or slightly higher than that of jennite.

Two additional peaks, overlapping some signals of the tobermorite-like structure, are also observed in the spectra (one of high intensity placed at $\sim 30° \ 2\theta$ and the other at $\sim 50° \ 2\theta$) which are associated to calcite Figure 2). Although phase synthesis and storage have been carried out in CO_2-free conditions, carbonation is an unavoidable process which can affect C-S-H gels to a great extent during their manipulation (e.g., for analysis).

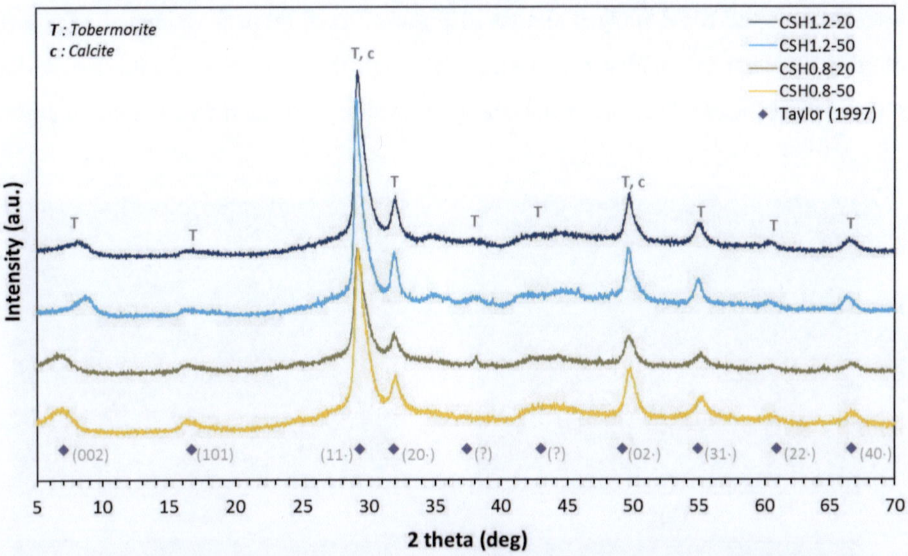

Figure 2: *X-Ray powder diffraction spectra of synthesised dried C-S-H samples. For comparison purposes, data reported by Taylor (1997) for C-S-H(I) has been included. Letters stand for tobermorite (T) and calcite (c).*

Sorption and desorption isotherms. BET surface area

Multipoint BET-N_2 (Brunauer et al., 1938) was employed to analyse the Surface Area (SA) of dried synthesised C-S-H gels with a Micromeritics ASAP 2010. The obtained SA values for each C-S-H formulation with associated errors are shown in Table 1. As can be seen, changes in either calcium-to-silica ratio or solid-to-liquid ratio affect solid surface area in a way which higher active surfaces are developed at lower C/S. Likewise, we can observe that there is an increase of the SA for the samples synthetised from the lowest S/L ratio suspensions.

Table 1: *Surface area data obtained from multipoint BET-N2 using C-S-H phases with different C/S and prepared from different S/L ratios.*

Samples	Suspensions S/L (g/L)	BET-N_2 (m^2/g)
C/S 0.8	50	173.72 ± 1.15
C/S 0.8	20	208.84 ± 1.14
C/S 1.2	50	39.02 ± 0.25
C/S 1.2	20	62.79 ± 0.38

SEM-SEI/EDX

SEM-SEI/EDX was carried out on C-S-H samples to observe surface morphology as well as to acknowledge the average C/S ratio for each sample. SEM-SEI images were obtained with a Field Emission Scanning Electron Microscope (FE-SEM, ZEISS Ultraplus) and the EDX analysis was performed with a X-Max EDX detector from OXFORD Instruments.

As can be seen through the SEM images shown in Figure 3 *a-d*, there is evidence of significantly diverse surface morphologies between the analysed samples. Although the presence of tobermorite-like structures is more evident in some cases, these structural differences cannot be attributed to changes in terms of S/C or S/L ratios.

Figure 3: *SEM images obtained for C-S-H gels:* ***(a)*** *Sample C/S 0.8 - S/L 50 at 10 kx;* ***(b)*** *Sample C/S 0.8 – S/L 20 at 20 kx;* ***(c)*** *Sample C/S 1.2 – S/L 50 at 5 kx;* ***(d)*** *Sample C/S 1.2 - S/L=20 at 10 kx.*

The EDX analysis performed on different spots of sample surfaces shows that the C/S ratio is a variable parameter which depends on the reagent dissolution and sample homogenisation, among others. Table 2 comprises the averaged EDX analysis (from 7 spots) in atomic percentage for each C-S-H sample studied and calculated C/S ratios. Besides sample C/S 0.8 – S/L 20 whose value has proven to be lower than expected, the rest of values are not far from the theoretical ones. This fact points out that a successful synthesis procedure has been carried out in which the reaction between lime and silica fume took place as expected.

Table 2: *EDX analysis (% atomic) on C-S-H samples at 700x.*

	Sample C/S 0.8 - S/L 50	**Sample C/S 0.8 – S/L 20**	**Sample C/S 1.2 – S/L 50**	**Sample C/S 1.2 – S/L 20**
Si (%atom)	18.40	18.25	15.35	18.55
Ca (%atom)	14.96	12.43	18.51	23.01
C/S	**0.81**	**0.68**	**1.21**	**1.24**

Conclusions and Future work

Pre-experimental calculations have allowed prediction of Mo limiting solid phases in cementitious media and their most probable oxidation state. Under highly alkaline conditions, the molybdate anion seems to be relatively soluble and thermodynamically stable even in a system ruled by a strong reducing potential.

Results also indicate that Powellite could be a solubility controlling phase under a relatively wide range of pH and Eh, although very sensitive to changes in calcium concentration; under an eventual de-calcification (leaching) of cement, the increase of dissolved Ca content would promote Powellite precipitation.

C-S-H gels have been successfully synthesized and targeted C/S ratios have been achieved. Characterization carried out herein points out that a higher surface area is developed at lower C/S and S/L ratios, and despite the different surface morphology observed between the solid samples, tobermorite-like structure is present in all of them.

On-going work includes the synthesis and characterization of other cementitious phases (AFt and AFm). Further experimental work will be devoted to the study of Mo solubility in pure phase porewaters as well as Mo adsorption mechanisms to pure cementitious phases.

Acknowledgement

The research leading to these results has received funding from the European Union's European Atomic Energy Community's (Euratom) Horizon 2020 Programme (NFRP-2014/2015) under grant agreement, 662147 – Cebama.

The authors are gratefully acknowledged to ANDRA from their co-financial support.

References

Brunauer, S., Emmett, P.H., Teller, E. (1938). Adsorption of gases in multimolecular layers. Journal of the American Chemical Society, 60(2), 309-319.

Glasser, F.P. (1993). Chemistry of cement-solidified waste forms. In: chemistry and microstructure of solidified waste forms, Spence, R.D., Ed.; Lewis Publishers, Boca Raton, FL.

Grangeon, S., Claret, F., Linard, Y., Chiaberge, C. (2013). X-ray diffraction: a powerful tool to probe and understand the structure of nanocrystalline calcium silicate hydrates. Acta Crystallographica Section B: Structural Science, Crystal Engineering and Materials, 69(5), 465-473.

Lindgren, M., Pettersson, M., Wiborgh, M. (2007). Correlation factors for C-14, Cl-36, Ni-59, Ni-63, Mo-93, Tc-99, I-129 and Cs-135, In operational waste for SFR 1. SKB Report, R-07-05.

Ochs, M., Mallants, D., Wang, L. (2016). Radionuclide and Metal Sorption on Cement and Concrete. Springer International Publishing, Switzerland.

Osborne, M.F., Collins, J.L., Lorenz, R.A., Norwood, K.S. (1984). Measurement and characterization of fission products released from LWR fuel. CONF-840914-28. TI85 007539.

Pointeau, I. (2000). Etude mécanistique et modelisation de la rétention de radionucléides par les silicates de calcium hdratés (CSH) des ciments Andra. PhD thesis. 'Université de Reims Champagne-Ardenne.

Taylor, H.F. (2002). Sulfates in Portland clinker and cement, International RILEM TC Workshop on Internal Sulfate Attack and Delayed Ettringite Formation. London, Thomas Telford.

Taylor, H.F. (1986). Proposed structure for calcium silicate hydrate gel, Journal of the American Ceramic Society, 69(6), 464-467.

Taylor, H.F.W. (1997). Cement chemistry. Thomas Telford.

Application of synthesis protocols of AFm phases

Nicolas Marty[1*], Sylvain Grangeon[1]

[1] BRGM, Bureau de Recherches Géologiques et Minières (FR)

* Corresponding author: n.marty@brgm.fr

Abstract

This document reviews the most efficient AFm synthesis protocols that were tested by the BRGM in the frame of the WP2 package from the CEBAMA project. AFm having interlayer Cl could be successfully synthesised. All synthesis protocols for AFm having interlayer OH led to products that contained impurities of synthetic katoite and of synthetic portlandite. Synthesis protocols from AFm having interlayer SO_4 and Cl were also tested, but the purity of the product is still under investigation at the present stage of the study.

Introduction

In the frame of workpackage 2 (WP2) from the European research project Cement-based materials (CEBAMA), BRGM aims at contributing to a better understanding of anion retention by cement materials. This retention is expected to be influenced, if not controlled, by AFm phases, a family of phases having a layered double hydroxide (LDH) structure that is being built of layers of atoms separated from each other by a hydrated interlayer space. In the case of AFm, layers are built of Ca and Al atoms in a 2:1 ratio, each being coordinated by oxygen atoms to form polyhedra that are connected to one belonging to a cation of the other type (i.e., two atoms of the same type – Ca or Al – are never directly adjacent). This layer structure induces a positive permanent charge, which is compensated by anions, which are located in the interlayer space, these latter also containing water molecules. Anions that are typically reported to be part of AFm interlayer space are OH^-, Cl^- and SO_4^{2-}. These anions, present in the interlayer space, are exchangeable with those present in the solution in which AFm are present. As such, AFm have a sorption capacity for radionuclides that will be in an anionic form. Determining cement sorption capacity with regards to anionic radionuclides of interest passes through a determination of the sorption capacity of AFm phases. However, each interlayer anion, including those initially contained in AFm interlayer space, has its own affinity for AFm interlayer sites. As a consequence, understanding radionuclides retention by AFm (and thus cements) requires the quantification of this affinity for each interlayer anion. BRGM will focus on the determination of the capacity of AFm phases to uptake Mo and

Se (stable forms will be employed and their behaviour will be assumed analogous to that of the radioactive isotopes). In this context, BRGM will study the relative affinities of Cl, Se and Mo for AFm.

In addition, cement structures present in geological disposal sites will possibly be contacted by water originating from the surrounding geological materials (e.g., clayey rock). This will induce disequilibrium conditions with regards to AFm, possibly leading to the degradation of AFm phases. This may lead to transformations of the AFm structure that may induce a reduction (or an increase) of the capacity of AFm to uptake Mo and Se. In an effort to better understand the evolution of AFm structure and reactivity under disequilibrium conditions, BRGM will study AFm alteration mechanisms.

Both sorption and alteration experiments require that pure AFm phases are available. In this view, BRGM has reviewed AFm synthesis protocols, and selected those, which were most likely to lead to the formation of pure AFm phases. Although a synthesis and study of AFm having interlayer Cl will be the core contribution of BRGM, protocols were also tested leading to AFm phases having other interlayer anions, possibly to be used in sorption and (or) alteration experiments.

This document will review the most efficient synthesis protocols that were tested by BRGM during the first six months of the CEBAMA project. When pure phases were obtained, their XRD pattern is shown. When impurities were observed, they were identified and reported, but their pattern is not shown here.

AFm synthesis

AFm-Cl

All AFm-Cl syntheses were performed in a N_2 filled glove box, using ultra-pure degassed water (Milli-Q18 MΩ), to avoid carbonation process (e.g., Goñi and Guerrero, 2003). The synthesis protocol involves mixing C_3A and $CaCl_2 \cdot 2H_2O$ (1:1 molar ratio) at room temperature, following Balonis and Glasser (2009), and allowing the reaction to run for 15 days, with periodical shaking. After 15 days of maturation at room temperature, synthetized AFm-Cl were filtered (using a 0.45 µm Millipore HVLP membrane filters) and dried (using a lyophilizer Christ Beta 2-8). Fluid was retrieved and analysed. This product was found to be pure, as demonstrated by the successful Rietveld refinement of its XRD pattern using an AFm-Cl structure model (Figure 1).

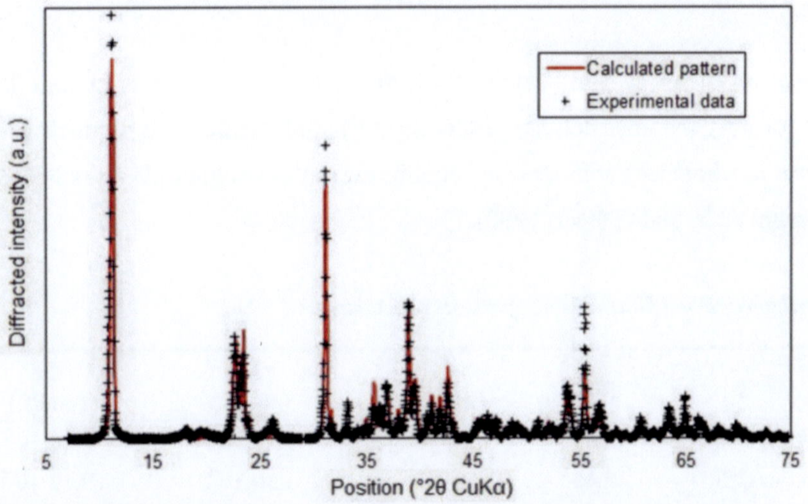

Figure 1: *Experimental (black crosses) and calculated (solid red line) XRD patterns of AFm-Cl.*

AFm-OH

The C_4AH_{13} ($Ca_4Al_2O_7 \cdot 13H_2O$), is called AFm-OH for convenience. Following the protocol describe in Matschei et al. (2007), it was synthetized from a mixture of C_3A into N_2 glove box and CaO (1:1 molar ratio) using cooled and degassed ultra-pure water (5°C). The solid obtained suspension has been sealed in box (to avoid carbonation process) and was maintained at 5°C during 15 days in a refrigerator. Solid suspensions were periodically shaken and after maturation, synthetized AFm was filtered (cut-off diameter of 0.45 µm) and dried before being characterized. Fluid was also retrieved and analysed to determine the saturation index of the solution with respect to the synthetized material. All syntheses led to products containing impurities of synthetic katoite and synthetic portlandite. Alternative syntheses protocols are going to be tested.

AFm-Cl/SO_4

The Kuzel's salt, also called AFm-Cl/SO_4 for convenience, has been synthetized in a N_2 glove box at room temperature, following the protocol described by Glasser et al. (1999). Respecting a water to solid ratio of 10, C_3A, $CaCl_2$, and $CaSO_4$ were mixed into degassed ultra-pure water. Solid suspensions were periodically shaken and after 15 days of maturation, synthetized AFm was filtered (cut-off diameter of 0.45 µm) and dried before to be characterized. Fluid was also retrieved and analysed to determine saturation indexes. The purity of the obtained product is still under investigation.

Saturation index

From the analyses of chemical data (Table 1) which are obtained on filtrated equilibrium solution (filter cut-off diameter of 0.1 µm), the ionic activity products (IAP) of all synthesized products can be computed. Note that AFm-Cl data are in agreement with reported equilibrium solid-solution pH values for various assemblages of Friedel's salt (Birnin-Yauri and Glasser, 1998).

Table 1: *Compositions of retrieved fluids during synthesis procedures.*

Exp.	pH	Cl (mol/L)	Al (mol/L)	Ca (mol/L)
AFm-OH(a)	12.41	$1.92 \cdot 10^{-3}$	$9.27 \cdot 10^{-5}$	$2.18 \cdot 10^{-2}$
AFm-OH(b)	12.43	---	$1.30 \cdot 10^{-4}$	$2.17 \cdot 10^{-2}$
AFm-Cl(a)	12.08	$8.46 \cdot 10^{-2}$	$6.93 \cdot 10^{-6}$	$5.86 \cdot 10^{-2}$
AFm-Cl(b)	12.11	$1.07 \cdot 10^{-1}$	$9.04 \cdot 10^{-6}$	$6.62 \cdot 10^{-2}$

AFm-OH and AFm-Cl are referenced as C4AH13 and Friedel_Salt in the THERMODDEM database (Blanc et al., 2012), respectively. The saturation index of these phases as well as of portlandite, C3AH6 (i.e., Katoite) and 2 forms of gibbsite (amorphous and well crystallized) have been reported in Table 2.

Table 2: *Saturation indexes (SI) of retrieved fluids during synthesis procedures.*

Exp.	$SI_{AFm\text{-}OH}$	$SI_{AFm\text{-}Cl}$	$SI_{Portlandite}$	$SI_{Gibbsite}$	$SI_{Gibbsite(am)}$	SI_{C3AH6}
AFm-OH(a)	-0.04	-1.76	-0.07	-1.40	-4.24	0.56
AFm-OH(b)	0.36	---	-0.04	-1.27	-4.11	0.92
AFm-Cl(a)	-2.94	-0.80	-0.38	-2.24	-5.08	-2.03
AFm-Cl(b)	-2.41	-0.14	-0.29	-2.16	-5.00	-1.59

PHREEQC calculations indicate that synthetized AFm materials are close to a thermodynamic equilibrium (i.e., $SI_{AFm\text{-}OH}$ and $SI_{AFm\text{-}Cl}$ close to 0 after the synthesis AFm-OH and AFm-Cl, respectively). Note that negative $SI_{AFm\text{-}Cl}$ (i.e., under-saturation) obtained at the end of AFm-Cl synthesis are inconsistent with the characterisation of the retrieved material (i.e., pure AFm-Cl, Figure 1); calculated negative values may be due to uncertainties on both thermodynamic constant reported in THERMODDEM and/or on solution analyses. SI reported in Table 2 for AFm-OH syntheses indicate that katoite (C3AH6) and portlandite could also have precipitated during synthesis if they were initially oversaturated, as modelling indicates SI values close to 0 in final solution. The presence of katoite has been confirmed by XRD analyses.

Conclusions and Future work

The objective of this work is to determine dissolution kinetics and sorption capacity of synthesised AFm. Even if solutions have been systematically analysed after each synthesis, solubility constants cannot be derived; their determination should be tackled by dedicated works (e.g., Birnin-Yauri and Glasser, 1998). However, solution analyses give reliable data on the degree of purity of synthesised materials and the nature of possible accessory phases.

After additional characterisations, sorption study will focus on the determination of the capacity of AFm to uptake Mo and Se using batch reactors at room temperature.

The alteration of AFm-Cl phases will be also studied with flow-through experiments performed at room temperature and at various pH conditions (i.e., $9.2 < pH < 13$). Flow-through experiments combined with mineralogical (XRD) and chemical (EPMA, TEM, ICP-AES, pH measurements) analyses will allow evaluating AFm alteration.

Acknowledgement

The research leading to these results has received funding from the European Union's European Atomic Energy Community's (Euratom) Horizon 2020 Programme (NFRP-2014/2015) under grant agreement, 662147 – Cebama.

References

Balonis, M. and Glasser, F.P. (2009). The density of cement phases. Cement and Concrete Research, 39, 733-739.

Birnin-Yauri, U.A. and Glasser, F.P. (1998). Friedel's salt, $Ca_2Al(OH)_6(Cl,OH)\cdot 2H_2O$: its solid solutions and their role in chloride binding. Cement and Concrete Research, 28, 1713-1723.

Blanc, P., Lassin, A., Piantone, P., Azaroual, M., Jacquemet, N., Fabbri, A., Gaucher, E.C. (2012). Thermoddem: A geochemical database focused on low temperature water/rock interactions and waste materials. Applied Geochemistry, 27, 2107-2116.

Glasser, F.P., Kindness, A., Stronach, S.A. (1999). Stability and Solubility Relationships in AFm Phases, Cement and Concrete Research, 29, 861-866.

Matschei, T., Lothenbach, B., Glasser, F.P. (2007). The AFm phase in Portland cement. Cement and Concrete Research, 37, 118-130.

Understanding the behaviour of safety relevant radionuclides in cementitious systems: uptake of radium and technetium by CSH, AFm and AFt

Steve Lange [1*], Martina Klinkenberg [1], Guido Deissmann [1], Dirk Bosbach [1]

[1] Forschungszentrum Jülich GmbH, Institute of Energy and Climate Research: Nuclear Waste Management and Reactor Safety (IEK-6) (DE)

* Corresponding author: s.lange@fz-juelich.de

Abstract

In this paper, we report initial results of the uptake of ^{226}Ra and ^{99}Tc by single model phases representative for phases present in hydrated cements, namely CSH, ettringite, and monosulphate (AFm-SO_4). The phases were synthesized under argon atmosphere, using well established procedures (e.g., Atkins et al., 1991, 1992; Baur et al., 2004). Sorption and uptake kinetics of $^{226}Ra^{2+}$ and $^{99}TcO_4^-$ by the model phases were studied in static batch experiments under anoxic conditions. Experiments performed for up to 40 days indicate a fast and strong uptake of radium by CSH phases, whereas the interaction in systems with AFt and AFm is significantly weaker. In contrast, only a negligible uptake of technetium, present as ^{99}Tc(VII) in solution, by the various model phases was observed. Microanalytical investigations of the radionuclide distribution in the solids and hardened cement pastes by TEM and ToF-SIMS are in progress.

Introduction

Cement-based materials are widely used in nuclear waste management, for example for the solidification of low and intermediate level wastes, as barrier materials in nuclear waste repositories, or in certain waste containers as buffer, for example, in the Belgian Supercontainer. The migration behavior of radionuclides in cementitious materials is controlled by radionuclide solubility phenomena, diffusion, interface processes such as surface complexation, or incorporation of radionuclides into solid phases, including the formation of solid solutions. Within the framework of CEBAMA, we study the uptake of selected safety relevant long-lived fission and decay products such as ^{129}I, ^{99}Tc, ^{79}Se, and ^{226}Ra in cementitious systems, using advanced micro analytical and spectroscopic tools. The objective of these investigations is to enhance the mechanistic understanding of the radionuclide uptake and retention in cementitious materials and to evaluate the relevance of cement alteration processes, such as carbonation, on the solid speciation of radionuclides in aged concrete. Within this context, a bottom-up approach is being pursued, studying the radionuclide interaction with synthesized phases present in cementitious materials (model phases), such as calcium silicate hydrates (CSH) with different Ca/Si ratios, monosulphate (AFm) and ettringite (AFt) phases on the one hand, and hardened cement pastes with different compositions (e.g., CEM I and CEM V) on the other. Here, we report first results on the uptake of ^{226}Ra and ^{99}Tc on CSH, ettringite, and AFm-(SO_4) as determined in batch sorption experiments.

Materials and Methods

Synthesis of model phases

Phase synthesis and sample preparation was carried out in an argon glove box (< 10 ppm CO_2). The products were washed by centrifugation (30 min at 4 500 rpm) and resuspended in water three to five times, followed by vacuum filtering using a Büchner funnel and a paper filter (Whatman). Drying was carried out over $CaCl_2$ using a desiccator under continuous evacuation. Calcium silicate hydrates (CSH) were prepared according to the procedure of Atkins et al. (1992). To achieve CSH with different CaO/SiO_2 ratios, various amounts of CaO, typically 1.8 g up to 3.4 g (weighted after 2 hours of calcination at 1 000°C), were mixed with SiO_2 and suspended in water to achieve a water/solid ratio of 20. Ettringite ($Ca_6Al_2(SO_4)_3(OH)_{12}{\cdot}26H_2O$) synthesis was carried out according to Atkins et al. (1991) and Baur et al. (2004), respectively. CaO was suspended in water and cooled to 4°C and a stoichiometric solution of $Al_2(SO_4)_3{\cdot}18H_2O$ (Ca/Al = 3) and mixed with the CaO suspension. The suspension was stirred for 4 hours at approx. 4°C and cured for one month at room temperature. C_3A ($3CaO{\cdot}Al_2O_3$), used as a reagent, was synthesized according to the method described by Atkins et al. (1991). Stoichiometric amounts (Ca/Al = 1.5) of CaO and Al_2O_3 were mixed and fired at 1 000°C for 2 hours in a Pt/Au (5% Au) crucible and ignited at 1 450°C for one hour. The product was crushed and well ground using a mortar and ignited in a second cycle at 1 450°C for 48 hours. Monosulphate ($Ca_4Al_2(SO_4)(OH)_{12}{\cdot}6H_2O$) synthesis was done according to Baur et al. (2004). Stoichiometric amounts (Ca/Al = 2) of freshly prepared C_3A and $CaSO_4{\cdot}2H_2O$ were mixed and suspended in water at 4°C and stirred for 4 hours. Afterwards the product was cured for 5 months at room temperature.

Table 1: *List of details for chemical substances that have been used.*

Chemical	Product description	Provider
CaO	puriss., 96 - 100.5% ex ignited substance	Sigma Aldrich
$Al_2(SO_4)_3{\cdot}18H_2O$	AnalaR normapur	VWR chemicals
SiO_2	fumed silica 395 m²/g	Sigma-Aldrich
Al_2O_3	99.5% purity based on trace metals analysis	Sigma-Aldrich
$CaSO_4{\cdot}2H_2O$	analytical grade, EMSURE	Merck Millipore

Phase characterization

Purity and composition of the final products was confirmed by XRD (D8 Advance with a θ-θ geometry and a D4 Endeavour with a θ-2θ geometry (CuK_{α}, λ = 0.15406 nm) from Bruker AXS GmbH) and by XRF-analyses conducted by an external service company (Terrachem GmbH). Phase morphology and sample microstructure was studied by SEM analyses performed with a Quanta 200F from FEI equipped with a field emission cathode. Samples were prepared on sticky carbon pads which were glued to an aluminium sample holder. SEM analysis was carried out under low vacuum (60 Pa), where a coating of the samples was not required.

Sorption experiments

Batch sorption experiments were carried out in two different ways, using 50 mL LDPE bottles and a L/S-ratio of 200. In the first set of experiments, weighted amounts of dried solids were equilibrated with deionized water prior to use. The obtained suspension was stored at room temperature for up to 14 days under anoxic conditions to achieve equilibrium between aqueous and solid phases. After equilibrium was reached, the liquid was separated from the solids by filtration. Weighted amounts of fresh solids was added to this solution and stored for seven days before the respective radiotracer was added for sorption studies.

In situ uptake experiments were performed with the radiotracers directly added to the suspended phases at the end of the curing time, using series of batches of samples produced under identical conditions at the same time. After synthesis, one of the batches was used for the characterization of the synthesized solid and the corresponding liquid phase, whereas the others were spiked with the radiotracer and studied with respect to the radionuclide uptake. The radiotracer concentrations were identical in both sets of experiments, namely $c(^{226}Ra^{2+}) = 10^{-8}$ mol/L and $c(^{99}TcO_4^-) = 10^{-7}$ mol/L. The activity concentrations in the liquid phase were determined using gamma-ray spectroscopy and liquid scintillation counting, respectively. The sorption of radionuclides to reaction vessels and filters was tested and found to be negligible. The radionuclide uptake by the respective solids is described in terms of the distribution between solid and liquid phase according to:

$$R_d = \frac{A_{ini} \cdot A_t}{A_t} \cdot \frac{V}{m} \qquad \textbf{\textit{Eq. 1}}$$

with

A_t: activity at time t in liquid phase (Bq)

A_{ini}: initial activity in liquid phase (Bq)

V: volume of liquid phase (L)

m: mass of solid phase (kg)

Gamma-ray spectroscopy

The γ-line of ^{226}Ra (186 keV) was measured using a 500 µL sample volume in a 2 mL borosilicate glass placed close to a high-purity germanium coaxial N type detector system (type: EGC 35-195-R), purchased from Eurisys Mesures, Lingolsheim, France, equipped with the spectrometer system obtained from EG & G Ortec, Munich, Germany. Analysis of the spectra was performed with the GammaVision® Modell A66-B32 software (version 5.20).

Liquid scintillation counting

Liquid scintillation counting (LSC) measurements of ^{99}Tc was performed using a 1 220 Ultra low level Quantulus™ LSC device, purchased from Perkin Elmer, Rodgau-Jügesheim, Germany, equipped with the Winq software (version 1.2). Aliquots of 50 µL were diluted in a 20 mL polyethylene Vial with 15 mL Hionic-Fluor or Ultima Gold™ LSC-cocktail, both obtained from Perkin Elmer, Rodgau-Jügesheim, Germany.

Results

The characterization of the solids by XRD and SEM revealed a high purity of CSH, ettringite, and AFm-SO_4 obtained by the synthesis route employed, since no impurities could be detected by the characterization methods applied. SEM measurements of the synthesized phases revealed that the microstructure and size of the CSH and AFm phases are comparable in size to phases in hydrated cement pastes. However, the crystal size of the synthesized single phase ettringite was found to be smaller (< 1 µm) compared to ettringite needles formed in CEM I paste hydrated for 6 days.

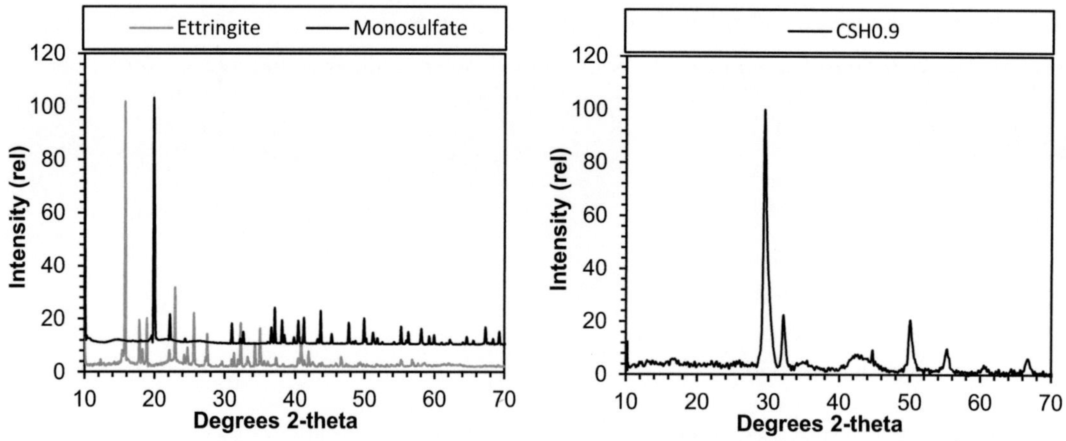

Figure 1: *XRD pattern of the synthesized ettringite, monosulfate and CSH 0.9.*

Uptake experiments performed for up to 40 days indicate a fast and very strong uptake of radium by CSH phases (R_d about 25 000 L/kg), whereas the interaction in systems with AFt and AFm is significantly weaker (Table 2). The uptake of radium by CSH0.9 reached equilibrium after about 15 to 30 days; the solution pH was found to be constant throughout these experiments at pH 11.9. A variation of the liquid/solid ratio in the range from 50 to 500 had no significant impact on the observed radium uptake by CSH. The radium uptake by AFm-SO_4 was found to be in equilibrium after about 15 to 30 days, similar to CSH0.9, however the R_d–value for ^{226}Ra for AFm -SO_4 was only about 0.2 L/kg.

Table 2: *Uptake kinetics of ^{226}Ra by different model phases under anoxic conditions ($c(^{226}Ra^{2+}) = 10^{-8}$ M; room temperature; L/S = 200).*

	$R_d(^{226}Ra)$ in L/kg		
time [d]	**CSH 0.9**	**Ettringite**	**Monosulphate**
0.04	-	0.02	0.002
0.05	439	-	-
1	1 025	0.07	0.03
3	2 408	0.01	0.04
7	6 783	0.03	0.15
23	-	0.01	0.18
33	25 584	-	-

The *in situ* experiments of radium uptake onto CSH0.9 (pH = 11.7) and CSH1.4 (pH = 12.4) phases showed that the uptake reaches equilibrium within 10 days. The R_d-value obtained for CSH0.9 was similar to the batch uptake experiments described above, however it was found that $R_d(^{226}Ra)$ decreases by a factor of about 10 with increasing the Ca/Si ratio up to 1.4.

In contrast to ^{226}Ra, only a negligible uptake of technetium, present as $^{99}TcO_4^-$ in solution, by the various model phases was observed. The $R_d(^{99}Tc^{VII})$ was found to be below 0.02 L/kg for all, CSH0.9, ettringite and AFm-SO_4.

Conclusions and Future work

Single model phases representative for phases in hydrated cements were synthesized and characterized. First sorption experiments addressing the uptake of radium and technetium by CSH, ettringite, and AFm-SO_4 under anoxic conditions were performed. The uptake of radium by CSH was found to be fast and strong, and equilibrium was reached within 30 days. In contrast, the uptake of radium on AFm-SO_4 and AFt was rather small. The uptake kinetics in the *in situ* experiments are in agreement with literature data from Tits et al. (2006). The sorption/uptake of technetium by CSH0.9, ettringite, and AFm-SO_4 was found to be negligible, in agreement with the expectations.

Work intended in the near future comprises sorption experiments on additional model phases like AFm-OH, AFm-CO_3, CSH1.4 and hydrotalcite. The synthesis of these phases is in progress. Studies addressing the uptake of molybdenum (present in solution as molybdate) on single phases started recently. Sorption studies using artificial cement waters representative for young alkali-rich cement pore waters (pH about 13.3) are proceeding as scheduled. Microanalytical investigations of the radionuclide distribution in the solids using e.g., SEM-EDX, TEM, and ToF-SIMS are under preparation. The results of this ongoing study will lead to a better understanding of the uptake and retention mechanisms of safety relevant radionuclides in cementitious barriers and materials.

Acknowledgement

The research leading to these results has received funding from the European Union's Horizon 2020 Research and Training Programme of the European Atomic Energy Community (EURATOM) (H2020-NFRP-2014/2015) under grant agreement n° 662147 (CEBAMA).

References

Atkins, M., Glasser, F.P., Kindness, A. (1992). Cement hydrate phases: solubility at 25°C. Cement Concrete Research, 22, 241-246.

Atkins, M., Macphee, D., Kindness, A., Glasser, F.P. (1991). Solubility properties of ternary and quaternary compounds in the calcia-alumina-sulfur trioxide-water system. Cement Concrete Research, 21, 991-998.

Baur, I., Keller, P., Mavrocordatos, D., Wehrli, B., Johnson, C.A. (2004). Dissolution-precipitation behaviour of ettringite, monosulfate, and calcium silicate hydrate. Cement Concrete Research, 34, 341-348.

Tits, J., Iijima, K., Wieland, E., Kamei, G. (2006). The uptake of radium by calcium silicate hydrates and hardened cement paste. Radiochimica Acta, 94, 637-643.

Carbon-14 transport parameters through CEM II mortar

Crina Bucur[1*], Ionut Florea[1]

[1] Institute for Nuclear Research Pitesti (RO)

* Corresponding author: crina.bucur@nuclear.ro

Abstract

In the frame of CEBAMA project, RATEN ICN proposes to understand the influence of concrete ageing on ^{14}C and ^{226}Ra retention on cement based materials. To achieve this, sorption and diffusion experiments will be carried out on fresh and aged hardened cement pastes (HCP). The fresh samples are prepared in the RATEN ICN while the aged ones consist of HCP cured for 7 years in saturated Portlandite water (pH = 12.4) and were received from ARMINES.

The experience of RATEN ICN on radionuclide sorption and diffusion on cement based materials was accumulated in the frame of Romanian R&D programme for near surface disposal. In this programme sorption and diffusion experiments were carried out on mortars samples based on CEM II, the cement to be used in the waste immobilisation as well as for disposal cells and disposal containers building.

This paper summarize the experiments previously performed in RATEN ICN on ^{14}C and ^{137}Cs sorption and diffusion and the literature review regarding ^{14}C and ^{226}Ra sorption on hardened cement paste and mortars carried out in the frame of CEBAMA project.

Introduction

In Romania, the long-lived radioactive waste (LL-ILW) and CANDU spent fuel are foreseen to be disposed of in the future geological disposal.

The main source of Carbon-14 in the LL-ILW to be disposed of in the Romanian geologic repository is the spent ion exchange resins generated mainly in the Moderator System and in the Primary Heat Transport System of the two units of Cernavoda NPP. The dominant Carbon-14 species in spent ion exchange resins is inorganic carbonate $^{14}CO_3^{2-}$, more than 90% of Carbon-14 is bound on the anion resin fraction (Park et al., 2008). Other LL-ILW to be geological disposed of consists in the pressure tubes and calandria tubes that could contain Carbon-14 both in inorganic and organic forms.

^{226}Ra is one of the major daughter nuclides of the ^{238}U that could be found in the fuel-contact spent ion exchange resins but also in other waste streams.

In form of dissolved carbonate or bicarbonate, inorganic Carbon-14 is strongly immobilized in cemented waste forms. Under strongly alkaline conditions, the dominant species of inorganic Carbon-14 is $^{14}CO_3^{2-}$. The carbonate ions can either precipitate as $CaCO_3$ or sorb onto cement phases.

Information on the speciation of Carbon-14 associated with organic compounds which might be released from the various waste forms, and their potential immobilization by cementitious materials is limited.

Literature data regarding Ra(II) sorption on fresh and aged HCP show that distribution coefficient is less than 0.1 m^3/kg for fresh HCP and less than 0.4 m^3/kg for aged HCP (Wieland, 2014). The uptake of Ra by HCP could be interpreted in terms of Ra binding to C-S-H phases (Tits et al., 2006a).

There is reported in the literature results of numerous studies on C-14 sorption on hardened cement paste and mortars (Allard et al., 1981; Bayliss et al., 1988; Bradbury and Sarott, 1994; Ochs et al., 2016; Wang et al., 2009) and some on Ra-226 sorption (Bayliss et al., 1989; Bayliss et al., 2000; Holland and Lee, 1992; Tits et al., 2006; Wieland and Van Loon, 2002; Wieland, 2014). Experimental data regarding the effect of the cement degradation in the disposal conditions on radionuclide sorption are limited and the experiments to be carried out in RATEN ICN in the frame of CEBAMA - WP2 will be specifically oriented to assess the influence of HCP degradation on C-14 and Ra-226 sorption.

Previous sorption experiments on cement based materials

Batch sorption experiments were carried out in the frame of Romanian R&D programme for low and intermediate level waste disposal. These wastes are generated mainly from Cernavoda NPP operation and decommissioning and are intended to be disposed of in a near surface repository located on Saligny site.

For a near surface repository, around 95 weight per cent of the material consists of concrete. Radionuclide sorption on concrete represents one of the most important retardation mechanisms in the disposal vaults. For this reason experiments to evaluate the ^{14}C sorption on concrete matrix (based on standard Portland cement) were performed in the frame of Romanian near-surface disposal R&D programme.

The most common type of cement used in Romania for radioactive waste conditioning is the Standard Portland Cement, CEM II type while for the disposal modules and vaults construction the cement type has not been decided yet. For these reasons, the CEM II OPC was used in solid matrix preparation for sorption experiments.

The radionuclide distribution coefficients were determinate by laboratory batch tests. This method is commonly used in many laboratories and it consists in contacting a certain volume of tracer solution with a known mass of crushed solid material (EPA, 2008). After the equilibration the two phases are separated (by filtration or by centrifugation) and the tracer concentration in solution is measured.

The sorption tests were performed on two cementitious materials. One of them (noted S1) was prepared only from Portland cement powder and water (water: cement ratio of 0.26), in order to get information about the affinity of the radionuclides for calcareous material, and the other one (S2) had additional in composition quartz sand with granulation between 1 and 3 mm (cement – 42.5%, sand – 42.5%, water – 15%).

The batch sorption tests were run for a solid to liquid ratio of 1 to 25 (1 gram of crushed cementitious materials to 25 mL of tracer solution), at room temperature (23 ± 3ºC), in 25 mL centrifuge tubes. During the equilibration period, the test tubes were horizontally placed on an orbital shaker at 70 rpm. After equilibration the tubes were centrifuged at 4 000 rpm for 30 min and the supernatant was separated and measured for ^{14}C concentration by liquid scintillation spectrometry using a 2100 TRI-CARB Packard model.

^{14}C sorption was evaluated for two different contact time - 7 days and 100 days - in order to see the influence of contacting time on carbon sorption (Bucur et al., 2005).

To get the sorption isotherms, the batch tests were carried out for four different concentrations of ^{14}C in the contacting solutions, ranging between 10^2 Bq/L and 10^6 Bq/L. These isotherms were obtained by plotting the amount of contaminant removed by the solid matrix during the batch tests versus the contaminant concentration in solution. The amount of the contaminant sorbed on the solid matrix was computed using the following mass balance Equation 1:

$$A_{eq} = \left(C_0 - C_{eq}\right)\cdot\frac{V}{m} \qquad \textbf{\textit{Eq. 1}}$$

where: A_{eq} is the quantity of contaminant sorbed on the solid matrix (Bq/g), C_0 is the initial concentration of contaminant in solution (Bq/mL), C_{eq} is the contaminant concentration in solution after the sorption equilibrium was attained (Bq/mL), V is the volume of tracer solution (mL) and m is the mass of loss sample used in the sorption tests (g).

Each sorption test was performed in duplicate and one control sample with only tracer solution was run in the same conditions to check the potential sorption of radiocarbon on the test tubes and the stability of the tracer solution. Also, a blank sample was run to measure the ^{14}C background.

The sorption isotherms obtained for ^{14}C uptake on the two cementitious samples (S1 and S2), at the two contacting times, are reported in Figures 1 and 2.

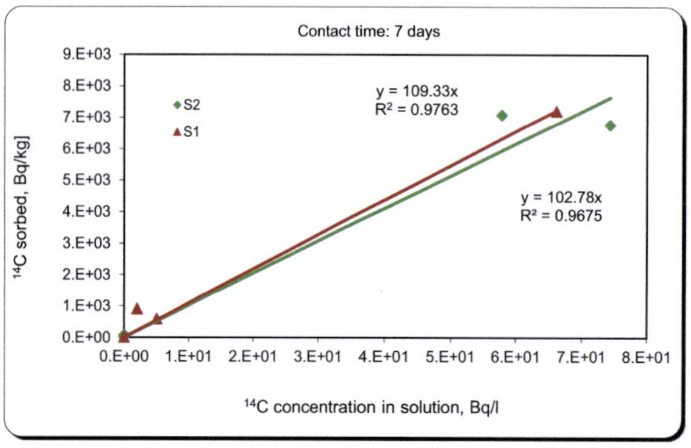

Figure 1: *^{14}C sorption isotherm after 7 days.*

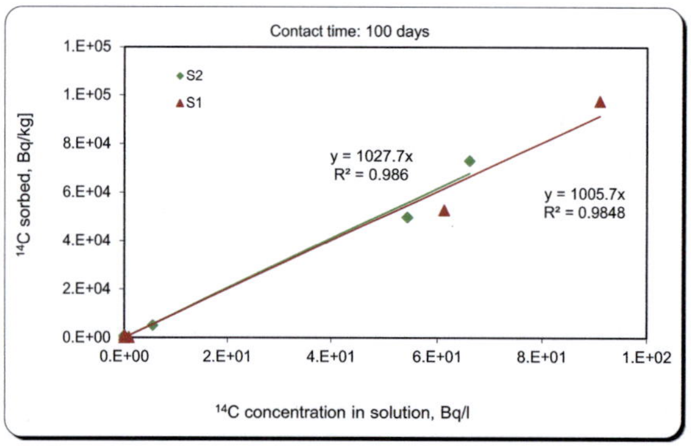

Figure 2: *^{14}C sorption isotherm after 100 days.*

The experimental data showed that ^{14}C sorption is linear, but the removal of ^{14}C from aqueous phase is slower than we previous obtained for cesium, with a K_d around 1 000 L/kg after a contacting time of 100 days. Generally, very high sorption values for ^{14}C on cementitious materials have been reported (~ 10^3 L/kg or even higher). A characteristic of such experiments is the long times required to reach a steady state sorption value (Bayliss et al., 1987). As it was expected, similar results were obtained both on HCP (S1 sample) and on mortar (S2 sample). Most cementitious systems contain $CaCO_3$. Since the cement powder has a finely ground nature, the $CaCO_3$ is also presented in a finely divided form. The pore water in a hardened concrete is saturated with respect to $CaCO_3$ and the concentration of CO_3^{2-} ions is pH dependent and influenced by the Ca^{2+} concentration arising from portlandite. The dominant removal mechanism of $^{14}CO_3^{2-}$ from solution is isotope exchange with CO_3^{2-} ions presented in the finely divided solid $CaCO_3$ phase (Wieland, 2014).

Previous diffusion experiments on cement based materials

Matrix diffusion is an important component to the radionuclide transport. In the mortar or concrete matrices, diffusion is the main mechanism for transport of ions through the pore water inside of concrete.

To assess the diffusion coefficients for different radionuclides, through-diffusion experiments were carried out using diffusion cells schematically represented in Figure 3. The experimental set-up consists of two reservoirs (inlet and outlet) of liquid separated by a cement wafer. The inlet reservoir was filled with a solution spiked with tritium as tritiated water ($1.3 \cdot 10^9$ Bq/m^3) and ^{137}Cs ($2.5 \cdot 10^8$ Bq/m^3) in 10 mM NaOH. The outlet reservoir was filled with tracer-free solution of 10 mM NaOH and more solution was continuously flushed through the outlet reservoir at very slow flow rate. The pressures between the two chambers were kept equal to minimize the advective flow and ensuring that main mechanism for the contaminants movement from the inlet reservoir through the cement pellet into the outlet reservoir is diffusion only.

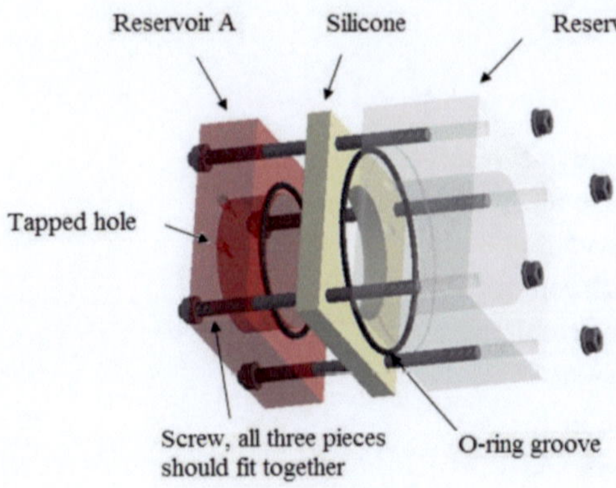

Figure 3: *Flow-through diffusion cell.*

The concentration of contaminant is measured in the flow of tracer-free solution leaving the outlet reservoir and the change of concentration over time is used to estimate the tracer apparent diffusion coefficient in cement.

Diffusion experiments were run on two types of cementitious materials as those previously presented at sorption experiments (S1 – a HCP sample and S2 a mortar based on CEM II). The cement wafers used in diffusion experiments were cut off from the large cement slabs, after 30 days from their preparation (Bucur et al., 2010).

Diffusion experiments were run in saturated conditions and to estimate diffusion coefficients, it was assumed that the tracers moved according to one-dimensional diffusive transport through the concrete wafers:

$$\frac{\partial c}{\partial t} = D_a \frac{\partial^2 c}{\partial x^2} = \frac{D_e}{R} \frac{\partial^2 c}{\partial x^2} \qquad \textit{Eq. 2}$$

Where c is the tracer concentration (Bq/m^3), D_a is the apparent diffusion coefficient (m^2/s), D_e is effective diffusion coefficient (m^2/s), R is the retardation factor (is 1 for non-sorbing elements such as tritium), x is the distance (m) and t represents the time (s).

Although analytical solutions to this partial differential equation exist for simple boundary conditions (Carslaw, 1959), the time-dependent concentration boundary conditions at the inlet and outlet reservoirs in the diffusion cell experiments demand a numerical solution. Thus, Equation 2 was solved using an implicit finite-difference technique. The equations describing the tracer concentrations in the inlet and the outlet reservoirs (the first and last finite difference nodes), respectively, are (Jenson et al., 1977):

$$\frac{\partial c_i}{\partial t} = \frac{\phi \pi r^2 D_e}{V_i} \left. \frac{\partial c}{\partial x} \right|_{x=0}$$ ***Eq. 3***

$$\frac{\partial c_o}{\partial t} = - \frac{\phi \pi r^2 D_e}{V_o} \left. \frac{\partial c}{\partial x} \right|_{x=L} - \frac{Q}{V_o} c_o$$ ***Eq. 4***

where,

c_i = tracer concentration in inlet reservoir, Bq/m^3
c_o = tracer concentration in outlet reservoir, Bq/m^3
V_i = volume of inlet reservoir, m^3
V_o = volume of outlet reservoir, m^3
Q = flush rate of outlet reservoir, m^3/s
ϕ = porosity of matrix cement
r = radius of rock wafer, m
L = thickness of rock wafer, m

The numerical solution for Equations 2, 3, and 4 was obtained using Diffcell Software developed at LANL.

The breakthrough curve for tritium as well as the theoretical fit to the experimental data are presented in Figure 4. Based on mercury intrusion porosimetry the total porosity of mortar sample used in these experiments was 0.6.

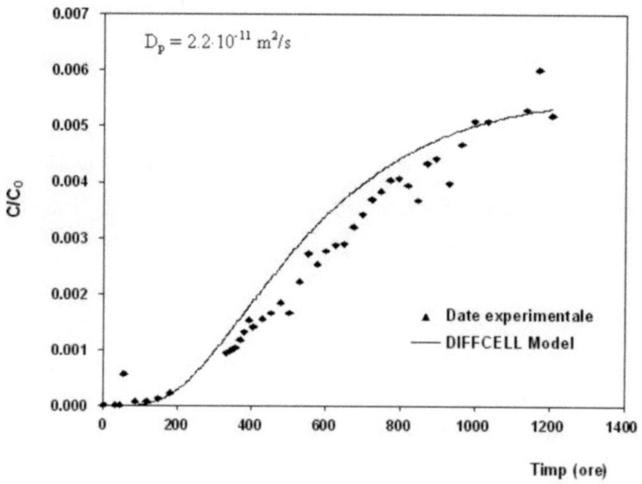

Figure 4: *^{3}H breakthrough curve and the theoretical one get for $D_e = 2.2 \cdot 10^{-11}$ m^2/s.*

Conclusions and Future work

The experimental procedures and sets-up previously used for radionuclide sorption and diffusion in cement based materials will be used to carry out the experimental programme we proposed to perform in the frame of CEBAMA project.

The experiments will consists of sorption/desorption tests and diffusion experiments for ^{14}C and ^{226}Ra by which the interaction of these radionuclides with high pH cementitious materials will be investigated. The sorption and diffusion experiments will be carried out on fresh HCP and aged ones, based on CEM I and CEM V cement types (Figure 5).

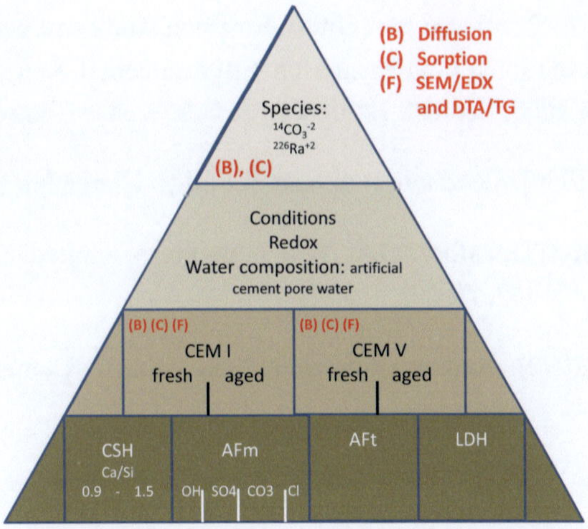

Figure 5: *Studied systems.*

The experimental data will be correlated with the concrete structure and composition obtained by SEM/EDX and DTA/TG.

Acknowledgement

The research leading to these results has received funding from the European Union's European Atomic Energy Community's (Euratom) Horizon 2020 Programme (NFRP-2014/2015) under grant agreement, 662147 – Cebama.

References

Allard, B., Torstenfelt, B., Andersson, K. (1981). Sorption studies of $H^{14}CO^{3-}$ on some geological media and concrete. MRS Proceedings 3, 465-473.

Bayliss, S., Ewart, F., Howse, R., Smith-Briggs, J., Thomason, H., Willmott, H. (1987). The solubility and sorption of lead-210 and carbon-14 in a near field environment. MRS Proceedings, 112, 33-42.

Bayliss, S., Ewart, F.T., Howse, R.M., Lane, S.A., Pilkington, N.J., Smith-Briggs, J.L., Willimams, S.J. (1989). The solubility and sorption of radium and tin in a cementitious near-field environment. MRS Proceedings, 127, 879-885.

Bayliss, S., Howse, R.H., McCrohon, R., Oliver, P., Smith-Briggs, J.L., Thomason, H.P. (2000). Near-field sorption studies. AEAT/ERRA-0073, Harwell.

Bradbury, M.H. and Sarott, F.A. (1994). Sorption databases for the cementitious near-field of a L/ILW repository for performance assessment. PSI Report, 95-06, and Nagra Technical Report, NTB 93-08.

Bucur, C., Olteanu, M., Pavelescu, M. (2010). Radionuclide transport through cement matrices. Revista de chimie, 5, 458-461.

Bucur C., Olteanu M., Olteanu C., Pavelescu M. (2005). Sorption studies of cesium and carbon on concrete, in proceedings of the 10th International Conference on Environmental Remediation and Radioactive Waste Management, September 4-8, 2005, Scottish Exhibition & Conference Centre, Glasgow, Scotland.

Carslaw, H.S. and Jaeger, J.C. (1959). Conduction of heat in Solids. Clarendon Press, Oxford.

EPA (2008). Fate, Transport and Transformation Test Guidelines. United States Environmental Protection Agency Report, EPA 712-C-08-009, 20-39.

Holland, T.R. and Lee, D.J. (1992). Radionuclide getters in cement. Cement and Concrete Research, 22, 247-258.

Jenson, V.G. and Jeffreys, G.V. (1997). Mathematical Methods in Chemical Engineering. 2nd ed., London Academic Press, 267.

Ochs, M., Dirk, M., Wang, L. (2016). Radionuclide and Metal Sorption on Cement and Concrete, Springer International Publishing Switzerland.

Tits, J., Iijima, K., Wieland, E., Kamei, G. (2006). The uptake of radium by calcium silicate hydrates and hardened cement paste. Radiochimica Acta 94, 637-643.

Wieland, E. and Van Loon, L.R. (2002). Cementitious near-field sorption data base for performance assessment of an ILW repository in Opalinus clay. PSI report, 03-06 and Nagra Technical Report, NTB 02-20.

Wieland, E. (2014). Sorption Data Base for the Cementitious Near Field of L/ILW and ILW Repositories for Provisional Safety Analyses for SGT-E2. Nagra Technical Report, NTB 14-08.

Investigation of the diffusion properties of inorganic ^{14}C species (dissolved and gaseous) through partially saturated hardened cement paste: Influence of water saturation conditions

Catherine Landesman[1*], Bernd Grambow[1]

[1] Armines/Subatech (FR)

* Corresponding author: Catherine.Landesman@subatech.in2p3.fr

Abstract

This report presents a description of the state of the art as well as a description of Armines/Subatech contribution for CEBAMA. It aims at giving a rapid overview of some scientific and experimental basis necessary to tackle the main topics of the project: water saturation of cement paste, carbonation process, aqueous and gas diffusion in cement, reactivity of inorganic carbon-14 in cement environment.

Introduction

The existence of partially saturated states in rocks, materials (engineered barriers, backfill materials) and surface environments is a common situation to all disposal sites. In France, this concerns either the current surface waste disposal facilities (very-low-level waste repository, low and intermediate-level short lived waste repository) or the future deep or sub-surface geological repositories such as CIGEO project for high-level waste (HLW) and intermediate-level long-lived waste (IL-LL waste) and FAVL project for low-level long-lived wastes. The unsaturated conditions are very variable depending on the type of disposal (surface or underground) and also on the origin of the gas which substitutes water inside the materials.

Under surface and sub-surface conditions or during the construction and the operation phase of a deep geological repository, the ventilation of underground galleries or shafts would induce, by gas exchanges with the atmosphere, a partial dehydration of the different materials of the disposal facilities. In porous media, partially saturated conditions denote the co-presence of an aqueous phase and a gaseous phase within the pore network. The degree of saturation (S_w) macroscopically defines the ratio between the pore volume filled with water to the total pore volume. The term "unsaturated" denotes situations where the degree of saturation is less than 100% (Drouet, 2010).

As an example, during the operational phase of the CIGEO project, the saturation of the Callovo-Oxfordian claystones will range from 70 - 95% in the excavation damaged zone (EDZ) around the galleries to over 95% in the intact clay rock. In a repository, cement materials play an important role either as engineered barriers or confinement matrices for some IL-LL wastes. Depending on their uses, saturation values of cement materials may range from 30 - 40% (backfill) to 80 - 90% (sealings). Moreover, some cemented IL-LL wastes may produce gases such as H_2, CO_2 and CH_4 that will be diluted and discharged by ventilation (operational period). During the post-closure period, hydrogen production due to the corrosion of metal components (HLW overpacks or IL-LL metallic wastes,...) and the radiolysis of some IL-LL wastes (polymers, bitumen) and/or of water (cemented wastes, salts) constitute the main sources of gas. This means that cemented wastes and cement based repository architectures (plugs, seals, linings...), in disposal cells, may remain partially unsaturated for many thousands of years before the completed resaturation by water of the repository (after 100 000 years) (Andra, 2014).

Carbon-14 has been identified as one of the important radionuclides in the inventory of radioactive waste due to its rather long half-live (5 730 years) (Andra, 2005). Results from previous and current European collaborative projects such as CARBOWASTE and CAST show that carbon-14 could be released by the corrosion of irradiated metals (zirconium alloys or stainless steels) and by the degradation of irradiated graphite and ion exchange materials. Thus the issue of carbon-14 in wastes is common to all disposal facilities (HLW, IL-LLW, LL-LLW). The speciation of carbon-14 covers various chemical forms: organic and/or inorganic species dissolved or gaseous. In its volatile form, carbon-14 may be present as CO_2 (inorganic species) or CH_4 and C_2H_6 (organic species) (CARBOWASTE, 2013; CAST, 2015; Wieland et al., 2015). Both the degree of saturation and the speciation of an element influence its sorption and transport properties. For carbon-14, inorganic species (CO_2/CO_3^{2-}) are the most reactive species toward cement materials due to their ability to enter in the carbonation process.

The carbonation process has been intensively studied for decades due to its importance in civil engineering (Suzuki et al., 1985; Groves et al., 1990; Cowes et al., 1992; Thiery et al., 2005; Borges, 2010). Depending on water saturation conditions, the reactivity of inorganic carbon with cement is due to the diffusion of either dissolved carbonate ions (saturated condition) or gaseous CO_2 (unsaturated condition) (Drouet, 2010). In partially saturated conditions, both gaseous and dissolved species may be transportable. The main consequences of the presence of carbonate are: i) a decrease of pH in the cement pore water, ii) a change in cement mineralogy (dissolution of Portlandite, progressive decalcification of C-S-H phases) and iii) a precipitation of calcite in the open porosity (Auroy, 2015). The formation of calcium carbonate inside the porosity may have an impact on the transfer properties of carbon gaseous species including $^{14}CO_2$ and thus contributes to the potential entrapment of carbon-14 species released from wastes.

The objective of the project is to understand the behaviour of inorganic carbon-14 species (dissolved and gaseous) by describing their diffusion properties (diffusion coefficients) in hardened cement paste (HCP) for different saturation conditions and different water degradation stages.

Carbon reactivity in cement materials under partially saturated conditions

Description of the carbonation process

Numerous studies describe the carbonation process of cement materials as a two-step diffusive process: diffusion of gaseous species (CO_2) in the unsaturated open porosity of the cement matrix and diffusion of aqueous carbonate ions in solution (Bary et al., 2004; Drouet, 2010; Auroy, 2015; and references inside). Carbon dioxide dissolves in alkaline solution with the following acid-base Reaction 1:

$$CO_2(g) + 2OH^- \leftrightharpoons CO_3^{2-} + 2H_2O \qquad (1)$$

The pKa values of the two acid-base couples are 6.37 and 10.33 (T = 298 K) (Cowes et al., 1992).

This dissolution potentially leads to the neutralization of the cement pore water and induces reactions with cement hydrated minerals.

❖ *Portlandite*

Portlandite ($Ca(OH)_2$) is the most sensitive hydrate for this reaction. The neutralization of hydroxide ions in solution leads to the precipitation of calcium carbonate following the Reaction 2.

$$Ca(OH)_2(s) + CO_2(g) \leftrightharpoons CaCO_3(s) + H_2O \qquad (2)$$

Due to the strong buffering effect of Portlandite, the pH value of the pore water is stabilized around 12.5 until its complete dissolution.

❖ *Calcium silicate hydrated (C-S-H)*

C-S-H minerals also react with carbon dioxide following a specific mechanism (Suzuki et al., 1985) (Reaction 3).

$$CaO_xSiO_{2y}H_2O_z + xCO_2(g) \leftrightharpoons xCaCO_3(s) + ySiO_2.tH_2O + (z-t)H_2O \qquad (3)$$

- The precipitation of calcium carbonate is concomittant to a noncongruente dissolution of C-S-H leading to a progressive decrease of the calcium to silica ratio (C/S) in the C-S-H.
- This decalcification induces the release of silicate ions in solution and in case of a complete carbonation, leads to their polymerization into silica gel.

Other cement hydrates such as hydrated calcium sulfoaluminate (ettringite, AFm phases) may react with carbon dioxide and form calcite, gypsum and hydrated alumina. An example is given for ettringite (Reaction 4):

$$3CaO.Al_2O_3.3CaSO_4.32H_2O + 3H_2CO_3 \leftrightharpoons 3CaCO_3 + 3(CaSO_4.H_2O) + Al_2O_3.xH_2O + (26-x)H_2O \qquad (4)$$

The total carbonation of cement hydrates may modify its microstructure. Due to their different molar volumes, the transformation of one mole of portlandite (V_{molar} = 33 cm^3/mol) into one mole of calcite (V_{molar} = 35 cm^3/mol) induces only a slight increase of the solid phase volume and as a consequence, a slight decrease of the porous space (Auroy, 2015). The pore space distribution is also affected by the carbonation process. Many studies report a decrease of the microporosity (< 100 nm) and a formation of a macroporosity (> 100nm). The changes in porosity are reported to be much more important for Portland cement (CEM I) than for Blast Furnace Slag cement (CEMV) (Houst et al., 1994; Thiery et al., 2005).

Influence of Relative Humidity (RH)

Because of the carbonation process is based on the diffusion of both gaseous (CO_2) and dissolved (carbonate ions) species, water saturation of the cement material has an impact on this process. Indeed, capillary porosity (open porosity) controls the rate of CO_2 transfer and as far as it is not saturated, carbon dioxide transfer is mainly due to a gas transfer. In partially saturated conditions, the diffusion rate of CO_2(g) (in the gaseous phase of the pore network) is 10 000 times higher than that of carbon aqueous species (in the liquid phase) (Bourbon, 2013). Auroy reports a synthesis of the evolution of the carbonation level for different cement materials as a function of relative humidity (Auroy, 2015).

The carbonation level is maximal for a relative humidity ranged between 55 - 65% of RH depending on the materials (cement paste, concrete, w/c ratio…) and on the carbonation protocol (natural or accelerated carbonation using $P_{CO2} > P_{CO2(atm)}$). This optimal value denotes the existence in the pore network of both a continuous unsaturated path through which CO_2 may diffuse and a continuous liquid phase in which CO_2 may dissolve and react.

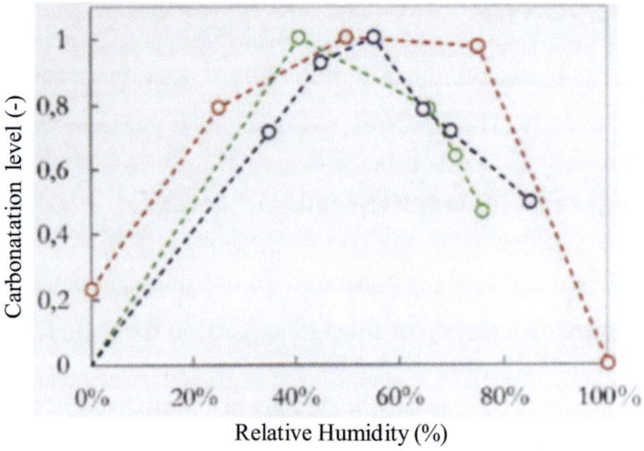

Figure 1: *Evolution of the carbonation level vs relative humidity for different cement materials (from (Auroy, 2015 and references herein): red (Verbeck et al., 1958), blue (Papadakis et al., 1991), green (Thiery et al., 2005).*

Moreover, at high saturation condition (> 80% RH), the percolation of the gaseous phase in the porous network is more and more difficult and the effective diffusion coefficient of gaseous species diminishes drastically in order to achieve very low value 10^{-14} - 10^{-13} m^2/s similar to those of aqueous species (Sercombe et al., 2007). It means that, for high water saturation conditions, a transition from a gas diffusion process to an aqueous diffusion process probably exists. Nevertheless, very few experimental data are available for these conditions.

Reactivity of inorganic carbon-14 in cement environment

Inorganic carbon-14 released from waste is expected to behave like stable carbon. Results of wet chemistry experiments (batch) performed on non carbonated hardened cement pastes show that $R_d(^{14}C)$ values (distribution ratio defined as the ratio of ^{14}C activity in the solid to ^{14}C activity in the corresponding equilibrium solution) range from 1 to 10 m^3/kg with a very slow kinetic of uptake (Allard et al., 1984; Bayliss et al., 1988).

These results were interpreted as a coupling of two mechanisms: first, the co-precipitation of $^{14}CO_3^{2-}$ with calcium carbonate and then the isotopic exchange of $^{14}CO_3^{2-}$ with $^{12}CO_3^{2-}$ and $^{13}CO_3^{2-}$ species.

For more complex cement materials (mortar, concrete) containing calcareous aggregates, Bradbury and Sarott (Bradbury, 1995) have developped a retention model based on the assumption that $^{14}CO_3^{2-}$ uptake relies only on the accessible quantity of (stable) carbonate in the solid and the solubility of calcite in solution. R_d expression is then given by Equation 1:

$$R_d = \alpha \times [\text{Carbon}]_{solid}/[\text{Carbon}]_{solution}\ (m^{-3}/kg^{-1}) \qquad \textit{Eq. 1}$$

With;

α: accessibility factor taking in account the fraction of accessible carbonate in aggregates.

α equals $0.17\times(1-\log\phi)$, where ϕ is the mean diameter of calcareous aggregate (Pointeau et al., 2002).

Diffusion in cement materials in partially saturated conditions

As previously explained, transfer properties of carbon (and so, of carbon-14) in cement materials are dominated by diffusion processes in aqueous or gas phases. Nevertheless, in literature very few studies are dedicated to the acquisition of transfer parameters in partially saturated conditions because of experimental difficulties to distinguish between these separate pathways. In the following paragraphs, basis of diffusion mechanism both in aqueous and gaseous phases will be described with a focus on results obtained in partially saturated conditions.

Generally speaking, diffusion process refers to the transport of species (molecular or ion) in an environment (solid, liquid or gas) due to a gradient of chemical potential. A proportional relationship exists between the magnitude of this gradient and the rate of transfer of the species expressed by the first Fick law. The proportionality coefficient is the diffusion coefficient (D_0). This description is valid also for the diffusion process applied to (radioactive) tracers in which the gradient of chemical potential is replaced by the activity gradient.

Based on Powers model (Powers, 1958), initially developed for CEM I cement, the microstructure of a cement paste can be described with two distinct classes of porosity:

- Capillary porosity (> 125-150 nm): It is a vestige of the inter granular space in the initially hydrated paste.
- Hydrate porosity (< 125 nm): porosity forms inside the hydrated phases mostly C-S-H phases

The microstructure depends on the nature of the cement, on the water/cement ratio (w/c) used for the hydration and on the age of the paste. Classically, the capillary porosity (and the maximal size of the pores) tends to diminish with a decrease of w/c and with time due to the progress of hydration reactions (ageing).

Consequently, the jointly decrease of the capillary porosity and the total porosity leads to a decrease of the porous volume accessible to the diffusion process.

Moreover for CEM V cement, the microstructure of the paste contains less capillary porosity than in a Portland paste. At a given w/c ratio, the total porosity value is then similar but the pore diameters are smaller than in a CEM I paste.

For the diffusion process, the most important parameter is not the total porosity but the interconnectivity of the capillary porosity. Results from Bentz and Garboczi (1991) show that, below a certain limit (percolation threshold), the interconnectivity of the capillary porosity falls to zero. The value is around 18% for a CEM I paste and do not depend on w/c ratio (Figure 2).

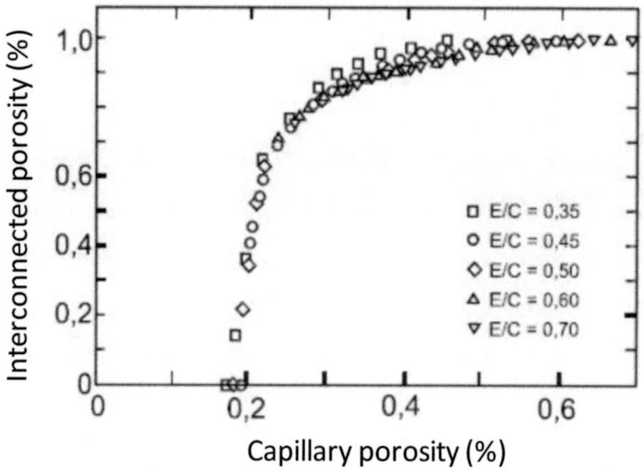

Figure 2: *Evolution of the fraction of interconnected porosity vs the fraction of capillary porosity (Bentz and Garboczy, 1991).*

Aqueous diffusion

In a porous media such as a cement paste, the aqueous diffusion of molecular or ion species (solute) takes place in the cement pore water through a volume of porous material. In saturated conditions, solute diffusion in cement-based materials is entirely described by the mass-balance equation. Thus, the first Fick law can be expressed by the following Equation 2:

$$J_e = -D_e\left(\frac{\partial c}{\partial x}\right) \qquad \textbf{\textit{Eq. 2}}$$

With;

J_e : flux of species through the surface ($mol/(m^2 \cdot s)$)

D_e : effective diffusion coefficient (m^2/s)

$\delta c/\delta x$: gradient of concentration (mol/m^4)

In a fully saturated material, the effective diffusion coefficient of a species can be related to its diffusion coefficient in solution (D_0) and to the geometry of the porous network (constrictivity and tortuosity) by the Equation 3:

$$D_e = \varepsilon \times D_0 \times \chi/\tau^2 \qquad \textbf{\textit{Eq. 3}}$$

With;

ε : water accessible porosity (-)

χ: constrictivity of the pore network

τ: tortuosity of the pore network.

Unfortunately, these two last parameters are not experimentally accessible.

If the species has no interaction with the material, it is considered as a "non reactive species". In cement material, this is often the case for tritiated water (HTO) which is considered as a tracer of water behaviour. Nevertheless, some results show that isotopic exchange with hydrated minerals takes place during the diffusion process (Tits et al., 2003).

On the contrary, reactive tracers are species which can interact with the materials following different chemical or physical processes (sorption, incorporation, precipitation, isotopic exchanges,...). These interactions are expressed by the rock capacity factor, α. For those species, the diffusion process is described by the second Fick law (Equation 4):

$$\frac{\partial C(x,t)}{\partial t} = \frac{D_e}{a} \times \frac{\partial^2 C}{\partial x^2} \qquad \textbf{\textit{Eq. 4}}$$

With;

α: rock capacity factor ; $\alpha = \varepsilon + \rho \times \frac{\partial [species]_{solid}}{\partial [species]_{solution}}$

ρ : bulk material density (kg/m^3)

$\frac{\partial[species]_{solid}}{\partial[species]_{solution}}$ stands for the chemical interaction of the species with the material. If this term is constant, it is equal to the distribution ratio (R_d, in kg/m^3). If not, it could be described by an isotherm for sorption processes (Langmuir or more complex).

Note : The ratio D_e/a expresses the apparent diffusion coefficient, D_a.

Classically, the effective diffusion coefficient of a tracer is measured with a through-diffusion set-up in which a water saturated cement sample is sandwiched between two reservoirs. In the upstream reservoir, the tracer is injected and maintained at a constant concentration. The diffusion process through the sample is then driven by the gradient created by the difference of the tracer concentration in the two reservoirs. The measurement of the tracer in the downstream reservoir allows to determine the steady state and to calculate the effective coefficient diffusion by applying the first Fick law.

Influence of water saturation (S_w)

There are very few studies dealing with the influence of water saturation (S_w) on the aqueous diffusion in cement materials and none dealing with radioactive tracers in HCP. A precise review of this topic for concrete has been done by Bourbon et al. (2013). The main results of this study are reported below.

First of all, as the aqueous diffusion takes place in the pore water (total porosity), transfer of species is possible only if the pore water constitutes a continuous phase meaning that water saturation should be higher than a critical threshold (S_c). Based on data obtained from impedance spectroscopy on concrete samples, Bourbon concluded that the effective diffusion coefficients for chloride ions tend to zero for S_w values below 0.36. This value may then be considered as the critical threshold S_c for concrete materials.

A recent study from Dridi and Lacour (2014) performed with a new specific experimental set-up (half-cell) confirms these results for CEM I paste (a drastic decrease of D_e below S_w= 0.3 - 0.5). The principle of this method consists in placing two samples of the material (source and target) into contact with one another in a sealed cell (Figure 3). The source sample is uniformly pre-doped with tracer (here lithium ions), while the target sample is tracer free. Bonding with fresh cement paste improves the continuity of the porous network between the source and the target during the diffusion test.

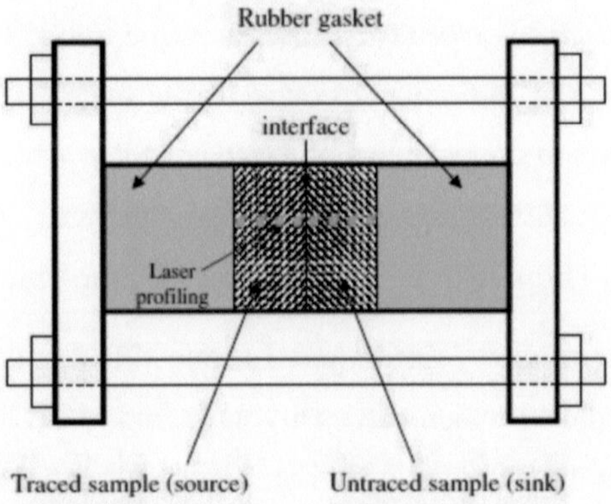

Figure 3: *Sketch of a half-cell diffusion test (Dridi and Lacour, 2014).*

After a certain diffusion time, concentration profiles of lithium within the samples are measured by elemental mapping (Laser Induced Breakdown Spectroscopy technique, LIBS). Diffusion profile of Li ions are ajusted by the analytical solution of a 1D problem in a semi-infinite environment (Crank, 1975) and lithium effective diffusion coefficients are then calculated by inverse analysis. Figure 4 summarizes results obtained for different S_w values.

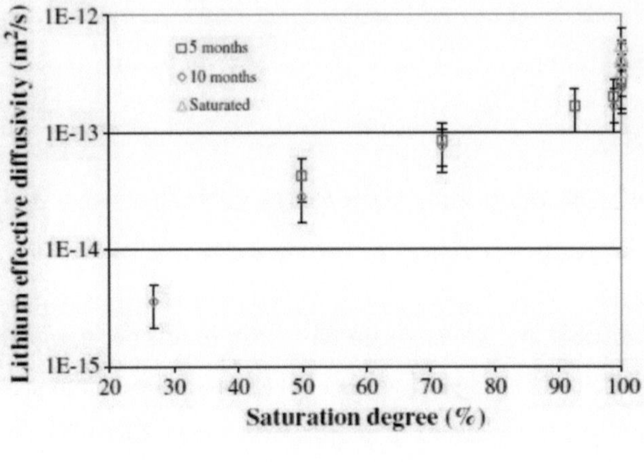

Figure 4: *Evolution of the effective diffusion coefficient of Li vs water saturation in a CEM I paste with w/c=0.4 (Dridi and Lacour, 2014).*

The authors interpret the drop of D_e(Li) values for S_w values ranged between 0.3 and 0.5 as the disappearance of a connected liquid phase in the capillary pores as well as in hydrates. The data obtained in the study are then consistent with the interpretation given for concrete materials (Bourbon, 2013).

Bourbon (2013) and Mercado et al. (2012) reported a comparison between CEM I and CEM V concretes based on impedance spectroscopy data (Figure 5). For CEM I concrete, the results show that, at high water saturation ($S_w > 0.75$), the decrease of D_e values with water saturation is very limited but becomes more important for $S_w < 0.75$ while for CEM V concrete, the D_e values are continuously decreasing on the all range of

S_w values. Below 0.75, the two materials seem to follow the same trend. These different behaviours have been interpreted as due to the difference in the microstructure of the materials. The evolution of D_e value at high saturation in CEM I material, is due to the presence of capillary pores which are progressively drained while, in CEM V material, those pores being absent, smaller pores (from hydrates) are drained. Below 0.75 (S_w), in both material, the same type of porosity (from hydrates) is concerned by the draining process.

From the experimental point of view, Dridi and Lacour (2014) reported several technical difficulties (e.g., contact between the two samples) which can easily have impact on the results. They have even proposed some important changes in the set-up (removal of the diffusion cell and use of a desiccator with constant RH). This means that even if this experimental set-up seems to be very attractive for acquiring data in partially saturated conditions, its realization appears to be touchy and has never been used for radioactive tracers.

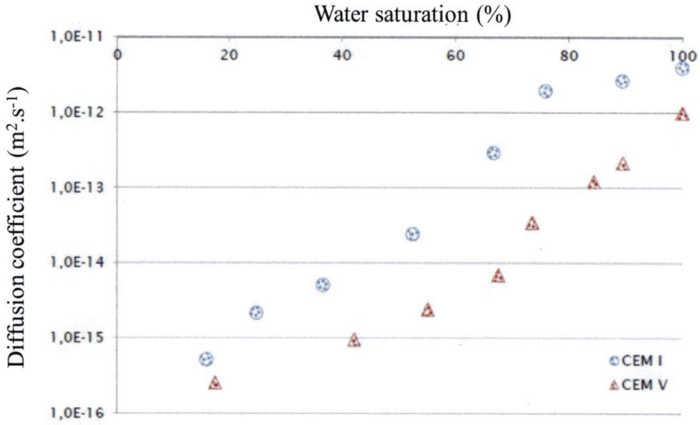

Figure 5: *Evolution of diffusion coefficient vs water saturation for CEM I (w/c=0.4) and CEM V (w/c=0.43) concretes (Bourbon, 2013).*

Due to these experimental difficulties, we have chosen to test, in this project, another experimental set-up for acquiring aqueous diffusion coefficient at high S_w value. This new experimental set-up is based on Savoye et al. (2010) work dedicated to the study of diffusion of radioactive tracers through partially saturated argillite clay stone. In this study, the diffusion parameters are determined using modified through-diffusion cells in which the water saturation of a sample is generated by the osmosis process. In fact, osmosis allows to control the water suction potential in a sample while maintaining contact with a chemical solution. This technique which is used by soil scientist and geotechnical engineering, is an alternative to the classical supersaturated salt solution technique (Zur, 1966; Delage et al., 1998). The suction is then generated by an osmosis process between the pore water and a highly concentrated solution of polymer (polyethylene glycol, PEG). The porous solid sample is separated from the PEG-solution by a semi-permeable membrane, which is permeable to all dissolved species except PEG. The exclusion of the PEG from the clay sample results in a osmotic-potential imbalance between the pore water and the water in the reservoirs of the diffusion cell. This osmotic suction has the effect of keeping the porous solid sample partially saturated. The value of the imposed suction depends on PEG concentration in solution. Delage et al. (1998) reported the evolution of the suction *vs* PEG concentration (Figure 6).

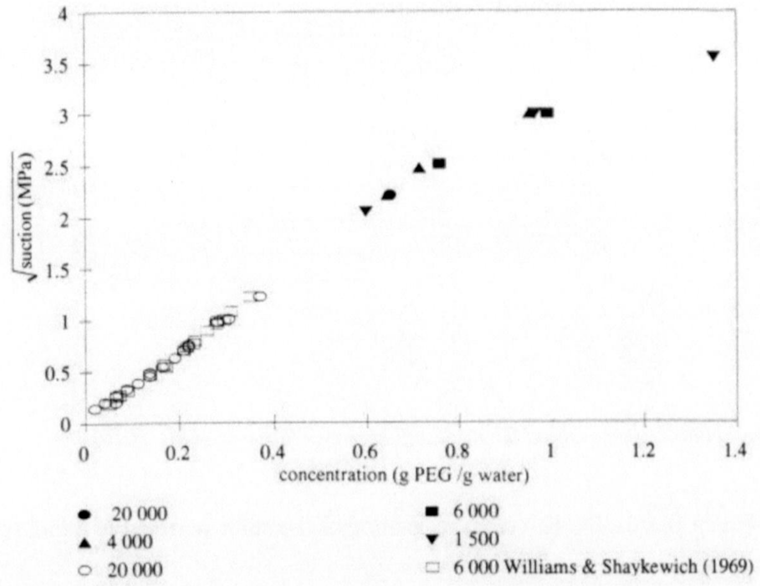

Figure 6: *Evolution of the suction vs PEG concentration with various molar mass (Delage et al., 1998).*

All data below 4 MPa fit the following parabolic relation between the suction (in MPa) and the concentration of PEG (g of PEG/g of water)

$$\Psi = 11\times[PEG]^2 \qquad \textit{Eq. 5}$$

Above 4 MPa, suction values are lower than those expected from Equation 9. Nevertheless, suctions up to about 10 MPa can be reached by using very highly concentration PEG solution (> 1 g of PEG/g of water). Previous geomechanical studies on Callovo-Oxfordian clay stone showed that suctions up to 10 MPa lead to a water saturation of about 80%. It is clear that the relation between suction and water saturation strongly depends on the nature of the material and more precisely on the microstructural properties (porosity, geometry of the pore network, permeability,…). The application of this technique to cement pastes will need first to determine such a relationship for these materials.

A sketch of the modified through-diffusion cell used for the diffusion experiments is given in Figure 7.

The cut-off of the membrane has to be adapted with the molar mass of PEG molecule. For example, with PEG (6 kDa), a membrane with a cut-off of 3.5 kDa has been used. With this set-up, Savoye et al. (2010) succeeded in measuring effective diffusion coefficients of molecular (HTO) and ion radionuclides (Na^+, Cs^+, Sr^{2+} and I^-) for water saturation ranged from 0.8 to 1 in partially saturated clay stone samples (Savoye et al., 2012).

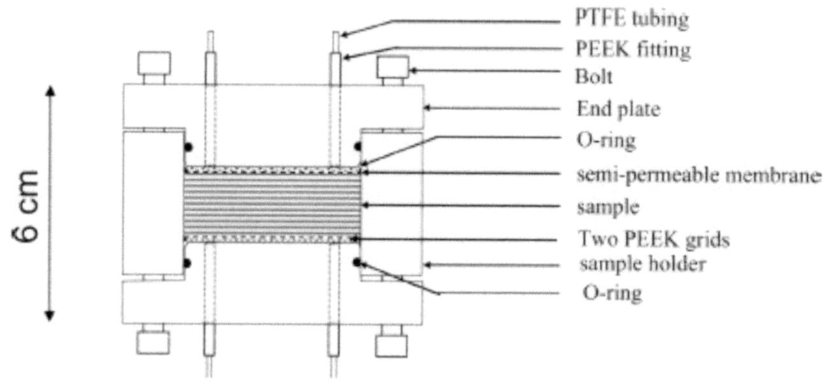

***Figure** 7: Schematic cross-section of a modified through-diffusion cell (Savoye et al., 2010).*

For application to cement pastes, the chemical resistance of the semi-permeable membrane to highly alkaline solution will have to be tested.

In conclusion, literature review shows that the osmosis technique is probably the easiest technique which can be adapted for the measurement of diffusion parameters of radionuclides in partially saturated conditions. Nevertheless, preliminary specific studies have to be done before performing experiments with cement pastes (checking the Ψ = f[PEG] relationship, testing the resistance of the membrane to high alkaline solutions,..).

Gas diffusion

This paragraph deals with the diffusion of gas molecule (in air) contained in the pore of an unsaturated porous media. Two diffusion mechanisms have then to be considered (Sercombe et al., 2007; Bourbon, 2013):

- Molecular (or ordinary) diffusion occurs predominantly when molecule-molecule collisions dominate over molecule-pore wall collisions. The molecule of different species move under the influence of concentration gradients and the diffusive transfer follows the Fick law.
- Knudsen (or free-molecule) diffusion occurs predominantly when molecule-molecule collisions can be ignored compared to molecule-pore wall collisions. The molecule of different species move entirely independently from each other.

The prevalence of ordinary or Knudsen diffusion depends first, on the mean free path of the gas molecule (approximately 100 nm for a gas molecule at atmospheric pressure and T = 293 K) depending itself on global parameters (total pressure, temperature, nature of the gas) as and second, on the geometry of the porous media (pores size, degree of connectivity of the unsaturated pores). As the pore sizes in cement pastes are widely distributed from nm to mm scale, it is difficult *a priori* to state what is the relevant mechanism for a given paste.

Experimental set-ups designed for studying gas diffusion in cement paste as a function of water saturation are scarce and experiments are performed with hydrogen or inert gas (helium or xenon). For these light gases (H_2 or He), the effective diffusion coefficient value is around $7 \cdot 10^{-7}$ m^2/s. Classically, the set-up consists in a diffusion cell, vacuum pumps, pressure sensors, gas flow lines and a gas chromatography system. As an example, the set-up used by Sercombe et al. (2007) is shown Figure 8.

To our knowledge, no use of gas diffusion of set-up has been already reported for radioactive gas experiment.

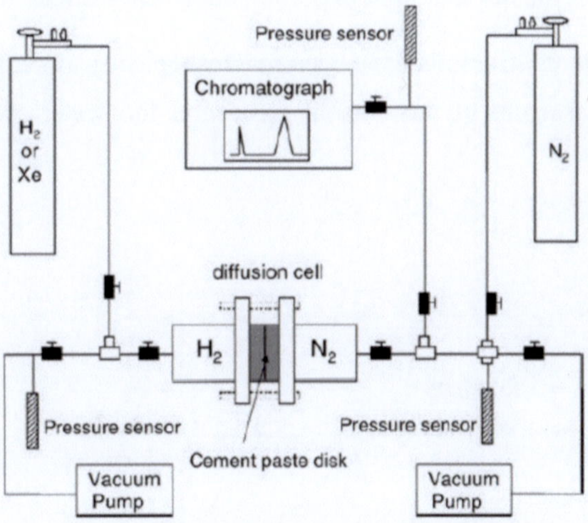

Figure 8: *Sketch of the experimental gas diffusion set-up (Sercombe et al., 2007).*

For CEM I paste, the evolution of effective diffusion coefficient *vs* water saturation is reported in Figure 9.

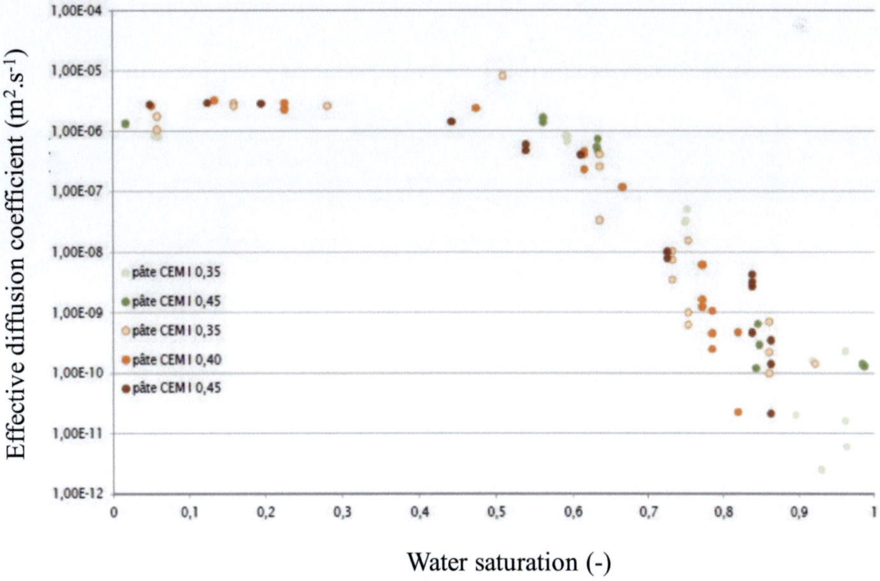

Figure 9: *Evolution of helium effective diffusion coefficient vs water saturation for CEM I pastes for different w/c ratio (Sercombe et al., 2007; Frizon and Gallé, 2009; Vu, 2009).*

For $S_w < 0.4$, D_e(He) values are constant at 10^{-6} m²/s. The interpretation is that the continuity of the gaseous phase is such that even great variations of water saturation have no impact on the diffusion. For $0.4 < S_w < 0.75$, D_e(He) values decrease linearly of almost four orders of magnitude ($D_e(He) = 5\cdot10^{-10}$ m²/s) and do not depend on w/c ratio. This linear drop is interpreted as the progressive closure of the percolation path due to the filling of the porosity by water and then the apparition of a continuous aqueous phase which prevents the progression of gas in the porosity. For $S_w > 0.75$, the dispersion of experimental is such that no trend can be given.

As a first modelling approach, these results have been macroscopically modelled with a simplified formalism taking into account the microstructure of the cement paste and the water saturation following Equation 6 (Bourbon, 2013).

$$D_e = \frac{De^{\text{º}}}{p^b}(1 - S_W)^a \qquad \textbf{\textit{Eq. 6}}$$

With;

p: total porosity (-) with p = 0.35 ± 0.05

D_e^0: He diffusion coefficient at dry state with $D_e^0 = (3.0 \pm 1.5)\cdot10^{-6}$ m²/s

S_W: water saturation (-)

This model is not applicable to S_w values close to 1 because it is unable to account for the simultaneous existence of a continuous water path and a discontinuous gas path in fully saturated materials.

The best fit of the experimental data has been obtained with a = 5.25 ± 1.25 and b = 2 and is reported Figure 10.

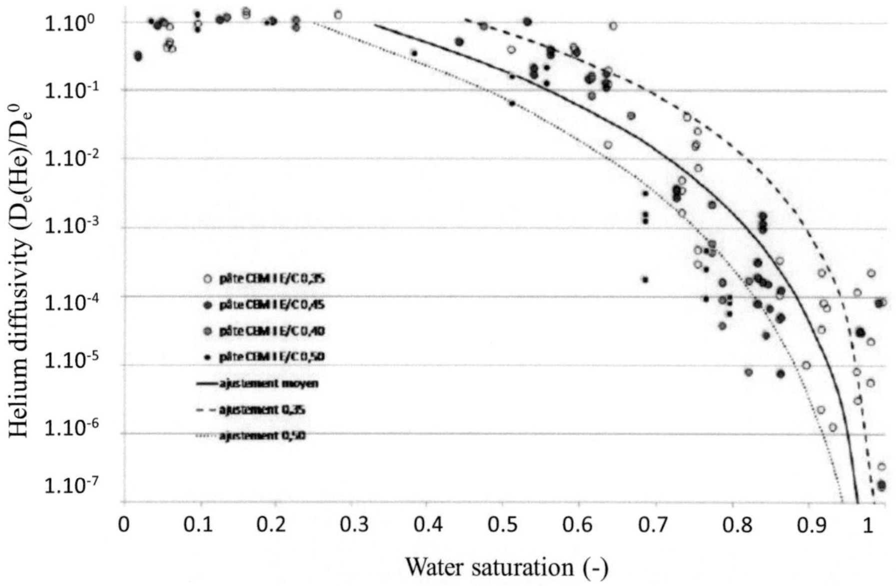

Figure 10: *Fit of helium diffusivity (D_e/D_e^0) vs water saturation for CEM I paste with various w/c ratios.*

The same fitting approach can be done with data acquired on CEM V based material with p = 0.35 ± 0.05 and $D_e^0 = (1.1 \pm 0.5)\cdot 10^{-7}$ m²/s. The best fit of the experimental data has been obtained with a = 3.1 ± 1.0 and b = 1.67 (Figure 11). In this type of material (CEMV), the presence of a very dense porosity due to the hydrates smoothes the difference between paste and concrete samples.

Moreover, Sercombe et al. have compared hydrogen diffusion coefficient data obtained for CEM I and CEM V pastes (Figure 12). The behaviour of CEM V paste is different from this of CEM I one. First, for S_w values ranges from 0.1 to 0.6, the hydrogen diffusion coefficient, is one order of magnitude lower than this of CEM I paste ($D_e(H) = 2$ to $8\cdot 10^{-8}$ m²/s) and second, the evolution of the diffusion coefficient *vs.* water saturation shows two steps:

For $0.6 < S_w < 0.85$, a very slow decrease (by one order of magnitude), then for $S_w > 0.85$, a sharp decrease (three orders of magnitude) is registered. This sharp decrease of the diffusion coefficient indicates a discontinuity (like a percolation threshold) of the capillary pore system in the CEM V cement paste. It shows that above a given saturation (here $S_w = 0.9$), the pore network of the CEM V paste accessible to gas species becomes highly discontinuous. This behaviour is not observed on CEM I paste which shows a continuous evolution of the diffusion coefficient in the same range of S_w.

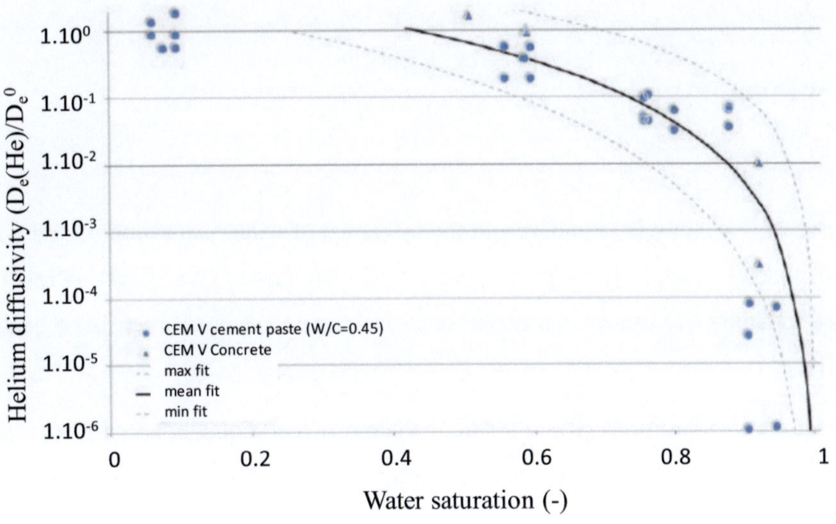

Figure 11: *Fit of Helium diffusivity (D_e/D_e^0) vs water saturation for a CEM V paste (w/c = 0.45) and CEM V concrete (Bourbon, 2013).*

This difference probably originates from a more uniform pore size distribution in CEM V paste, centered on smaller pore diameters. The diffusion threshold would then be related to the saturation of an important fraction of pores of similar diameter.

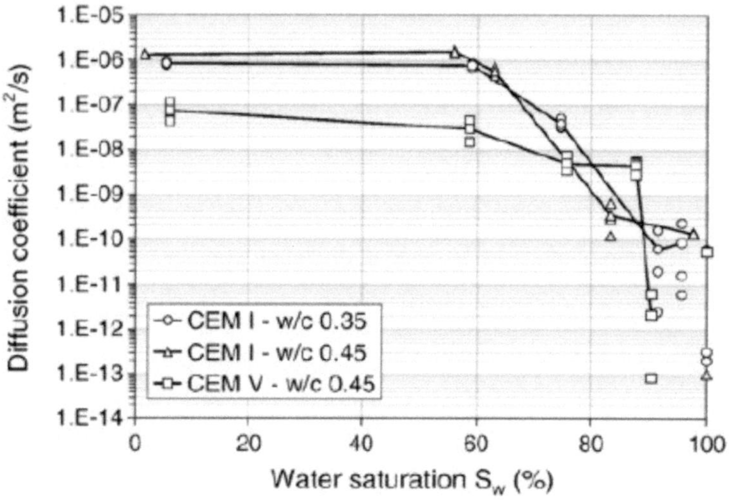

Figure 12: *Hydrogen diffusion coefficient vs water saturation for CEM I (w/c = 0.35 and 0.45) and CEM V (w/c = 0.45) cement paste (Sercombe, 2007).*

The wider pore size distribution in CEM I paste leads to a smoother evolution of the diffusion coefficient since the saturation (or desaturation) of the porosity occurs progressively (Sercombe, 2007).

Description of the program

Overal objectives

The first objective of this study is to understand the influence of water saturation on the diffusion profiles of inorganic carbon-14 species in partially saturated hardened cement paste. Two ranges of relative humidity (RH) will be investigated inducing the use of two different experimental set-ups which have been chosen from the state of the art review.

For high water saturation conditions (RH > 90%), the osmotic technique, adapted from Savoye et al. (2010), will be applied to impose the water saturation in HCP samples. The diffusion profile (and diffusion parameters) of aqueous carbon-14 (as carbonate anion) will be thus determined in in-diffusion experiments.

For intermediate water saturation conditions (70% < RH < 90%), water saturation will be imposed by the "oversaturated salt solution" technique and a classical gas diffusion experiment will be used to determine the diffusion profiles of gaseous carbon-14 ($^{14}CO_2$).

The second objective is to investigate the influence of water degradation of the HCP on the carbon-14 diffusion profiles.

Fresh cement pastes with high pH conditions will be studied: one in presence of alkaline ions (Na, K) and pH = 13.6 and the second, in alkaline free solution (Portlandite buffer pH = 12.4). These high pH conditions are representative of the two first stages of water degradation of a cement paste under repository conditions.

Depending on the results obtained on fresh samples, aged cement paste (obtained after a leaching stage) might be studied (optional study).

Description of the work to be performed (tasks)

The cement material used is a CEM V HCP (Calcia from Rombas factory) cast with a standard water/cement ratio equal to 0.4 and cured during 18 to 24 months. HCP samples are disks with 50 mm in diameter and 2 - 3 mm in thickness (sliced with a diamond wire saw).

Armines/Subatech contribution is divided in 4 tasks.

Task 1: Application of the osmotic technique to a cement paste

The osmotic technique described in paragraph 2.2 will be applied to HCP sample in order to reach high water saturation conditions (RH > 90%). The diffusion parameters in partially saturated samples will be determined using diffusion cells in which the (water saturation (suction) will be generated by the osmosis process between the pore water (present in the pores of the sample) and a highly concentrated solution with large-sized molecules of polyethylene glycol (PEG) separated by a semi-permeable membrane. The sketch of the experimental set-up is given Figure 7. Three preliminary tasks have to be done before starting diffusion experiments.

- Modification of the home-made diffusion cell (Figure 13) to the osmotic technique (modifications of sample-holder)
- Test of the chemical resistance of the semi-permeable membrane (regenerated cellulose) and of PEG macro molecule in high alkaline solution. If these tests are non conclusive, other membrane and/or macromolecule will have to be tested.

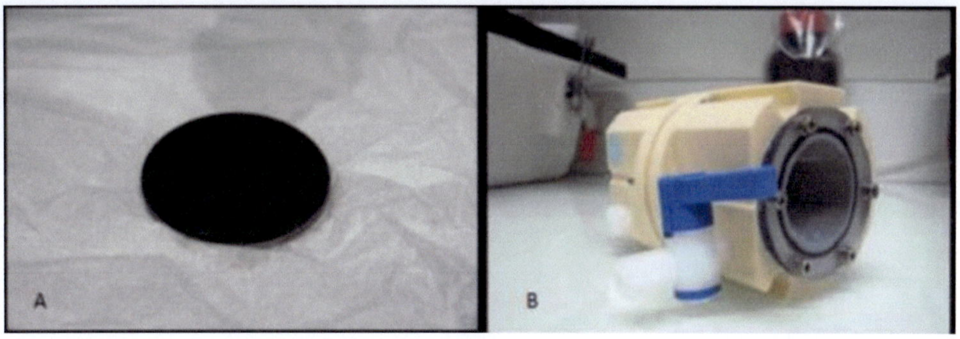

Figure 13: *Classical diffusion cells used at Subatech (right) and sliced HCP.*

- Desaturation procedure: HCP samples will be desaturated in a dessicator with a oversaturated NaCl solution (RH = 75% at 20°C) before being resaturated with the osmotic technique.

The first diffusion experiment will be performed with tritiated water (HTO) in order to

i) test the set-up (absence of leak),

ii) to acquire diffusion data (effective diffusion coefficient) for HTO in partially saturated sample,

iii) to compare them with literature data (Bejaoui et al., 2006).

Task 2 :Acquisition of effective diffusive coefficients data for inorganic aqueous carbon-14 species under partially saturated conditions (RH > 90%)

For these experiments, carbon-14 will be used as carbonate ions. In-diffusion experiments will be performed on two different fresh HCP samples (high pH conditions) in order to study the influence of the calcium concentration:

- Fresh cement paste with alkaline ions (Na, K) and pH value equal to 13.6. In these conditions, Ca concentration is around 2 mmol/L (equilibrium with Portlandite).
- Fresh cement paste without alkaline with a pH value equal to 12.4 (Portlandite buffer)

Optionally, if it is experimentally achievable, an aged cement paste will also be studied (leached sample; pH $\approx$ 11).

For tasks 1 and 2, radionuclides (HTO and C-14) activities will be measured by direct liquid scintillation (SL). ICP-MS and ion chromatography analysis will also be performed for determining the other aqueous compounds. Carbon-14 diffusion profiles will be analysed by beta-autoradiography and or micro-abrasion in order to obtain a more precise localization of C-14 in the solid. Additional Scanning Electron Microscopy (SEM/EDX) analysis and/or ICP-MA (laser ablation) will also be performed for the elemental mapping of the samples.

Task 3 : Acquisition of transport parameter for volatile carbon-14 species (CO2) in partially saturated HCP samples (70% < RH < 90%)

This task is dedicated to the study of CO_2 transport in partially saturated HCP samples. The experimental program is based on classical gas diffusion experiments. The experimental set-up, adapted from Sercombe et al. (2007) is given Figure 14.

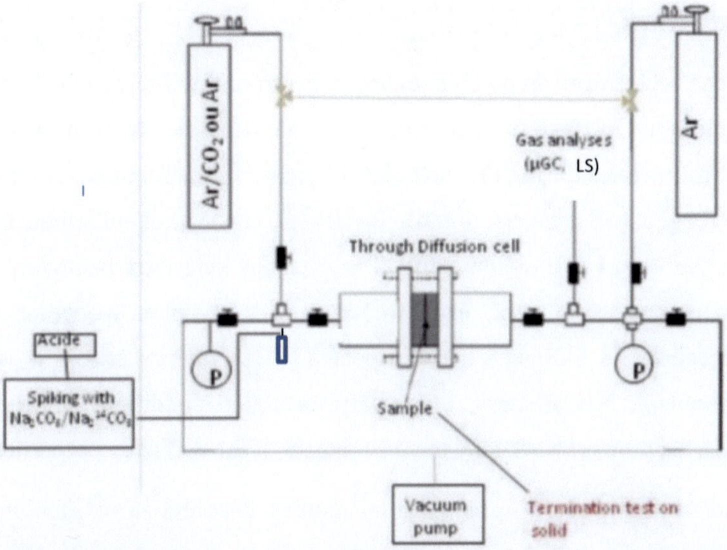

Figure 14: *Sketch of the gas-diffusion set-up.*

Gas diffusion experiment is described as the following. Centimetric HCP cylinders (36 mm in diameter and 25 mm thick max.) partially saturated, by oversaturated solution technique (NaCl, RH = 75%) in dessicator, are put on a specifically designed gas diffusion cell. A mixture of inert gas (argon) and CO_2 labelled with C-14 will be injected in the upstream reservoir. The mixture of gas, in the downstream reservoir, is then analyzed either by liquid scintillation (after bubbling in a NaOH solution) for $^{14}CO_2$ or by micro gas chromatography for stable CO_2 in order to follow the diffusion of gas in the sample. At the end of the experiment, autoradiography technique will be used to measure the penetration depth of carbon-14 into the samples and to determine the shape of the diffusion profile. Moreover, one experiment with enriched $^{13}CO_2$ gas mixture instead of $^{14}CO_2$ is also planned in order to use other technique such as nano SIMS cartography (Secondary Ions Mass Spectrometry) for solid characterization. The key-point of this experimental program is the *in situ* generation of the $^{14}CO_2/CO_2$ from an acidified solution of $^{14}CO_3^{2-}/ CO_3^{2-}$.

Task 4: Coupled geochemical and transport modelling

The extent of the task will depend strongly on the experimental results which will be obtained in the experimental tasks. Geochemical modelling will be used to interpret the diffusion profile of carbon-14 in the solid, taking into account the different chemical processes (sorption isotopic exchange, incorporation,…) which may occur during the diffusion and the partial saturation of the solid (two-phase system). The modelling task will be based on PhreeqC using the 1D transport module.

Conclusion and future work

Unsaturated conditions will prevail in surface and/or underground waste disposal for a very long period of time (100 000 years) due to the interaction of materials with atmospheric condition during the operation phase or to the release of gas from wastes during the post-closure phase. Cement materials which are used for several purposes in a storage (engineered barriers, matrix for IL-LL wastes,...) will then be affected. Due to its relatively long half life, carbon-14 is a radionuclide of interest for long-term behaviour studies. It is now well established that carbon-14 can be released from wastes with a complex speciation (organic or inorganic, dissolved or gaseous species). As inorganic carbon species ($^{14}CO_2$ and its bases) is very reactive species in saturated cement environment. Nevertheless, in partially saturated conditions, the presence of two phases (gas/solution interface) in the pore network may have an impact of the diffusive properties of carbon-14.

This rapid overview shows that if some (scarce) studies describe the diffusion properties of gas in unsaturated cement materials as a function of water saturation, none is related to CO_2. Moreover, due to its strong reactivity in cement pore water, carbonate ions are not studied as a potentially diffusive species. In conclusion, as far as we know, no data is available on the diffusive properties of carbon-14 in partially saturated cement materials.

From the experimental point of view, only few experimental set-ups are available for studying the diffusive parameters of radionuclides (or tracers) under partially saturated conditions. Depending on water saturation (S_w), either a modified through-diffusion set-up (high $S_w > 0.8$) (Savoye et al., 2010) or a more classical gas diffusion set-up (intermediate S_w, 0.5 - 07) (Sercombe et al., 2007) will be then tested in this project.

A 2-year post-doctoral position is now open at Subatech for the realization of this scientific program.

Acknowledgement

The research leading to these results has received funding from the European Union's Horizon 2020 Research and Training Programme of the European Atomic Energy Community (EURATOM) (H2020-NFRP-2014/2015) under grant agreement n° 662147 (CEBAMA).

References

Allard, B., Eliasson, L., Hoglund, H., Anderson, K. (1984). Sorption of Cs, I and actinides in concrete systems. SKB Technical report, TR-84-15.

Andra (2005). Dossier Argile – tome Evolution phénoménologique du stockage géologique. Juin 2005, www.andra.fr (site accessed 15 april 2016).

Andra (2014). Programme détaillé du Groupement de Laboratoires: Comportement chimique et transfert dans des environnements / ouvrages insaturés. ANDRA report, DRD Z.NT.ASTR.14.0017.A.

Auroy, M. (2015). Impact of the carbonation on the water transport properties in cementitous material. PhD thesis, Paris Est University.

Bary, B. and Sellier, A. (2004). Coupled moisture-carbon dioxide –calcium transfer model for carbonation of concrete. Cement and Concrete Research, 34, 1859-1872.

Bayliss, S., Ewart, F.T., Howse, R.M., Smith-Briggs, J.L., Thomason, H.P., Willmott, H.A. (1987). The solubility and sorption of lead-210 and carbon-14 in a near-field environment. MRS Proceedings, 122, 35-42.

Bejaoui, S., Bary, B., Nitsche, S., Chaudanson, D., Blanc, C. (2006). Experimental and modelling studies of the link between microstructure and effective diffusivity of cement pastes. Revue Européenne de Génie Civil, 10, 1073-1106.

Bentz, D.P, Garboczi, E.J, Stutzman, P.E. (1992). Computer modeling of the interfacial zone in concrete. RILEM Proceeding: Interfaces in cementitious composites, 18, 107-116.

Borges, P.H.R., Costa, J.O., Milestone, N.B., Lynsdale, C.J., Streatfield, R.E. (2010). Carbonation of CH and C–S–H in composite cement pastes containing high amounts of BFS. Cement and Concrete Research, 40, 284-292.

Bourbon, X. (2012). Référentiel des matériaux d'un stockage de déchets de haute activité et de moyenne activité à vie longue – Tome 2 : les matériaux cimentaires. Andra report, CG.RP.ASCM.12.001.

Bradbury, M. and Sarott, F.A. (1995). Sorption databases for the cementious near-field of L/ILW repository for performance assessment. PSI report, 95-06.

Toulhoat, N., Narkunas, E., Zlobenko, B., Diaconu, D., Petit, L., Schumacher, S., Catherin, S., Capone, M., von Lensa, W., Piña, G., Williams, S., Fachinger, J., Norris, S (2015). CAST project, Carbon-14 Source Term. Work Package 5, Review of Current Understanding of Inventory and Release of C14 from Irradiated Graphite (D5.5).

Grambow, B., Norris, S., Petit, L., Petit, L., Blin, V., Comte, J., de Visser-Tynova, E. (2013). CARBOWASTE project. Work Package 6, Disposal Behaviour of Irradiated Graphite & Carbonaceous Wastes -Final Report.

Cowie, J. and Glasser, F.P. (1992). The reaction between cement and natural waters containing dissolved carbon dioxide. Advances in Cement Research, 4, 119-134.

Crank, J. (1975). The Mathematics of diffusion. Oxford Univ. Press.

Delage, P., Howat, M.D., Cui, Y.J. (1998). The relationship between suction and swelling properties in a heavily compacted unsaturated clay. Engineering Geology, 50(1-2), 31-48.

Dridi, W. and Lacour, J.L (2014). Experimental investigation of solute transport in unsaturated cement paste. Cement and Concrete Research, 63, 46-53.

Drouet, E. (2010). Impact of temperature on the carbonation process of hardened cement paste. PhD thesis, Ecole Normale Supérieure de Cachan.

Frizon, F. and Gallé, C. (2009). Experimental investigations of diffusive and convective transport of inert gas through cement pastes. Journal of Porous Media, 12(3), 221-237.

Groves, G.W., Rodway, D.I., Richardson, I.G. (1990). The carbonation of hardened cement pastes. Advances in Cement Research, 3, 117-125.

Houst, Y.F. and Wittmann, F.H. (1994) Influence of porosity and water content on the diffusivity of CO_2 and O_2 through hardened cement paste. Cement and Concrete Research, 24, 1165-1176.

Mercado, H., Lorente, S., Bourbon, X. (2012). Chloride diffusion coefficient: a comparaison between impedance spectroscopy and electrokinetic test. Cement and Concrete Composites, 34(1), 68-75.

Papadakis, V.G., Vayenas, C.G., Fardis, M.G. (1991). Fundamental modeling and experimental investigation of concrete carbonation. ACI Materials Journal, 83(4), 363-373.

Pointeau, I., Coreau, N., Reiller, P. (2003). C-14 carbonate uptake by limestone in concrete buffer environments. Nagra Workshop.

Powers, T.C. (1958). Structure and physical properties of hardened cement paste. Journal of the American Ceramic Socciety, 41(1), 1-6.

Savoye, S., Page, J., Puente, C., Imbert, C., Coelho, D. (2010). New experimental approach for studying diffusion through an intact and unsaturated medium: a case study with Callovo-Oxfordian argillite. Environmental Science & Technology, 44(10), 3608-3704.

Savoye, S., Beaucaire, C., Fayette, A., Herbette, M., Coelho, D. (2012). Mobility of Cesium through the Callovo-Oxfordian claystones under partially saturated conditions. Environmental Science & Technology, 46(5), 2633-2641.

Sercombe, J., Vidal, R., Gallé, C., Adenot, F. (2007). Experimental study of gas diffusion in cement paste. Cement and Concrete Research, 37, 579-588.

Suzuki, K., Nishikawa, T., Ito, S. (1985) Formation and carbonatation of C-S-H in water. Cement and Concrete Research, 15, 213-224.

Thiery, M. (2005). Modelling of the atmospheric carbonation process in cement material: influence of kinetic effects and microstructural and hydric changes. PhD thesis, Ecole Nationale des Ponts et Chaussées.

Tits, J., Jakob, A., Wieland, E., Spieler, P. (2003). Diffusion of tritiated water and $^{22}Na^{+}$ through non-degraded hardened cement pastes. Journal of contaminant Hydrology, 61, 45-62.

Verbeck, G. (1958). Carbonation of hydrated Portland cement. ASTM Special Publication, 205, 17-36.

Vu, T.H., Frizon, F., Lorente, S. (2009). Architecture for gas transport through cementitious materials. Journal of Physics D: Applied Physics, 42(10), 105501.

Wieland, E. and Hummel, W. (2015). Formation and stability of ^{14}C-containing organic compounds in alkaline iron-water systems: preliminary assessment based on a literature survey and thermodynamic modeling. Mineralogical Magazine, 79(6), 1275-1286.

Zur, B. (1966). Osmotic control the matrix soil water potential. Soil Science, 102, 394-398.

Procedures for assessing radionuclide retention in cementitious systems and single mineral phases

David Read[1,2*], Matthew Isaacs[1,3], Mónica Felipe-Sotelo[1]

[1] Department of Chemistry, University of Surrey (UK)
[2] National Physical Laboratory (UK)
[3] Institute of Energy and Climate Research: Nuclear Waste Management and Reactor Safety (IEK-6), Forschungszentrum Jülich GmbH (DE)

* Corresponding author: d.read@surrey.ac.uk

Abstract

This paper presents an overview of the experimental procedures that will be used as part of the contribution of the University of Surrey to the CEBAMA project. The solubility of long-lived fission and decay products, Ra, Tc(IV), I(-1, V), Se(IV, VI) and Cl, in cementitious media will be determined from both the over- and under-saturation directions. A series of batch studies investigating uptake of the radionuclides of interest will be carried out with 5 different cementitious mixtures designed either for conditioning and packaging of radioactive waste, in construction, or as backfill material, as well as with specific mineral phases present in the cement matrices (CSH, AFm, AFt). These sorption studies will be complemented with through-diffusion experiments. Examples of the potential solid phase characterisation techniques are given.

Introduction

Cementitious materials are used widely in radioactive waste management, for example in the solidification of low and intermediate level wastes, or as construction and barrier materials in underground and surface repositories. The retention of radionuclides in cements is controlled by solubility phenomena, diffusion, adsorption and/or incorporation into solids, including the formation of solid solutions. Within the framework of CEBAMA WP2, we are studying the solubility of selected radionuclides in conditions relevant to a geological disposal facility (GDF). The uptake of long-lived fission and decay products (Ra, Tc, I, Se, Cl) in cementitious materials, the diffusion of radionuclides through hardened cement paste and the radionuclide distribution between and within various cement phases on the micro-scale are also being investigated, using advanced micro-analytical and spectroscopic tools. The objective of these investigations is to enhance mechanistic understanding of their uptake and retention and to assess the relevance of chemical alteration processes, such as carbonation, in aged concrete. In this context, a bottom-up approach is being pursued using synthesised cement phases (model phases) on the one hand and hardened cement pastes with different compositions on the other.

Numerous studies have been carried out investigating: 1) the solubility of the radionuclides in solution under alkaline conditions analogous to those expected in a GDF (Baston et al., 1997; Heath et al., 1998; Felipe-Sotelo et al., 2014), 2) batch sorption studies with cement admixtures (Glasser et al., 1985; Beaudoin et al., 1990; Solem-Tishmack et al., 1995; Bonhoure et al., 2006; Tits et al., 2006; Mace et al., 2007; Tanabe et al., 2010) and 3) diffusion studies in cement using different experimental protocols (under equilibrium conditions, i.e., in-, out- or through–diffusion, or non-equilibrium, i.e., leaching; Atkinson and Nickerson, 1984; Tallent et al., 1987; Brodda and Mingxia, 1988; Gilliam et al., 1990; Sarott et al., 1992; Mattigod et al., 2001; Chida and Sugiyama, 2008; Felipe-Sotelo et al., 2014; van Es et al., 2015). A full review of these topics can be found in the CEBAMA WP2 state of the art report (Lange et al., 2016). Several of the radionuclides addressed in this study can exhibit a range of oxidation states under anticipated GDF conditions (Se, Tc, I). Selenium will be considered as both selenite and selenite whereas iodine will be considered as iodide and iodate; these are believed to be the dominant species likely under GDF conditions (Thoenen et al., 2014). Technetium will be considered as Tc(IV) as ambient conditions will be reducing and interaction between pertechnetate and solids is expected to be limited.

Methodology

Materials

The solids of interest for this work fall into two categories: hardened cement pastes (HCP) and single mineral phases. Five cement blends are included in the work plan, namely a CEM I, a ground granulated blast furnace slag: ordinary Portland cement blend (GGBS:OPC), a pulverised fuel ash blend (PFA:OPC), a backfill material (NRVB, Nirex Reference Vault Backfill) and a CEBAMA reference cement blend. The compositions used for the formation of the hardened cement pastes are detailed in Table 1. In addition to these freshly made cements, experiments will also be performed on aged NRVB, PFA:OPC and GGBS:OPC cements that have been stored submerged in cement equilibrated water for ~ 2 years.

Table 1: *Powder ratios of cement to be studied within the scope of this work.*

Cement Blend	OPC	PFA	GGBS	Hydrated Lime	Lime Flour	Silica Fume
CEM I	1	-	-	-	-	-
PFA:OPC	1	3	-	-	-	-
GGBS:OPC	1	-	9	-	-	-
NRVB	1	-	-	0.38	1.1	-
CEBAMA Reference mix	1	-	0.62	-	-	0.87

The individual phases to be studied include: calcium silicate hydrate (CSH), ettringite, monosulphate and hydrotalcite. These phases have been selected as they comprise the major components of a CEM I cement, excluding portlandite; no significant interaction of the latter with radionuclides is expected. CSH will be produced according to a direct synthesis method from Atkins et al. (1992) performed by mixing CaO with either silicic acid or silica fume in distilled, degassed water. Ca/Si ratios of between 0.9 and 1.4 will be obtained using this method, with the lower ratios being applicable to an aged cement. Ettringite is synthesised using a method from Atkins et al. (1991) in which a slurry of water containing CaO is prepared at 5°C (w/s ratio = 10) and added to a solution containing $Al_2(SO_4)_3 \cdot 16H_2O$. This is stirred for 24 hours followed by aging at 25°C. Excess gypsum is removed by repeatedly re-dispersing in water. The ettringite formed will be used to synthesise monosulfate via a reaction with tricalcium aluminate (C_3A) over a 2 week period. Hydrotalcite will be prepared via solid state reaction (Long et al., 2014), in which aluminium sulphate, magnesium sulphate and sodium carbonate are ground along with polyethylene glycol-400, and stored at 80°C for 3 hours to give a phase pure product after washing with deionised water, anhydrous ethanol and drying. Phase purity will be confirmed via X-ray diffraction (XRD).

Solubility

Solubility studies will be performed using under-saturation and oversaturation methodologies. They will use water that has been equilibrated with cement for a minimum of 28 days and will be performed under N_2. Differences between the results of over- and under-saturation tests can arise from several sources. Incipient precipitation in the case of the former may generate colloidal particles that pass filtration, leading to over-estimation of solubility; to minimise this effect, two filtration procedures will tested; ultrafiltration with 30 kDa MWCO (molecular weight cut off) regenerated cellulosed and with fine (0.22 μm) PES (polyester sulfone). Conversely, the dissolution kinetics of solids used in under-saturation measurements may be sufficiently slow that equilibrium is not reached, even on longer timescales (~ 1 year in this work), biasing the data to lower values. Finally, there may be compositional or textural differences in the solids formed by each method. In order to address the latter, structural techniques will employed to highlight any changes in particle size or crystal form that could have a bearing on phase solubility. Concentration of elements in solution will be measured by inductively-coupled plasma optical emission spectroscopy (ICP-OES, Se, I, Cl), liquid scintillation counting (LSC, ^{36}Cl, ^{79}Se, ^{99}Tc, ^{129}I, ^{226}Ra). XRD will be utilised to determine solubility limiting phases.

Sorption/ Desorption

Sample preparation will be carried out in a glove box (N_2 or Ar atmosphere) to avoid carbonation. Hardened cement pastes and individual mineral phases are prepared, weighed and allowed to equilibrate in deionised (DI) water for 28 days. The element of interest is added from a stock solution to give an appropriate initial concentration, i.e., $5 \cdot 10^{-3}$ mol/dm^3 for Se (Bonhoure et al., 2006) or $1.5 \cdot 10^{-8}$ mol/dm^3 for ^{226}Ra (Tits et al., 2006). A liquid to solid ratio of 100:1 is used in the batch studies and expected to run for a month with daily sampling in the first week followed by weekly sampling for the remainder of the experiment. Where applicable, radioactive tracers will also be added. Samples are allowed to reach equilibrium and aliquots of the supernatant are taken and analysed by ICP-OES (Se, I, Cl) or LSC (^{36}Cl, ^{79}Se, ^{99}Tc, ^{129}I, ^{226}Ra). Residual solids are filtered and some analysed via XRD, energy dispersive X-ray spectroscopy, scanning electron microscopy, extended X-ray absorption fine structure (XAFES) to provide insight to the distribution of radionuclides on the phases present. Desorption studies are performed via the addition of cement/phase-equilibrated water to the solids. Samples are again allowed to reach equilibrium and concentration in the supernatant is monitored.

These results should allow the determination of uptake and release kinetics, as well as the partitioning of the radionuclide between solid and solution, quantified in terms of a distribution ratio (R_d):

$$R_d = \frac{c_s}{c_{aq}} = \left(\frac{c_{tot} - c_{aq}}{c_{aq}}\right) \cdot \left(\frac{V}{m}\right) \qquad \textbf{\textit{Eq. 1}}$$

where c_s and c_{aq} denote analyte concentrations measured on the solid (mol/kg) and in solution (mol/dm^3), respectively, at equilibrium. c_{tot} denotes the total analyte concentration added. V and m denote the volume of the suspension (m^3) and mass of solid phase (kg) respectively.

Through-Diffusion

Through diffusion experiments will be performed in the manner described by Felipe-Sotelo et al. (2014, 2016) and van Es et al. (2015). For these experiments, cylindrical blocks of hardened cement paste (40 mm diameter and a 45 - 50 mm height) are drilled longitudinally to create a well for the addition of radioisotopes. The well is drilled centrally and to a depth of 25 - 30 mm. The radionuclide of interest is spiked into the central well, which is then sealed and the block placed into equilibrated water. The movement of the radionuclide through the block can be monitored by measuring the equilibrated water surrounding the block for the presence of the radionuclide. After a specified time (~ 1 year) the block is removed from the equilibrated water, sawn axially and the distribution of radionuclides measured by digital autoradiography. These experiments will be performed on both fresh and archive samples.

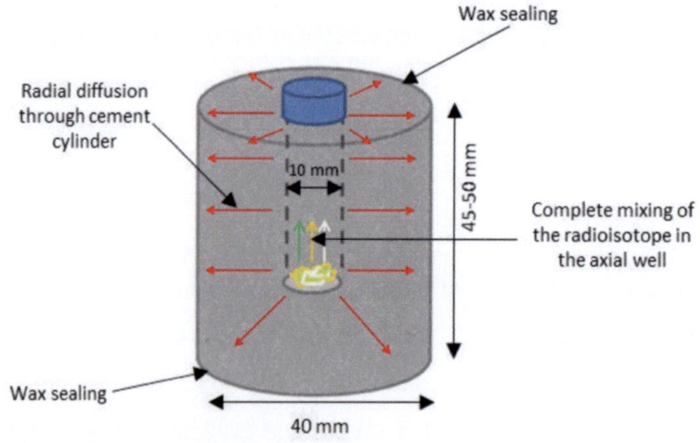

Figure 1: *Schematic of the radial diffusion experiments.*

Future work

The experiments planned during the course of the CEBAMA project aim to give an enhanced mechanistic understanding of the retention of Ra, Tc, I, Se, Cl in high pH cementitious environments. The experiments chosen investigate radionuclide solubility together with diffusion, adsorption and incorporation by hardened cement pastes and individual cement phases.

Acknowledgement

The research leading to these results has received funding from the European Union's European Atomic Energy Community's (Euratom) Horizon 2020 Programme (NFRP-2014/2015) under grant agreement, 662147 – Cebama.

References

Atkins, M., Macphee, D., Kindness, A., Glasser, F.P. (1991). Solubility properties of ternary and quaternary compounds in the CaO-Al2O3-SO3-H2O system. Cement and Concrete Research, 21, 991-998.

Atkins, M., Glasser, F.P., Kindness, A. (1992). Cement hydrate phase: Solubility at 25°C. Cement and Concrete Research, 22, 241-246.

Atkinson, A. and Nickerson, A.K. (1984). The diffusion of ions through water-saturated cement. Journal of Materials Science, 19, 3068-3078.

Baston, G.M.N., Brownsword, M., Smith, A.J., Smith-Briggs, J.L. (1997). Further Near-field Solubility Studies. Nirex Report, NSS/R257.

Beaudoin, J.J., Ramachandran, V.S., Feldman, R.F. (1990). Interaction of chloride and CSH. Cement and Concrete Research, 20, 875-883.

Bonhoure, I., Baur, I., Wieland, E., Johnson, C.A., Scheidegger, A.M. (2006). Uptake of Se(IV/VI) oxyanions by hardened cement paste and cement minerals: An X-ray absorption spectroscopy study. Cement and Concrete Research, 36, 91-98.

Brodda, B.-G. and Mingxia, X. (1988). Leaching of chlorine, cesium, strontium and technetium from cement-fixed intermediate level liquid waste. MRS Proceedings, 127, 481-488.

Chida, T. and Sugiyama, D. (2008) Diffusion behavior of organic carbon and iodine in low-heat Portland cement containing fly ash. MRS Proceedings, 1124, 1124–Q10–15.

Felipe-Sotelo, M., Hinchliff, J., Drury, D., Evans, N.D.M., Williams, S., Read, D. (2014). Radial diffusion of radiocaesium and radioiodide through cementitious backfill. Physics and Chemistry of the Earth, Parts A/B/C, 70-71, 60-70.

Gilliam, T.M., Spence, R.D., Bostick, W.D., Shoemaker, J.L. (1990). Solidification/stabilization of technetium in cement-based grouts. Journal of Hazardous Materials, 24, 189-197.

Glasser, F.P., Rahman, A.A., Macphee, D., Angus, M.J., Atkins, M. (1985). Immobilization of radioactive waste in cement-based matrices. Department of the Environment, DOE-RW--85.063.

Heath, T., Pilkington, N.J., Tweed, C., Williams, S. (1998). Radionuclide solubility at high pH. In The Chemistry of Deep Disposal of Radioactive Waste (S. Baker, TG Heath and R. McCrohon, editors). Proceedings of the Royal Society of Chemistry/Nirex Symposium. Nirex Report S/98/008.

Hinchliff, J., Evans, N.D.M., Read, D. (2016). Solubility constraints affecting the migration of selenium through the cementitious backfill of a geological disposal facility. Journal of Hazardous Materials, 305, 21-29.

Long, Q., Xia, Y., Liao, S., Li, Y., Wu, W., Huang, Y. (2014). Facile synthesis of hydrotalcite and its thermal decomposition kinetics mechanism study with masterplots method. Thermochimica Acta, 579, 50-55.

Mace, N., Landesman, C., Pointeau, I., Grambow, B., Giffaut, E. (2007). Characterisation of thermally altered cement pastes. Influence on selenite sorption. Advances in Cement Research, 19, 157-165.

Mattigod, S.V, Whyatt, G.A., Serne, R., Martin, P.F., Schwab, K.E., Wood, M.I. (2001). Diffusion and leaching of selected radionuclides (iodine-129, technetium-99 and uranium) through Category 3 waste encasement concrete and soil fill material.

Sarott, F.A., Bradbury, M.H., Pandolfo, P., Spieler, P. (1992). Diffusion and adsorption studies on hardened cement paste and the effect of carbonation on diffusion rates. Cement and Concrete Research, 22, 439-444.

Solem-Tishmack, J.K., McCarthy, G.J., Docktor, B., Eylands, K.E., Thompson, J.S., Hassett, D.J. (1995). High-calcium coal combustion by-products: Engineering properties, ettringite formation, and potential application in solidification and stabilization of selenium and boron. Cement and Concrete Research, 25, 658-670.

Tallent, O.K., McDaniel, E.W., Cul, G.D. Del, Dodson, K.E., Trotter, D.R. (1987). Immobilization of technetium and nitrate in cement-based materials. MRS Proceedings, 112, 23-32.

Tanabe, H., Sakuragi, T., Yamaguchi, K., Sato, T., Owada, H. (2010). Development of new waste forms to immobilize iodine-129 released from a spent fuel reprocessing plant. Advances in Science and Technology, 73, 158-170.

Thoenen, T., Hummel, W., Berner, U., Curti, E. (2014). The PSI/Nagra Chemical Thermodynamic Databse 12/07. PSI report, 14-04.

Tits, J., Kamei, G., Iijima, K., Wieland, E. (2006). The uptake of radium by calcium silicate hydrates and hardened cement paste. Radiochimica Acta, 38, 637-643.

van Es, E., Hinchliff, J., Felipe-Sotelo, M., Milodowski, A.E., Field, L.P., Evans, N.D.M., Read, D. (2015). Retention of chlorine-36 by a cementitious backfill. Mineralogical Magazine, 79, 1297-1305.

Characterization of hydrated cement paste (CEM II) by selected instrumental methods and a study of ^{85}Sr uptake

Barbora Drtinová[1*], Jana Kittnerová[1], Dušan Vopálka[1]

[1] Department of Nuclear Chemistry, Czech Technical University in Prague (CZ)

* Corresponding author: barbora.drtinova@fjfi.cvut.cz

Abstract

In relation with the Czech program of radioactive waste disposal, the two cements of CEM II grade were chosen for studying. Both materials have been evaluated by standard and also advanced methods. The obtained characteristics of selected cements are quite different. All methods used for characterization of cements have not been applied to both materials yet. A system, consisting of crushed hydrated cement paste (CEM II / A-S 42.5), its leachate obtained at the phase ratio $m/V = 0.2$ kg/L (with the natural concentration of Sr $3.5\cdot10^{-4}$ mol/L), and a radioactive tracer ^{85}Sr, was studied in order to understand the interaction of hydrated cement with Sr as an analog of Ra. In a wider range of phase ratio, the equilibrium of tracer, described by means of distribution coefficient K_d, was reached after approx. 2 days. The value of distribution coefficient for m/V ratio in the interval of (0.01 - 0.1) kg/L was constant, while for higher values of m/V K_d increased linearly. This effect may be connected with the fact that the hydrated cement paste itself contains strontium and that the equilibrium concentration of Sr in both liquid and solid phases belonging to this "exchangeable Sr" could be influenced by conditions of the experiment.

Introduction

A significant part of low and intermediate radioactive wastes containing radium is stored in Czech Republic in the repository Bratrství. This repository is running out of space and will be closed in the near future. It is important to provide safety studies, collect necessary parameters and check safety of the repository before its closure. As barriers in this repository are based on cementitious materials and ^{226}Ra is there the main contaminant of interest, SURAO (the Czech Radioactive Waste Repository Authority) aims to study systems radium – hydrated cement materials.

As there are not many studies available dealing specifically with radium and all the necessary techniques are not at hand yet, we decided to perform an introductory methodological study with strontium, which can be considered in some aspect chemically similar to radium. This approach corresponds with that of other laboratories (e.g., Berner, 2002; Tits et al., 2006). Our aim is to gain experience and knowledge about optimal experimental conditions and procedures while working with hydrated cement pastes. The description of interaction of radionuclides with cementitious materials, which is necessary in modeling of radium transport in the repository environment, belongs to our intentions too.

Two cements of CEM II grade, A: CEM II / A-S 42.5 R (produced by Lafarge Cement, a.s.) and B: CEM II / B-M (S-LL) 32.5R (produced by Českomoravský cement, a.s. – Heidelberg Cement Group) were chosen for the study, based on recommendations of SURAO. The first cement (A) is made, according to an information of the producer, by grinding together silicate clinker, blast furnace granulated slag and gypsum. Slag adjusts the cement blend and decreases hydration heat and its cement content is max. 20%. Gypsum acts as a regulator preventing cement flash setting. The second cement (B) is Portland composite cement of the strength class made of clinker, granulated blast furnace slag (approx. 15%), of limestone (approx. 12%), and with low TOC content and anhydrite as retarder.

Both materials have undergone testing by instrumental methods available in our laboratory (determination of density and specific surface area, FTIR, XRD, AAS). Part of the results of this work became the basis for a bachelor thesis (Kittnerová, 2015). With the hydrated cement paste of cement A, a preliminary study of the interaction using ^{85}Sr tracer was carried out.

Preparation of cement pastes

For preparation of cement blocks with water/cement ratio w = 0.667 with time of hydration 19 days in co-operating laboratory in ÚJV Řež cement A was used. A relatively short time used for hydration was chosen. In our laboratory blocks of cement paste from cement B were prepared with water/cement ratio w = 0.667 (a mixture of water and ice was used) with time of hydration in humid atmosphere of for almost 3 months.

For some experiments the blocks of hydrated cement paste were crushed and sieved. For interaction experiments and for characterizations (determination of density and specific surface) the fraction <0.71 mm was used. For the purpose of IR analysis and during leaching larger pieces and blocks were applied.

Characteristics of cement pastes

Standard methods

Values of density of hydrated cement were determined by a pycnometric method on crushed materials. For cement A, a value of 2 177 ± 44 kg/m^3 was measured, and 1 998 ± 29 kg/m^3 for cement B (the uncertainty represents an estimate of the standard deviation; throughout the whole text).

The rapid dynamic flow method for determinations of single-point B.E.T. (Quantachrome Monosorb MS-22 device) was utilized to measure specific surface areas of both materials. Obtained values were 20.1 ± 0.3 m^2/g and 48.5 ± 0.3 m^2/g for cements A and B, respectively.

Advanced instrumental methods

FTIR spectra were measured for two blocks of cement B by a Nicolet iS 50 FT-IR spectrometer using the Attenuated Total Reflection (ATR) technique in the scan range of 400 - 4 000 cm^{-1}. One of these blocks (sample W) was placed for 3 months in a closed vessel submerged in distilled water (cement-to-water ratio 0.1 kg/L). The second one (reference sample R) was left dry in sealed flask without contact with a liquid phase. From the results on Figure 1, it can be seen that leaching of hydrated cement caused only a very small change of the FTIR spectrum.

An attempt of the interpretation of the character of obtained spectra, based on the database of the device, is presented in Table 1.

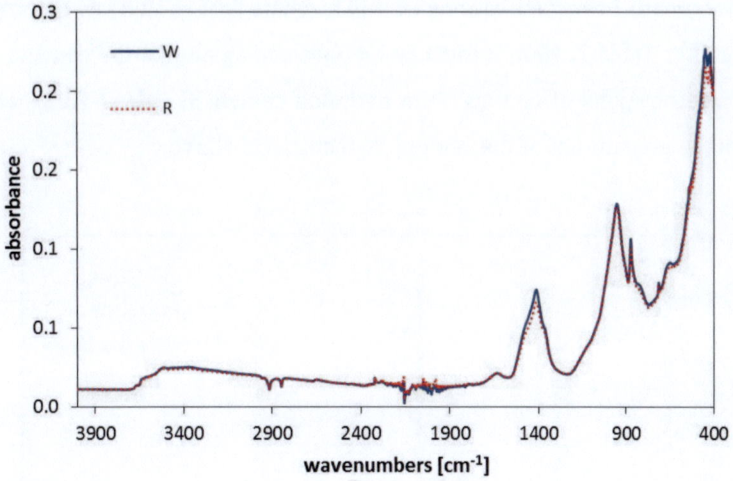

Figure 1: *FTIR spectra of two blocks of cement B, W: in contact with water for 3 months, R: no contact with water after hydration.*

Both cementitious materials (cement A and B) were studied by X-ray diffraction (Rigaku Mini Flex 600). As an illustration, XRD spectrum of cement A is presented on Figure 2. The comparison of measured spectrum with database ICDD PDF-2 (version 2013) of the measuring system enabled to identify four mineral phases, namely calcite $CaCO_3$, portlandite $Ca(OH)_2$, hydrotalcite $Mg_6Al_2CO_3(OH)_{16}:4(H_2O)$ and ettringite $Ca_6Al_2(SO_4)_3(OH)_{12}:26H_2O$.

It should be noted here that in XRD spectra of cement B (samples W and R), only calcite and portlandite were identified. From a comparison of spectra of samples R and W, it could be proven that the content of $Ca(OH)_2$ decreased significantly in the bulk of the cementitious block by leaching of hydrated cement in water.

***Table 1:** Possible assignment to some peaks observed in FTIR spectra.*

Wave number (cm^{-1})	Vibration	Compound
3 600 – 3 300	(OH)	H_2O or $Ca(OH)_2$
1 414	$(CO_3)^{2-}$	$CaCO_3$
1 100 – 950	(OH) and (Si-O)	$Ca(OH)_2$ and SiO_2
870, 420, 300, 250	$(CO_3)^{2-}$	$CaCO_3$
650	(Si-O)	SiO_2
450	Ca-Si	×

Sorption experiments with ^{85}Sr

Preparation of working solution

The crushed hydrated cement A was contacted with distilled water for 1 month (m/V equaled to 0.2 kg/L) to prepare a working solution (an approximation to the actual conditions considering the different m/V ratios in samples). The concentrations of important cations in this leachate (pH = 12.8) as determined by AAS (Varian AA 240 FS) are presented in Table 2. Here it must be emphasized again that the cement A itself contains Sr. In comparison, leachate concentrations of cations from hydrated cement B (pH = 12.2) were also shown. These samples originated from the preparation of the sample W mentioned above.

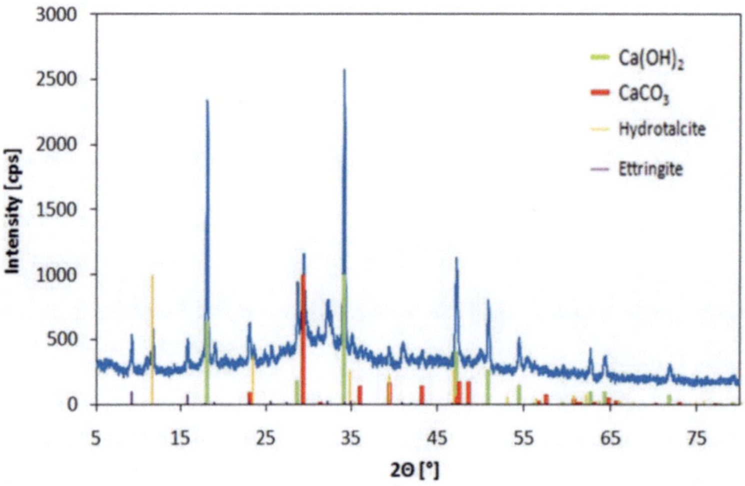

***Figure 2:** XRD spectrum of the hydrated cement A.*

Table 2: *Concentrations of important cations in leachates of hydrated cements A and B.*

cation	cement A, V/m = 0.2 kg/L, contact time 19 days		cement B, V/m = 0.1 kg/L contact time 3 months	
	C (mmol/L)	σ_C (mmol/L)	C (mmol/L)	σ_C (mmol/L)
Na^+	5.88	0.01	1.37	0.04
K^+	23.8	0.3	4.08	0.05
Ca^{2+}	13.5	0.1	3.1	0.2
Mg^{2+}	0.0016	0.0002	< 0.001	
Sr^{2+}	0.350	0.002	n.d.	

Kinetic experiments

In a set of kinetic experiments, crushed hydrated cement A (diameter of grain ≤ 0.71 mm) was contacted with a cement leachate (prepared as described above). Strontium (10^{-7} mol/L) traced by radioactive isotope ^{85}Sr was added as $SrCl_2$. Its activity was monitored by an NaI(Tl) detector. The m/V ratio was in the range from 0.033 (1:30) to 0.33 (1:3) kg/L. The uptake of ^{85}Sr is presented in Figure 3.

The observed kinetics of ^{85}Sr uptake was relatively fast – after 2 days equilibrium was reached. So the duration of each experiment in the subsequent set of equilibrium experiments was set to 4 days.

Equilibrium experiments

A broad set of equilibrium experiments was performed in which the initial total concentration of strontium in the liquid phase was changed from 0.35 to 1 mmol/L. The lower limit of concentration range corresponds to the content of Sr in the working leachate (Table 2). Experiments were carried out for 6 different m/V ratios (from 1:100 to 1:3). The distribution coefficient K_d, obtained from the balance evaluation of ^{85}Sr activity in the liquid phase as the measure of the ^{85}Sr uptake, was used. No influence of the total initial Sr concentration on the ^{85}Sr uptake was observed. The spread of results obtained for different initial concentrations enabled to describe the uncertainty of determined K_d values.

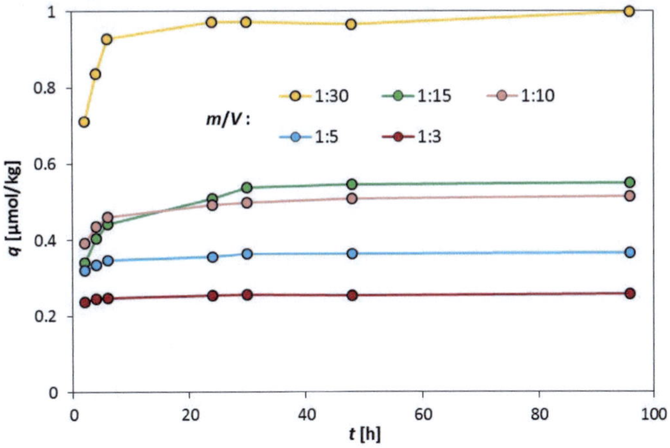

Figure 3: *Kinetics of the ^{85}Sr uptake on the hydrated cement A.*

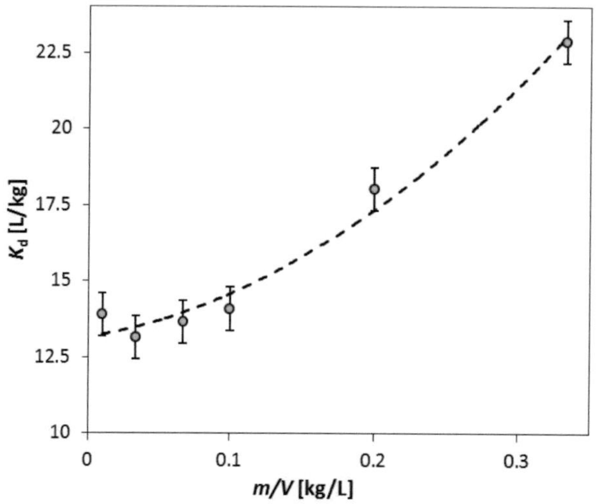

Figure 4: *Dependence of determined K_d values describing ^{85}Sr uptake on hydrated cement A on the phase ratio m/V.*

The obtained results are shown in Figure 4. The increase of K_d values with increasing phase ratio m/V is evident. This trend is unexpected if the ion-exchange is taken into account as an important mechanism of the uptake, as stated e.g., Wieland et al. (2008).

The presence of Sr in both working solution and solid phase at the beginning of sorption experiment needs to be taken into acount. The so-called exchangeable Sr in the solid phase q_0 (mol/kg) was determined according to Equation 1 (Davis and Curtis, 2003), in which C_0 and C are initial and equilibrium concentrations of total Sr in the liquid phase, m is mass of solid phase (kg), V volume of liquid phase (L) and K_d (kg/L) is the distribution coefficient determined from the balance of ^{85}Sr in the system.

$$q_0 = \frac{C(V + K_d m) - C_0 V}{m} \qquad \textbf{\textit{Eq. 1}}$$

The q_0 values were obtained for three selected m/V ratios (1:3, 1:10, 1:30) and for three initial concentrations ($3.5 \cdot 10^{-4}$, $0.9 \cdot 10^{-3}$ and $1.9 \cdot 10^{-3}$ mol/L). No significant influence of the phase ratio was observed, an ambiguous influence of initial concentration should be further studied. From nine obtained values the mean q_0 value (4.04 ± 0.26 mmol/kg) was determined.

Conclusions and Future work

Hydrated cement pastes were tested by techniques available in the laboratory of Department of Nuclear Chemistry, CTU in Prague. The samples stem from two different cements considered in the Czech program of disposal of ILW. For an introductory study of ^{85}Sr uptake on crushed hydrated cement, cement CEM II / A-S 42.5 R was selected. As the base for both kinetic and equilibrium experiments a leachate of hydrated cement (m/V = 0.2 kg/L, pH 12.8) was used, containing Sr in a significant concentration. The equilibrium was attained in approx. two days. An unexpected result was that the measured distribution coefficient K_d of ^{85}Sr increased with higher m/V ratios. This could be explained by the presence of Sr in the initial solid. As Sr and Ra are similar to some extent, the performed study of Sr uptake will be followed by the study of Ra uptake (the reopening of the question of above mentioned unusual behavior and work with other cementitious materials are planned).

Acknowledgement

The research leading to these results has received funding from the European Union's Horizon 2020 Research and Training Programme of the European Atomic Energy Community (EURATOM) (H2020-NFRP-2014/2015) under grant agreement n° 662147 (CEBAMA). This contribution is partially result of Radioactive Waste Repository Authority project "Research support for Safety Evaluation of Deep Geological Repository".

References

Berner, U. (2002). Project Opalinius Clay. Radionuclide concentration limits in the cementitious near-field of an ILW repository. PSI Report, 02-26.

Davis, J.A. and Curtis, G.P. (2003). Application of surface complexation modeling to describe uranium(VI) adsorption and retardation at the uranium mill tailings site at Naturita, Colorado. US Nuclear Regulatory Commission, NUREG/CR-6708.

Kittnerová, J. (2015). Cementitious materials in barriers of radioactive waste repositories. Thesis, Department of Nuclear Chemistry, CTU in Prague (in Czech).

Tits, J., Wieland, E., Möller, C.J., Landesman, C., Bradbury, M.H. (2006). Strontium binding by calcium silicate hydrates. Journal of Colloid and Interface Science, 300, 78-87.

Wieland, E., Tits, J., Kunz, D., Dähn, R. (2008). Strontium uptake by cementitious materials. Environmental Science and Technology, 42, 403-409.

Conceptual reactive transport model of low pH cement / clay interface processes

Vanessa Montoya[1*], Naila ait Mouheb[1], Thorsten Schäfer[1], Volker Metz[1]

[1] KIT-INE: Karlsruhe Institute of Technology, Institut für Nukleare Entsorgung (DE)

* Corresponding author: vanessa.montoya@kit.edu

Abstract

Reactive transport modelling activities described in this work are focused on the definition of the conceptual model to be applied on laboratory through diffusion experiments performed in KIT-INE (see contribution of ait Mouheb et al., these proceedings). The model will include different processes observed on through diffusion experiment of HTO, Cl^-, and I^- across the interface between bentonite porewater and low pH cement and will take into account porosity changes due to dissolution/precipitation reactions.

Introduction

The understanding of the evolution of a repository system over geological time scales requires a detailed knowledge of a series of highly complex coupled processes. By using small-scale laboratory experiments, under well-defined boundary conditions, numerical modelling can provide information to help in the repository design and predict radionuclide migration. The conceptual understanding of the different processes and their integration in numerical models will increase the reliability and predictive capabilities of the simulations.

The development of reactive transport models describing the interface low pH cement/clay processes and the impact on radionuclide migration in a deep geological disposal is a topic of importance in the Performance Assessment of a nuclear waste repository. In this work, the conceptual model to be used in the description of through diffusion experiments of HTO, Cl^- and I^- across the interface between bentonite porewater and low pH cement is described. The processes considered are diffusion transport in saturated media under isothermal conditions, sorption reactions, and both local equilibrium and kinetically controlled mineral dissolution-precipitation reactions. Update of the effective diffusion coefficient as a function of porosity changes is also considered. The reactive transport code where the model is implemented is iCP, which couples two different codes: the Finite Element code COMSOL Multiphysics V.5 (COMSOL, 2014) and the geochemical simulator PHREEQC v.3 (Parkhurst and Appelo, 2013).

Model description

The reactive transport model consists of a two dimensional (2D) fully water saturated isothermal (298 K) problem representing a laboratory through diffusion experiment of HTO, Cl^-, and I^- across the interface between bentonite porewater and low pH cement. Changes of porosity and diffusion coefficients will be coupled with the precipitation/dissolution of different solid phases happening in the interface. Information about the new phases precipitated in the system will be obtained experimentally. A schematic representation of the diffusion experiments is represented in Figure 1.

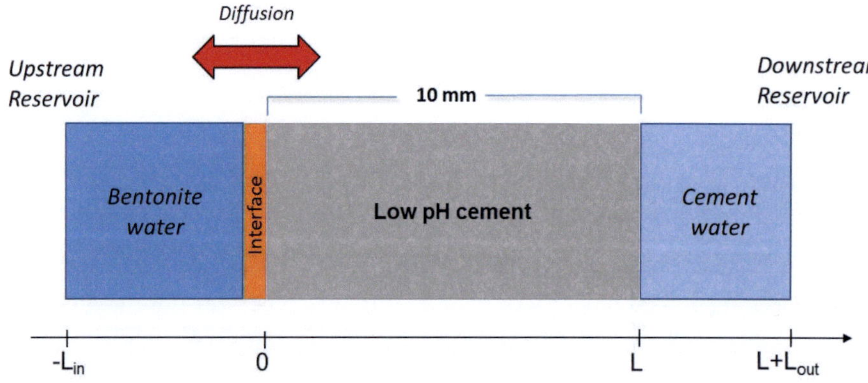

Figure 1: *Schematic representation of the diffusion experiments.*

Different assumptions/cases will be tested with increasing geochemical and geometrical complexity. The selection of the two modelling cases is summarized in Table 1 and details are provided below.

Table 1: *Short Summary of two different cases to be modelled.*

Cases	Geometry	Chemistry	Transport
1	2D	Thermodynamic	diffusion
2	2D	Kinetics	diffusion

Description of the software to be used (iCP)

The system studied is implemented in the newly developed code/interface iCP (Nardi et al., 2014). iCP is developed in Java® and the input data is given in a JSON format. One of the features of iCP is that the code is not in charge of the numerical calculations, and it is only used as an interface to couple and maximize the synergies between COMSOL and PHREEQC. This interface provides a numerical platform that can efficiently simulate a wide number of multiphysics problems coupled to geochemistry (i.e., liquid flow, solute and heat transport, elastic and plastic mechanical deformations and geochemical reactions).

Geometry and physical properties

Geometrical and transport parameters including the discretization of the system will be initially implemented in 2D in COMSOL Multiphysics. The cementitious solid phase is represented by a block of 10 mm put in contact with a bentonite porewater (see Figure 1). The mesh size and the time steps will be selected to ensure a satisfactory compromise between computation time and sufficient spatial resolution of the expected geochemical and transport processes, especially at the interface between the bentonite porewater and the low pH cement hydrated phases.

Considering the very low permeability of the low pH cement (10^{-10} - 10^{-14} m/s), mass transport will be considered diffusion driven only, following Fick's law.

$$J = -D\frac{\partial C}{\partial x} \qquad \textbf{\textit{Eq.1}}$$

where J is the substance flux (kg/m^2·s); $\frac{\partial C}{\partial x}$ is the concentration gradient (kg/m^4); and D is the diffusion coefficient (m^2/s). In porous media, the diffusion coefficient depends on the properties of the diffusing chemical species, the pore fluid and the porous medium, and it is called effective diffusion coefficient (D_e).

The transport behaviour of HTO, I^- and, Cl^- will be modelled. Porosity changes due to mineral precipitation/dissolution and feedback on transport will also be taken into account. Electrostatics may influence the transport of the Cl^- and I^- anions in the cement system (Chagneau et al., 2015). The importance of electrostatic surface interactions will be assessed taking into account experimental data obtained in the project (Leroy et al., these proceedings) and if needed this will be incorporated in the reactive transport model.

Mineralogical and chemical conditions

The geochemical conceptual model is implemented in the code PHREEQC v.3.1.7. (Parkhurst and Appelo, 2013). The mineralogical composition of the cement hydrated phases considered is representative of a low pH cement (pH ~ 11.0) and determined experimentally (see contribution of ait Mouheb, these proceedings). C-S-H phases with low Ca/Si ratio will be the main solid phases incorporated in the model. The porewater composition of the low pH cement will be defined in equilibrium with the hydrated solid phases present in the system.

The porewater composition of the clay is representative of the MX-80 bentonite described in the literature (Wersin et al., 2003).

Thermodynamic database

As a first approximation, chemical reactions at equilibrium will be simulated using the thermodynamic database ThermoChimie v.9. (https://www.thermochimie-tdb.com/, Giffaut et al., 2014), available in PHREEQC format. Precipitation of secondary phases to be formed in the cement/clay interface and dissolution of the major components of the low pH cement will be taken into account. Even though ThermoChimie v.9

includes the Specific Ion Interaction Theory (SIT) parameters for the activity correction, at present status, the Davies equation, valid for the ionic strength of the studied system (< 0.5 M) will be preferred to save computational time.

ThermoChimie v.9. database includes thermodynamic data for 50 major, radioactive and toxic elements. A significant data set is also available for the stability of cement hydrates, clay minerals (Blanc et al., 2015) and other aluminium-silicate phases, such as zeolites, although no thermodynamic data are currently available for C-S-H phases with low Ca/Si expected to be present in our system. For this reason, thermodynamic data available in recent studies will be integrated in ThermoChimie database if needed (Walker et al., 2016).

Kinetic rates

Rate equations of precipitation/dissolution of secondary/primary phases will be provided directly in the input files of PHREEQC only as a second modelling approximation if local equilibrium does not provide a good approximation to the experimental results.

The ThermoChimie v.9. database (Marty et al., 2015) contains a compilation of kinetic parameters for different minerals (albite, biotite, calcite, celestite, chlorite, C-S-H, dolomite, gibbsite, illite, kaolinite, portlandite, quartz, siderite and smectite) but probably this will need to be extended after a literature review of precipitation/dissolution kinetic parameters of C-S-H phases.

Sorption processes

Sorption reactions of $^{36}Cl^-$ and $^{129}I^-$ tracers into the low pH cement matrix will be considered as a thermodynamic mechanistic sorption model to be implemented in PHREEQC. Observations described in the recent review of Ochs et al. (2016) will be considered to select the most appropriate data, as well as data generated in the project with batch experiments of the studied radionuclides on the same solid materials.

Conclusions and Future work

Development of the conceptual model to be used is one of the most important parts of the modelling process. A parallel experimental program in a laboratory scale will support the modelling work Next steps include the compilation of the different parameters (porosity, initial chemical composition) needed to perform the reactive transport modelling exercise taking into account the information obtained in the laboratory. Predictive modelling will be used to select the most appropriate conditions to perform the lab experiments.

Acknowledgement

The research leading to these results has received funding from the European Union's Horizon 2020 Research and Training Programme of the European Atomic Energy Community (EURATOM) (H2020-NFRP-2014/2015) under grant agreement n° 662147 (CEBAMA).

References

Blanc, P., Vieillard, P., Gailhanou, H., Gaboreau, S., Marty, N., Claret, F., Madé, B., Giffaut, E. (2015). ThermoChimie database developments in the framework of cement/clay interactions. Applied Geochemistry, 55, 95-107.

Chagneau, A., Tournassat, C., Steefel, C.I., Bour, I.C., Gaboreau, S., Esteve, I., Kupcik, T., Claret, F., Schäfer, T. (2015). Complete Restriction of $^{36}Cl^-$ Diffusion by Celestite Precipitation in Densely Compacted Illite. Environmental Science & Technology Letters, 2 (5), 139-143.

COMSOL (2014). Comsol Multiphysics. Version 5.0, Available at: www.comsol.com.

Giffaut, E., Grivé, M., Blanc, P., Vieillard, P., Colàs, E., Gailhanou, H., Gaboreau, S., Marty, N., Madé, B., Duro, L. (2014). Andra thermodynamic database for performance assessment: ThermoChimie. Applied Geochemistry, 49, 225-236.

Marty, N., Claret, F., Lassin, A., Tremosa, J. Blanc, P., Madé, B., Giffaut, E., Cochepin, B., Tournassat, C. (2015). A database of dissolution and precipitation rates for clay-rocks minerals. Applied Geochemistry, 55, 108-118.

Nardi, A., Idiart, A., Trinchero, P., de Vries, L.M., Molinero, J. (2014). Interface COMSOL-PHREEQC (iCP), an efficient numerical framework for the solution of coupled multiphysics and geochemistry. Computers & Geosciences, 69, 10-21.

Ochs, M., Mallants, D., Wang, L. (2016). Radionuclide and Metal Sorption on Cement and Concrete. Springer International Publishing Switzerland.

Parkhurst, D.L. and Appelo, C.A.J. (2013). User´s guide to PHREEQC (version 2). A computer program for speciation, batch reaction, one dimensional transport and inverse geochemical calculations. U.S. Geological Survey Techniques and Methods, book 6, chap. A43.

Walker, C., Sutou, S., Oda, C., Mihara, M., Honda, A. (2016). Calcium silicate hydrate (C-S-H) gel solubility data and a discrete solid phase model at 25°C based on two binary non-ideal solid solutions. Cement and Concrete Research, 79, 1-30.

Wersin, P. (2003). Geochemical modelling of bentonite porewater in high-level waste repositories. Journal of Contaminant Hydrology, 61, 405-422.

Chemo-mechanical modelling of calcium leaching experiments in cementitious materials

Andrés Idiart[1*], Emilie Coene[1]

[1] Amphos21 Consulting, Barcelona (ES)

* Corresponding author: andres.idiart@amphos21.com

Abstract

In order to assess the long-term integrity and performance of cementitious barriers for their use in underground nuclear waste disposal, the evolution of the main physical properties over time needs to be well understood. These properties can be altered by chemical reactions as a result of clay-cement interaction or the effect of groundwater. Not only the transport properties need to be estimated, but also the mechanical long-term behaviour, to ensure the mechanical stability of the repository. Calcium leaching from cement paste is generally accepted to be one of the most relevant degradation processes in the long-term for nuclear waste repositories. It is well-known that this process leads to a more porous microstructure, negatively impacting the transport and mechanical properties of concrete. In this contribution, a reactive transport model is coupled to a micromechanical model to simulate a set of existing accelerated calcium leaching experiments. The impact of calcium leaching on the mechanical and transport properties is explicitly considered in the model. Preliminary results show how chemically-induced damage impacts the physical properties of cement paste.

Introduction

Degradation of cementitious materials due to mineral dissolution and precipitation processes can significantly impact their physical properties and thus negatively impair their barrier functions. One of the most important functions of cementitious materials to be used for disposal of radioactive waste is to limit and retard the ingress of deleterious species to the waste packages and the release of radionuclides in the case of failure. Transport properties such as the diffusion coefficient and the permeability govern these processes and are the subject of study within CEBAMA. In addition, the mechanical stability and long-term performance of the cementitious barriers also needs to be assessed to guarantee a correct functioning.

One of the main degradation processes expected to play a role in deep geological repositories is calcium leaching from the solid matrix of cementitious materials due to the chemical gradient between the barriers and the environment. Due to the fact that this process is very slow, calcium leaching experiments are typically performed using a more aggressive leaching solution than expected under repository conditions, e.g., using deionized water or different concentrations of ammonium nitrate (Adenot and Buil, 1992; Faucon et al., 1996).

The effect of calcium leaching on concrete properties is a result of the dissolution of a sequence of cement hydrates (portlandite, C-S-H gels, ettringite, etc.) leading to a more porous microstructure. As a consequence, the diffusion coefficient and permeability can significantly increase (e.g., Camps, 2008). Moreover, the mechanical properties such as the Young's modulus and the tensile and compressive strengths can substantially decrease (Gérard et al., 1998; Nguyen et al., 2007; Heukamp et al., 2005; Hu et al., 2014).

Brief description of the experimental data

Recently, a set of calcium leaching experiments using OPC hardened cement paste and concrete specimens has been conducted at Chalmers University of Technology in Sweden (Babaahmadi, 2015; Babaahmadi et al., 2015). In their setup, a 0.3 M ammonium nitrate solution is used coupled to an electrical migration through cement paste and concrete samples. In this way, the leaching process is substantially accelerated, enabling the use of relatively large specimens which can then be used for macroscopic characterization tests (permeability, diffusivity and mechanical properties).

The characterization of the experiments was performed over the reference (i.e., intact) and aged (i.e., leached) states of the same samples. It included microstructural aspects such as porosity and pore size distribution by MIP and gas sorption, as well as macroscopic mechanical (tensile and compressive strength, elastic modulus), hydraulic (gas and water permeability, water sorptivity) and (reactive) transport properties (diffusion coefficients of cations and chloride, sorption properties, qualitative mineral phase assemblages, and profiles of the Ca/Si ratio in the solid phase). Table 1 summarizes the characterization techniques. The accelerated leaching experiments were conducted until reaching a point where portlandite has been entirely depleted from the sample, while only partial degradation of the C-S-H gels was observed (change in Ca/Si ratios using LA-ICP-MS). All the specimens were casted with a diameter of 50 mm and a length of 75 mm, except the concrete specimens used for measuring the tensile strength (using the splitting test based on the American standard ASTM C49). The cement used is an OPC CEM I 42.5N MH/SR3/LA mixed with different water-to-cement ratios (Table 2).

Table 1: *Characterization techniques used in the experiments by Babaahmadi et al. (2015).*

Specimen type	Chemical properties				Transport properties						Mechanical properties			
	IC& PT	XRD	LA-ICP-MS	SEM-EDX	MIP & Gas Sorption	Freezable Water	Diffusion Cell Tests	Diffusion Adsorption Test	Gas Permeability	Water Permeability	Tensile Strength	Compressive Strength	Elastic Modulus	Resonant Frequency
hcp	✓	✓	✓	✓	✓	✓	✓	✓	✗	✗	✗	✓	✗	✓
Mortar	✓	✓	✓	✓	✓	✓	✗	✗	✗	✗	✗	✓	✗	✓
Concrete	✓	✓	✓	✓	✓	✗	✗	✗	✗	✓	✗	✓	✓	✓
	✓	✓	✓	✓	✓	✗	✗	✗	✓	✗	✓	✗	✗	✗

hcp = hardened cement paste; PT = Potentiometric Titration

Multi-scale chemo-mechanical model

A multi-scale chemo-mechanical model is implemented in iCP (interface Comsol-Phreeqc, Nardi et al., 2014) at the continuum scale, where the cementitious material is regarded as homogeneous. The model encompasses a classical reactive transport framework based on (1) solute transport and (2) chemical reactions, with (3) a non-linear mechanical model of the cementitious system, and (4) a multi-scale micromechanical model. The latter is used to describe the coupling between changes in the microstructure of cement paste and concrete (due to chemically-driven degradation) and the mechanical properties. The micromechanical model computes the elastic constants as a function of the volume fractions of all the minerals in concrete and is used to calculate the isotropic chemical damage, d_c (between 0 and 1). The non-linear mechanical constitutive law is based on Mazars' damage model (Mazars, 1986) and is used to calculate the isotropic mechanical damage, d_m (between 0 and 1). Both damage variables are scalar and coupling of chemical and mechanical damage is considered to be multiplicative (see e.g., Gérard et al., 1998). For these experiments, solute transport is limited to Fickian diffusion (i.e., using a single diffusion coefficient for every species to maintain electroneutrality of the pore solution), although both the diffusion coefficient and permeability are updated as a function of porosity in the model. Aqueous speciation and mineral dissolution/precipitation are calculated at chemical equilibrium using the PhreeqC format of the thermodynamic database CEMDATA07 (Lothenbach et al., 2008; Jacques, 2009). Changes in porosity due to chemical reactions are explicitly considered as well as the impact on transport properties.

The model developed in this work does not intend to reproduce the accelerated experiments in terms of the time needed to reach an aged sample. The goal has been to determine, by means of reactive transport modelling, the leaching acceleration factor obtained with the experiment when compared to the leaching of the same samples subjected to groundwater interaction. Furthermore, the outcome of the model in terms of chemical and physical properties of the aged samples can be compared to the experimental data to gain insight into the mechanisms of chemical degradation when using the accelerated leaching test.

Given that the experiments were accelerated by means of an imposed electrical current, a proper representation of this system should take into account the electrochemical migration effect, i.e., the Nernst-Planck equations, and adapted boundary conditions (e.g., Liu et al., 2012). This is not standard in currently available reactive transport codes and is not addressed in this study.

The multi-scale micro-mechanical model implemented in iCP is based on the work by Stora et al. (2009) and later refined in Bary et al. (2014). It uses homogenization theory to describe the effective properties at different scales based on information about the morphology of the structure at a lower scale and the volume fractions of the different constituents. Four different levels are defined, ranging from a few microns for low and high density C-S-H structures to the cm-scale for concrete (Table 3). Three homogenization schemes are used to better adapt the modelled microstructure at a given scale to experimental observations.

Table 2: *Summary of multi-scale micromechanical model based on Stora et al. (2009).*

Level	Homogenized domain	RVE size	Microstructure	Homogenization scheme	Components
I	High Density C-S-H (HD)	10 μm	matrix-inclusion type	Interaction Direct Derivative (IDD)	Porous C-S-H gel and other hydrates
I'	Low Density C-S-H (LD)	10 μm	matrix-inclusion type	Interaction Direct Derivative (IDD)	Porous C-S-H gel, capillary pores and other hydrates
II	Hardened cement paste (hcp)	100 μm	doubled coated spheres	Generalized Self-Consistent Scheme (GSCS)	Unhydrated clinker, HD & LD C-S-H
III	Mortar	mm-scale	doubled coated spheres	Generalized Self-Consistent Scheme (GSCS)	hcp, ITZ and sand particles
IV	Concrete	cm-scale	matrix-inclusion type	Mori Tanaka (MT)	mortar and larger aggregates

RVE: representative volume element; ITZ: interfacial transition zone

Input data and results

Diffusion-driven (initial effective diffusivity of $1.14 \cdot 10^{-12}$ m^2/s) leaching of cement paste samples is simulated in a 2D axisymmetric setup using a FE mesh with maximum element size of 2.5 mm and considering the mineralogical, porewater, and boundary groundwater compositions depicted in Table 3. The results of the simulation in terms of mineral volume fractions after 5 000 years and depletion of portlandite as a function of time are presented in Figure 1. It may be observed that the porosity increases substantially due to the leaching process even though with this groundwater composition significant calcite precipitation is expected in the long-term. Portlandite is completely depleted from the sample after 8 400 years, that is ~ 50 000 times slower than the accelerated leaching test, where ammonium nitrate and electrical migration are considered. Figure 2 shows the evolution of the physical properties over time at a point in the modelled domain and an example of stress-strain curves calculated with the mechanical damage model at three different levels of chemical damage.

Table 3: *Composition of the cement paste, porewater and groundwater used in the simulations.*

Component	Groundwater (mol/L)	Porewater (mol/L)	Cement phases	Volume fraction (-)
pH	8.64	12.975	Portlandite	0.176
Al	$1.21 \cdot 10^{-6}$	$5.54 \cdot 10^{-7}$	C-S-H (jennite)	0.314
C	$6.91 \cdot 10^{-4}$	$1.00 \cdot 10^{-5}$	Ettringite	0.029
Ca	$5.26 \cdot 10^{-4}$	$3.68 \cdot 10^{-3}$	Hydrogarnet (Si)	0.050
Cl	4.54·10-3	$5.55 \cdot 10^{-5}$	Hydrotalcite (OH)	0.012
K	$7.60 \cdot 10^{-5}$	$8.84 \cdot 10^{-2}$	Thaumasite	0.039
Mg	$1.48 \cdot 10^{-4}$	$2.53 \cdot 10^{-8}$	Unhydrated clinker*	0.014
Na	$4.79 \cdot 10^{-3}$	$2.70 \cdot 10^{-2}$	Initial porosity	0.366
S(VI)	$3.73 \cdot 10^{-4}$	$3.29 \cdot 10^{-3}$		
Si	$1.42 \cdot 10^{-4}$	$4.52 \cdot 10^{-5}$		

*considered unreactive

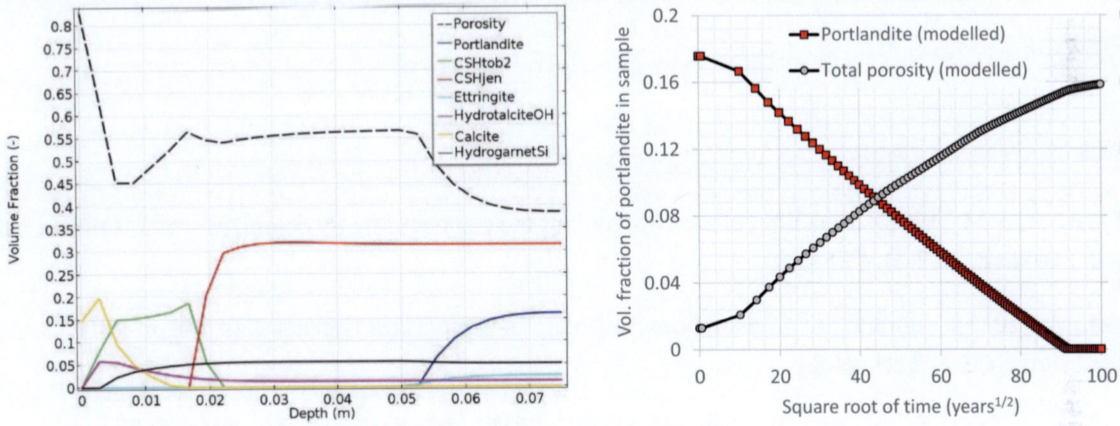

Figure 1: *Mineral volume fractions and porosity profiles after 5 000 years* ***(left)****, and portlandite volume fraction and total porosity in the entire modelled sample as a function of time* ***(right)****.*

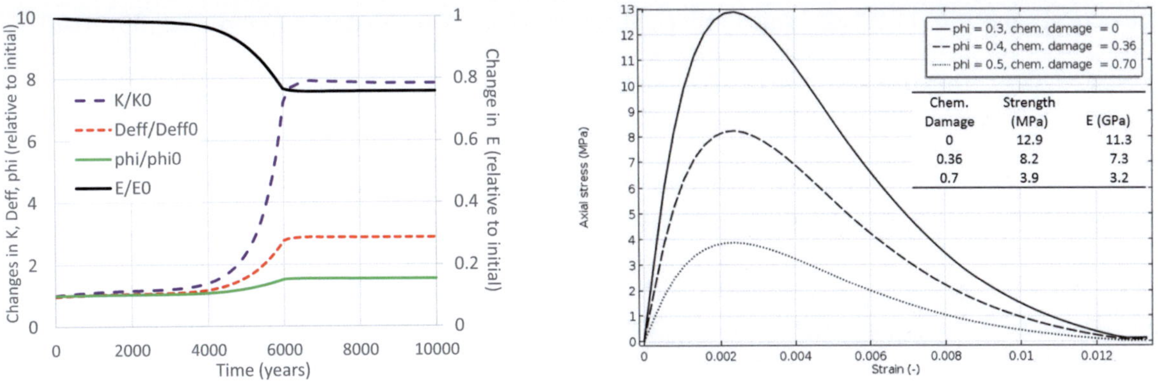

Figure 2: *Temporal evolution of physical properties at 16 mm from leaching surface* ***(left)****. Stress-strain curves for uniaxial compression at different levels of chemical damage* ***(right)****.*

Conclusions and Future work

A multi-scale chemo-mechanical model for cementitious materials has been implemented in iCP to simulate the degradation of transport and mechanical properties of cement paste and concrete as a function of the degree of calcium leaching from the samples. Preliminary results of the multi-scale chemo-mechanical model show the impact of long-term interaction with groundwater on the physical properties. On-going work aims at comparing the model results with the experimental data (when available) to check the validity of different model assumptions. Furthermore, the extension of homogenization methods to predict the transport properties will also be explored (see for instance Stora et al., 2009).

Acknowledgements

The research leading to these results has received funding from the European Union's European Atomic Energy Community's (Euratom) Horizon 2020 Programme (NFRP-2014/2015) under grant agreement, 662147 – Cebama.

References

Adenot, F. and Buil, M. (1992). Modelling of the corrosion of the cement paste by deionized water. Cement and Concrete Research, 22(2-3), 489-496.

Babaahmadi, A. (2015). Durability of Cementitious Materials in Long-Term Contact with Water. PhD Thesis, Chalmers University of Technology.

Babaahmadi, A., Tang, L., Abbas, Z., Mårtensson, P. (2015). Physical and Mechanical Properties of Cementitious Specimens exposed to an electrochemically derived accelerated leaching of calcium. International Journal of Concrete Structures and Materials, 9(3), 295-306.

Bary, B., Leterrier, N., Deville, E., Le Bescop, P. (2014). Coupled chemo-transport-mechanical modelling and numerical simulation of external sulfate attack in mortar. Cement and Concrete Composites, 49, 70-83.

Camps, G. (2008). Etude des interactions chemo-mécaniques pour la simulation du cycle de vie d'un élément de stockage en béton. PhD Thesis, Paul Sabatier University.

Faucon, P., Le Bescop, P., Adenot, F., Bonville, P., Jacquinot, J.F., Pineau, F., Felix, B. (1996). Leaching of cement: Study of the surface layer. Cement and Concrete Research, 26, 1707-1715.

Gérard, B., Pijaudier-Cabot, G., Laborderie, C. (1998). Coupling diffusion-damage modelling and the implications on failure due to strain localisation. International Journal of Solids and Structures, 35(31-32), 4107-4120.

Heukamp, F.H., Ulm, F.-J., Germaine, J.T. (2005). Does calcium leaching increase ductility of cementitious materials? Evidence from direct tensile tests. Journal of Materials in Civil Engineering, 17(3), 307-312.

Hu, D., Zhou, H., Zhang, F., Shao, J. (2014). Modeling of short- and long-term chemomechanical coupling behavior of cement-based materials. Journal of Engineering Mechanics, 140(1), 206-218.

Jacques, D. (2009). Benchmarking of the cement model and detrimental chemical reactions including temperature dependent parameters. Project near surface disposal of category A waste at Dessel. NIRAS-MP5-03 DATA-LT(NF) Version 1. NIROND-TR 2008–30 E.

Liu, Q.F., Li, L.Y., Easterbrook, D., Yang, J. (2012). Multi-phase modelling of ionic transport in concrete when subjected to an externally applied electric field. Engineering Structures, 42, 201-213.

Lothenbach, B., Le Saout, G., Gallucci, E., Scrivener, K. (2008). Influence of limestone on the hydration of Portland cements. Cement and Concrete Research, 38(6), 848-860.

Mazars, J. (1986). A description of micro- and macroscale damage of concrete structures. Engineering Fracture Mechanics, 25(5-6), 729-737.

Nardi, A., Idiart, A., Trinchero, P., de Vries, L.M., Molinero, J. (2014). Interface COMSOL-PHREEQC (iCP), an efficient numerical framework for the solution of coupled multiphysics and geochemistry. Computers & Geosciences, 69, 10-21.

Nguyen, V.H., Nedjar, B., Torrenti, J.M. (2007). Chemo-mechanical coupling behaviour of leached concrete: Part II: Modelling. Nuclear Engineering and Design, 237(20-21), 2090-2097.

Stora, E., Bary, B., He, Q.-C., Deville, E., Montarnal, P. (2009). Modelling and simulations of the chemo–mechanical behaviour of leached cement-based materials: Leaching process and induced loss of stiffness. Cement and Concrete Research, 39, 763-772.

Interfacial and petrophysical properties of low ph cements inferred from electrical measurements

Philippe Leroy[1*], Stéphane Gaboreau[1], Johannes Lützenkirchen[2], Sander Huisman[3], Egon Zimmermann[3], Francis Claret[1], Christophe Tournassat[1], Vanessa Montoya[2]

[1] BRGM (FR)
[2] KIT (DE)
[3] Forschungszentrum (DE)

* Corresponding author: p.leroy@brgm.fr

Abstract

Ions diffusion is the main transport process in cementitious materials because these materials have a large specific surface area and resulting low permeability. The electrochemical and petrophysical properties of cement pastes control their reactive transport properties and durability. However, the zeta potential at the low pH cement/water interface is not exactly known because of the surface electrical conductivity of calcium-silicate-hydrate (C-S-H). Furthermore, there is a need to develop non-intrusive methods to monitor evolution of the mineralogy, porosity or pore size distribution of cement pastes. Geophysical electrical methods such as streaming potential and induced polarization can be used to get information on the zeta potential and to non-intrusively monitor evolution of the mineralogical and petrophysical properties of low pH cement pastes. Laboratory streaming potential and spectral induced polarization experiments will be performed on hydrated low pH cement pastes. These experiments will be interpreted using surface complexation and transport models developed at the pore and sample scales. Surface electrical conductivity of C-S-H and its influence on zeta potential and spectral induced polarization measurements of low pH cement pastes will be thoroughly investigated.

Introduction

Hardened cement pastes have a very low permeability (typically $< 10^{-18}$ m^2; Powers, 1958) and high alkalinity (pH > 9; Bach, 2012) that make them a material of choice for the storage of nuclear wastes in geological formations (Berner, 1992). The C-S-H phase is one of the major phases of hardened cement pastes (Richardson, 2008). C-S-H forms lamellar crystallites of dimension typically $30 \times 60 \times 5$ nm^3 characterized by four layers separated by three interlayer spaces (Labbez et al., 2006; Haas and Nonat, 2015). Each layer is composed of three sheets, a "pseudooctahedral" calcium plane between two tetrahedral silicate chains (Viallis-Terrisse et al., 2001). The specific surface area of C-S-H crystallite is very high, between 100 and 300 m^2/g, and is higher than that of other minerals composing cement paste (such as Porlandite for instance) (Viallis-Terrisse et al., 2001; Jennings, 2008). The high specific surface area of C-S-H explains why the electrochemical properties of C-S-H dominate the reactive transport properties of hardened cement (Viallis-Terrisse et al., 2001; Labbez et al., 2006).

The zeta potential is the electrical potential at the shear plane between immobile and mobile water at the vicinity of the mineral surface (Hunter, 1981). The zeta potential of C-S-H particles suspended in water is usually deduced from their electrophoretic mobility measurements using the Smoluchowski equation (Labbez et al., 2006). Electrophoretic mobility measurements consist in measuring the velocity of the charged and suspended particle in water under the applied electric field (Delgado et al., 2007). The zeta potential is considered as a measured parameter representative of the surface charge of C-S-H to check the values of the surface electrical potentials predicted by electrostatic surface complexation models (Viallis-Terrisse et al., 2001). Zeta potential measurements and calculations are particularly important for estimating the electrochemical properties of C-S-H and cements because they are often the sole experimental data available to test the predictions of surface complexation models (Labbez et al., 2006). Complementary adsorption experiments can be performed to estimate for instance sorption equilibrium constants of radionuclides on cementitious materials (De Windt and Pellegrini, 2004), but the surface charge of these materials can't be estimated according to potentiometric titration experiments because of their high chemical reactivity (Labbez et al., 2006).

Two problems are inherent to the use of electrophoretic mobility (em) measurements to obtain the zeta potential of hardened cement pastes. Em measurements are performed for suspended C-S-H particles and not for hardened cement pastes, the latter being representative of cementitious materials under in-situ conditions. Furthermore, the Smoluchowski (S) equation is usually employed to infer C-S-H zeta potentials from their em measurements. Nevertheless, S equation does not consider surface conductivity effects of C-S-H particles on their em measurements. Surface conductivity effects may be strong for low pH cements (pH of pore water ~ 11; Bach et al., 2012) containing alpha-C-S-H (that has a low Ca/Si ratio, typically < 1; Viallis-Terrisse et al., 2001) because of the high specific surface area, negative surface charge and relatively dilute pore water of alpha-C-S-H (typically < 0.1 M; Bach et al., 2012). It results that the use of the S equation may lead to erroneous low zeta potentials of alpha-C-S-H (read Leroy et al. (2015) for the zeta potential of montmorillonite dispersions).

Laboratory streaming potential measurements can be used to estimate the zeta potential of hardened cement pastes. During these measurements, the sample is submitted to a water pressure difference, a shear plane is induced at the surface of the particles, and the measured resulting electrical potential difference, the streaming potential, is sensitive to the zeta potential (Delgado et al., 2007). However, as in the case of electrophoretic mobility measurements, the conversion of streaming potential into zeta potential is not straightforward, especially if the ionic strength of the pore water of low pH cement is low because of the effects of surface conductivity of C-S-H particles on measured streaming potential (read Li et al. (2016) for the effects of surface conductivity of calcite particles on streaming potential).

In addition, the spectral induced polarization (SIP) method may be very useful to non-intrusively monitor concrete durability in terms of mechanical, petrophysical, chemical and mineralogical evolution of concrete. This is due to the sensitivity of this method to conduction and polarization processes in the pore water and in the electrical double layer (EDL) surrounding particles (Leroy et al., 2016). One difficulty of this method concerns SIP measurements in very conducting materials such as cement pastes. In this case, a high level of accuracy is required for the measured phase shift between injected alternative current and measured electrical potential difference (Zimmermann et al., 2008).

Streaming potential combined with SIP experiments, interpreted by surface complexation and transport models are proposed to get information on interfacial, textural and mineralogical properties of hardened low pH cement pastes. SIP experiments will be interpreted using zeta potentials inferred from streaming potential experiments. *μ* on electrical measurements of low pH cements will be investigated.

Materials and methods

Low pH cements

Electrical measurements will be performed on low pH cements of water to cement ratio of 0.4 provided by Andra and VTT. Cebama reference mix design consists in a ternary binder composition with CEM I, granulated silica fume, and blast furnace slag (Table 1). Low pH cement provided by Andra contains 20% in mass of ordinary Portland cement, 32.5% of silica fume, and 47.5% of blast furnace slag.

Table 1: *Binder composition of Cebama reference mix design provided by VTT.*

Total binder content	280 kg/m^3
CEM I	105 kg/m^3
Silica Fume	110 kg/m^3
Blast furnace slag	65 kg/m^3
CaO/SiO_2 -ratio	0.61

Streaming potential measurements

During streaming potential experiments, induced water flow drags the excess counter-ions in the diffuse layer along the pores surface and creates a macroscopic current density, the streaming current and a macroscopic electrical potential difference, the streaming potential (Figure 1). The electrical field induced by the streaming potential is responsible for conduction currents in the bulk pore water and in the EDL around particles, i.e., of their surface conductivity (Hunter et al., 1981; Delgado et al., 2007). The surface conductivity of the particles increases the conduction current density, which is opposed to the streaming current density and hence decreases the magnitude of the measured streaming potential (Figure 1). It results that the use of the Helmholtz-Smoluchowski (HS) equation to convert streaming potential measurements into zeta potentials can lead to low apparent zeta potentials because this equation does not consider effects of surface conductivity on streaming potential. These effects are particularly important when the pore water is diluted (ionic strength typically < 0.1 M; Li et al., 2016), which can be the case of the pore water of low pH cement.

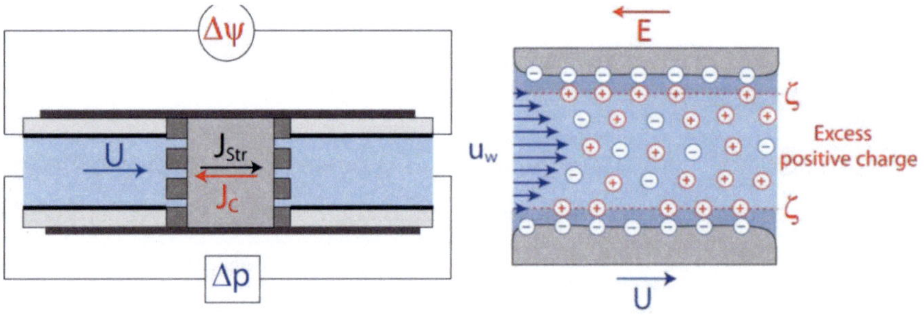

Figure 1: *Sketch of the streaming potential measurement (Li et al., 2016).*

Streaming potential experiments on low pH cement pastes will be performed at the laboratory of KIT-INE during the year 2016. The streaming potential device is the SurPASS Anton Paar Electrokinetic Analyzer, which measures the streaming potential but also the streaming current (not influenced by surface conductivity) and the electrical conductivity of the sample. The zeta potential of cement pastes will be constrained by the combination of streaming potential/current and electrical conductivity measurements. Two types of samples are considered:

- crushed cements of grain size ranging between 100 and 1 000 μm (0.5 kilogram maximum). Cement powders will also be used to prepare the pore water in chemical equilibrium with cement,
- 7 flat and rectangular cement pastes of dimensions (10 mm × 20 mm × 2 mm), to have slit channels of given thickness between two cement pastes of thickness 2 mm where water flow will occur. Water will be in chemical equilibrium with the cement paste.

Streaming potential experiments will be performed with or without contact with atmospheric CO_2. Apparent zeta potentials inferred from HS equation will be compared to zeta potentials corrected for surface conductivity and flow regime effects, following the approach developed by Li et al. (2016) for the streaming potential of calcite. Corrected zeta potentials will be used to constrain the parameters of the electrostatic surface complexation model that will be used to interpret spectral induced polarization measurements of low pH cements (Haas and Nonat, 2015).

Spectral induced polarization measurements

Frequency domain induced polarization measurements consist of imposing a sinusoidal current at a given frequency across the sample with two current electrodes and measuring the resulting electrical potential difference between two non-polarizing voltage electrodes (Figure 2). A phase shift is measured between injected current and measured electrical potential difference. This phase shift is associated with electromigration and diffusion currents in the bulk pore water and at the surface of non-conducting particles, hence to the interfacial, textural, and mineralogical properties of the sample, and also to its water content and chemical composition.

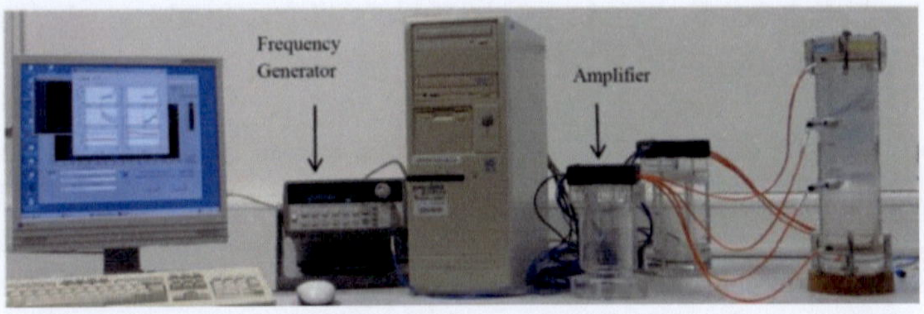

Figure 2: *Sketch of the spectral induced polarization experiment.*

Complex conductivity measurements will be performed in FJZ Jülich during year 2016 using the SIP ZEL apparatus (Zimmermann et al., 2008) on low pH cement pastes and will be interpreted using electrostatic surface complexation and transport models, following the work of Bücker and Hördt (2013) (membrane polarization model) and Leroy et al. (2016) for calcite (grain polarization model). Carbonation or Ca-leaching will be observed using SIP.

Conclusions and Future work

Streaming potential and spectral induced polarization experiments are proposed to describe the surface electrical, petrophysical, and mineralogical properties of low pH cement pastes. Measurements will be interpreted using electrostatic surface complexation and transport models at the pore and sample scales. Deduced zeta potential, porosity and mineralogical evolution can improve the predictions of the reactive transport properties of low pH cements for the storage of nuclear wastes in geological formations. The induced polarization method can also be used in the future to non-intrusively monitor durability of concretes in boreholes under in situ conditions.

Acknowledgement

The research leading to these results has received funding from the European Union's European Atomic Energy Community's (Euratom) Horizon 2020 Programme (NFRP-2014/2015) under grant agreement, 662147 – Cebama.

References

Bach, T.T.H., Coumes, C.C.D., Pochard, I., Mercier, C., Revel, B., Nonat, A. (2012). Influence of temperature on the hydration products of low pH cements. Cement and Concrete Research, 42, 805-817.

Berner, U.R. (1992). Thermodynamic Modeling of Cement Degradation - Impact of Redox Conditions on Radionuclide Release. Cement and Concrete Research, 22, 465-475.

Bücker, M. and Hördt, A. (2013). Analytical modelling of membrane polarization with explicit parametrization of pore radii and the electrical double layer. Geophysical Journal International, 194, 804-813.

Delgado, A.V., Gonzalez-Caballero, F., Hunter, R.J., Koopal, L.K., Lyklema, J. (2007). Measurement and interpretation of electrokinetic phenomena. Journal of Colloid and Interface Science, 309, 194-224.

De Windt, L., Pellegrini, D., van der Lee, J. (2004). Coupled modeling of cement/claystone interactions and radionuclide migration. Journal of Contaminant Hydrology, 68, 165-182.

Haas, J. and Nonat, A. (2015). From C-S-H to C-A-S-H: Experimental study and thermodynamic modelling. Cement and Concrete Research, 68, 124-138.

Hunter, R.J. (1981). Zeta Potential in Colloid Science: Principles and Applications. Academic Press, New York.

Jennings, H.M. (2008). Refinements to colloid model of C-S-H in cement: CM-II. Cement and Concrete Research, 38, 275-289.

Labbez, C., Jonsson, B., Pochard, I., Nonat, A., Cabane, B. (2006). Surface charge density and electrokinetic potential of highly charged minerals: experiments and Monte Carlo simulations on calcium silicate hydrate. The Journal of Physical Chemistry B, 110, 9219-9230.

Leroy, P., Li, S., Jougnot, D., Revil, A., Wu, Y. (2016). Modeling the evolution of spectral induced polarization during calcite precipitation on glass beads. Submitted to Geophysical Journal International.

Leroy, P., Tournassat, C., Bernard, O., Devau, N., Azaroual, M. (2015). The electrophoretic mobility of montmorillonite. Zeta potential and surface conductivity effects. Journal of Colloid and Interface Science, 451, 21-39.

Li, S., Leroy, P., Heberling, F., Devau, N., Jougnot, D., Chiaberge, C. (2016). Influence of surface conductivity on the apparent zeta potential of calcite. Journal of Colloid and Interface Science, 468, 262-275.

Powers, T.C. (1958). Structure and physical properties of hardened Portland cement paste, Journal of the American Ceramic Society, 41, 1-6.

Richardson, I.G. (2008). The calcium silicate hydrates, Cement Concrete Res., 38, 137-158.

Viallis-Terrisse, H., Nonat, A., Petit, J.C. (2001). Zeta-potential study of calcium silicate hydrates interacting with alkaline cations. Journal of the American Ceramic Society, 244, 58-65.

Zimmermann, E., Kemna, A., Berwix, J., Glaas, W., Munch, H.M., Huisman, J.A. (2008). A high-accuracy impedance spectrometer for measuring sediments with low polarizability. Measurement Science and Technology, 19.

Pore-scale reactive transport modelling of cementitious systems: concpet development based on the lattice-boltzmann method

Stephan Rohmen[1*], Babak Shafei[2], Andrés Idiart[2], Guido Deissmann[1], Dirk Bosbach[1]

[1] Forschungszentrum Jülich GmbH, Institute of Energy and Climate Research: Nuclear Waste Management and Reactor Safety (IEK-6) (DE)
[2] Amphos21 Consulting (ES)

* Corresponding author: st.rohmen@fz-juelich.de

Abstract

Pore-scale models represent an appealing approach for obtaining a more accurate and mechanistic description of physical and chemical processes in heterogeneous porous media such as cement-based materials. In this paper, the concept for the development of a coupled code, which is able to calculate reactive transport processes in porous media on the pore scale using the Lattice-Boltzmann approach, is outlined. The aim of this coupled code is to simulate chemically-driven alteration and degradation processes and solute transport in cement-based materials. The current development status of the code and the validation for simple diffusion problems against analytical solutions is described. Resulting from the comparison of analytical solution and numerical results it is concluded that solute transport by diffusion can be correctly simulated at time of writing.

Introduction

Cement-based materials have been widely adopted as engineered barriers for nuclear waste storage both for their radionuclide fixation and immobilization properties as well as for its low permeability and diffusivity. When hydrated, the hardened cement paste is composed of cement hydrates such as C-S-H phases, portlandite, or aluminate/ferrate compounds (AF_m/AF_t), porosity, and unhydrated clinker. The detailed composition of a given cementitious material and its microstructure (e.g., pore size distribution) depends on the cement mix formulation and the mixing and curing processes (Taylor, 1997).

Physical and chemical heterogeneities exist in a range of length scales (nm to dm) in cementitious systems. The macroscopic physical (including mechanical) and chemical properties of hardened concrete strongly depend on the pore structure evolution in the long-term. The microstructure and porosity will be modified as a result of chemical and/or physical degradation processes, e.g., leaching or sulfate attack.

Pore-scale models simulate similar physical and chemical processes than traditional continuum-scale models. The main difference is that the former models explicitly consider the pores and the solid as distinct domains, while water-rock interactions occur at the interface between these two. On the other hand, continuum-scale models are based on the definition of a representative volume element where the different phases (solid, liquid, and gas) are defined implicitly.

Therefore, pore-scale models may be regarded as an appealing tool to describe the pore space structure and non-uniform distribution of chemical species in a realistic manner (Zhang et al., 2014). The use of pore-scale models in other disciplines, such as oil and gas, is relatively well-established for natural rocks (Huber et al., 2014; Kang et al., 2014; Shafei et al., 2013). However, their use in cement science is very recent (Patel et al., 2014; Raoof et al., 2013). The advantage of these models over traditional continuum-scale models at the meso- and macro-levels is that they allow to explicitly capturing microscopic effects of chemical reactions on the physical properties in a mechanistic way. In particular, by this means it is possible to describe multi-phase interactions (e.g., fluid and solid) in the model in a more physically sound way. This is the main focus of this work, which is in line with the objectives of CEBAMA. In order to study the mass transport behavior and the impact of chemical degradation due to the interaction with different groundwater compositions on the transport properties of cement pastes, a pore-scale reactive transport model will be developed and validated in this work. To this end, two existing widely used tools will be coupled: a three dimensional advection-diffusion solver at the pore scale based on Lattice-Boltzmann Method (LBM), Palabos, and a geochemical software, PhreeqcRM. The main advantages of LBM compared to other methods such as pore network models are that concentration gradients within the pores can be represented, and that the pore space geometry can be discretized in a more physical manner. A recent review of pore-scale models based on LBM and application in different fields can be found in Yoon et al. (2015).

Lattice-Boltzmann Method (LBM)

The Lattice-Boltzmann method can be used to simulate the microscopic behavior of particles in a fluid in a discrete way. Space, time and velocity vectors are discretized. By choosing proper discretization steps, mesoscopic and macroscopic properties can be recovered (see Mohamad, 2011; Succi, 2001). The fundamental Lattice-Boltzmann Equation (LBE) is:

$$f_i(r + e_i\Delta t, t + \Delta t) = f_i(r,t) + \Omega \qquad \textit{Eq. 1}$$

In Equation 1, f_i is the particle distribution function at a particular discrete velocity direction i, r is the spatial vector and t is time. The left-hand side of the equation represents the particle probability distribution at the neighboring cell and next time step. On the right-hand side, the current local flow in the particular direction is corrected by a collision operator, Ω. The collision operator is the location in which the physical processes take place. By properly choosing collision terms and discretization magnitudes the macroscopic Navier-Stokes equation for fluid flow can be derived.

Macroscopic properties such as fluid density, mass flux and flow velocity can be recovered from the sum of the probability functions and product of probability functions and local velocities, respectively. Diffusion processes can be calculated in a similar manner as advection. In order to simulate chemical reactions such as dissolution and precipitation and their effect on porosity changes the collision operation has to be extended (Patel et al., 2014). Dissolution and precipitation reactions and consequent porous media geometry evolution need to be updated in the reaction term.

In some cases, non-Fickian multi-species diffusion has been shown to be important (Galíndez and Molinero, 2010). This addresses the issue of chemical species interactions, like ion-ion collision effects, represented by Nernst-Planck or Maxwell-Stefan diffusion equations, and which can be described in the LBM transport algorithm.

Neither the LBM transport algorithm nor the interface coupling implementation is limited to a particular LBM velocity vector scheme (D2Q5, D2Q9, D3Q19, etc.) or collision operators (BGK, TRT, MRT, etc.). This enables the possibility to extend the simulation code for the scope of other applications besides cement alteration modelling in which possibly high Péclet numbers are involved or complex phase interfaces are important.

Interface development

The pore-scale reactive transport model will be based on the existing software Palabos (FlowKit, 2011) and PHREEQC (Parkhurst and Appelo, 2013) for Lattice-Boltzmann calculations and for solving chemical reactions, respectively. Since both codes are open access the coupling interface, which will be developed within the framework of CEBAMA, will also be released on an open access platform.

As programming language C++ has been chosen. This opens the possibility for a trade-off between efficient overhead minimized programming capabilities, good scalability for multi-CPU techniques and encourages code maintainability at the same time. One of the goals is to design the coupling interface in an abstract way in order to allow changing the chemical and transport codes used in the simulations, if required in future.

The coupling scheme between transport and chemical code is based on the sequential non-iterative approach (SNIA). Therefore, the transport and chemical processes are split into two distinct operations (operator-splitting approach). This leads to a cyclic interaction between chemical and transport calculations. The solid phase compositions, as well as the geometry of the pore space are updated after each time step (see Figure 1, center).

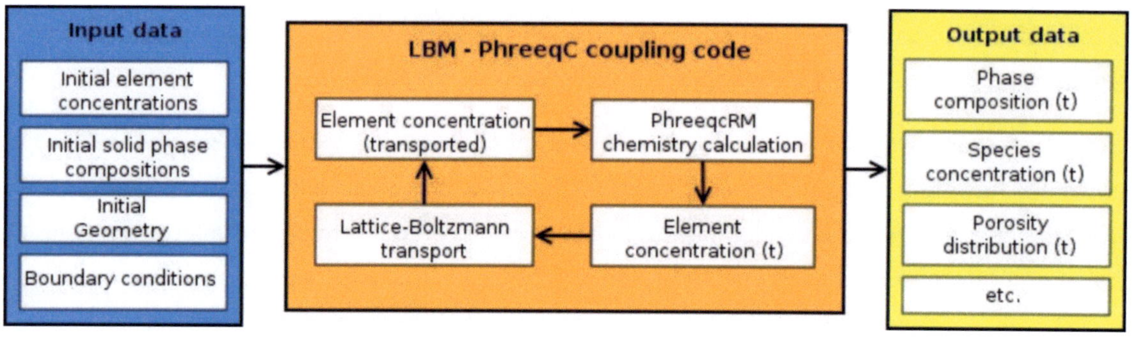

Figure 1: *Calculation loop and data input/output scheme.*

Particular input data like initial solute concentrations, solid phase compositions and geometric information of the simulation model have to be provided to satisfy the model description. Those can be defined either programmatically via a C++ configuration interface or as human readable JSON input file which can be imported by the coupling code input system. The current input file definition is preliminary and can be modified and extended to the needs of the end user (see Figure 1, left). In order to conduct scientific results, output data can be written with arbitrary frequency at each time step and do include for instance phase composition, chemical species concentration, altered porosity distribution and pH-value. Depending on the demands of the scientific research field or engineering problem the kind of output data can be extended to particular needs (see Figure 1, right).

Palabos has already incorporated the capability of shared and distributed memory multi CPU, i.e., OpenMP and MPI. The software can be executed on common CPU based hardware. Due to the nature of dynamic locality of the LBM algorithm it is possible to develop an interface that is easily parallelizable on multiple CPUs. This assures good scalability with increasing number of calculation units as CPU cores.

A modified version of PHREEQC will be used, named as PhreeqcRM (Parkhurst and Wissmeier, 2015), which is suitable for coupling with external codes. PhreeqcRM implements a C++ interface to the PHREEQC core which can be used programmatically in order to prevent most of the overhead due to parsing text based input files. Another advantage is the possibility to use multi-CPU capabilities via multi-threading or MPI to reduce computing time. It is planned that the ThermoChimie database as implemented in PHREEQC format will be used (Giffaut et al., 2014), contains relevant data on cement phases.

At the time of writing a first preliminary version of the interface coupling code was programmed. This code is currently able to synchronize concentrations between PhreeqcRM and Palabos. Thus, it is already possible to calculate diffusive transport of conservative tracers. In order to achieve cluster computing capability, the interface code was developed with multi-CPU supporting features (MPI).

Validation of the LBM approach for analytical solvable problems

In order to validate the current version of the coupling between the LBM transport code and PHREEQC, a simple 1D diffusion-based setup was studied. It consists of a NaCl solution domain surrounded by semi-infinite regions with zero concentration on both sides. No source or sink terms are considered, since the NaCl concentration (1 mol/L) is low enough to prevent precipitation processes at the selected temperature (25°C). Therefore, precipitation was neglected in the PHREEQC setup. The total width of the high concentration domain is set to 10 cm. This NaCl solution domain is surrounded by 45 cm pure water domain on each side and zero concentration boundary condition was imposed at the edge of the system. Thus, the complete system length was 1 m and was discretized in 1 000 Lattice-Boltzmann nodes which led to 1 mm cell resolution. The diffusion coefficient in the modelled domain is set to 10^{-9} m^2/s, representative for diffusion of aqueous solutes in free solution. This system can be solved analytically with the following expression:

$$C(x,t) = \frac{1}{2} C_0 \left(erf\left(\frac{h-x}{2\sqrt{Dt}}\right) + erf\left(\frac{h+x}{2\sqrt{Dt}}\right) \right) \qquad \textbf{\textit{Eq. 2}}$$

Equation 2 shows the position (x) and time (t) dependent concentration function for this problem. *C* is the concentration, C_0 the initial concentration in the domain, *D* is the diffusion coefficient, h is the half width of the domain, while *erf* is the so called *error function.* The detailed derivation of the analytical solution can be found in (Crank, 1975).

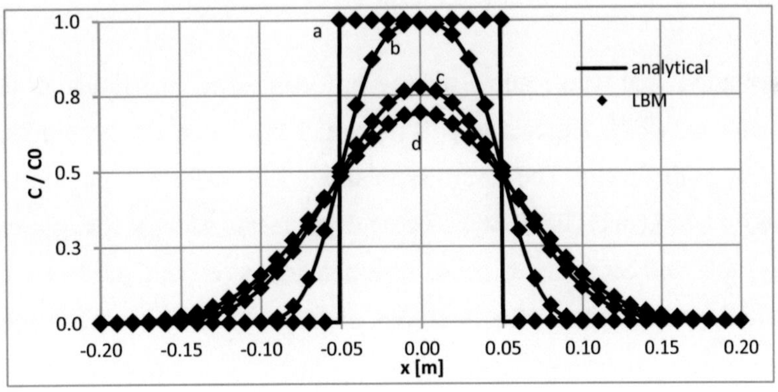

Figure 3: *Comparison of LBM diffusive transport results (dots) to analytical solution (solid lines). Results for four time steps are conducted: a=0 s, b=166 667 s, c=858 333 s and d=1 216 667 s.*

The concentration profiles were calculated for four different times with the LBM code and compared to the analytical solution (Figure 3). The initial concentration profile (a) is obviously a step function. With increasing elapsed time, the solute diffuses to the boundaries of the model domain, while the maximum concentration in the center decreases (b - d). It can be seen that the resulting concentration values determined by the numerical LBM scheme and those from the analytical solution are in very good agreement. Minor variations could result from the chosen spatial discretization in the LBM simulations.

Conclusions and Future work

At the current state a preliminary version of the coupling code does already exist. A first validation with a diffusion-only setup, dealing with a conservative tracer, has shown good agreement with analytically determined values. This demonstrates that the current implementation correctly transfers solute concentrations between Palabos and PHREEQC. Additionally, this code is already capable of multi-CPU computing features via MPI. The actual development state and the computing scalability were tested on local code development workstation and by utilization of an in-house computer cluster.

In the next development steps of the coupling code, the transport will be extended in order to support Fickian diffusion of several species. Afterwards, a set of benchmarks will be modelled to validate the code. These benchmarks will be simple enough to be either solvable analytically or well-known by other benchmarking experiments. The cases include (but are not limited to) Fickian diffusion, coupled advection-diffusion transport, coupled diffusion-reaction problems (dissolution and precipitation) and reactive transport benchmarks of flow through material involving porosity effects. The pore size distribution can be either generated numerically, e.g., by CEMHYD3D (Bentz, 2005), or taken from 2D/3D imaging techniques such as μCT or SEM from real samples.

Additionally, code-to-code benchmarks can be performed among CEBAMA WP3 project partners. However, the pore-scale reactive transport model for the comparison is not fixed yet and will be further discussed when the coupling software modules are capable to solve the corresponding simulation models with respect to CEBAMA.

Once the implementation and verification phases are complete, experimental data produced within CEBAMA will be simulated, which requires microstructural information on the pore space of the hardened cement pastes used in the experiments. The future simulations will address in particular the CEBAMA WP1 experiments performed by USFD and USURREY. These experiments address changes in transport parameters (porosity, permeability) and cement mineral phase assemblages in various hardened cement pastes due to interaction with different groundwaters. The leaching experiments will be performed with cementitious materials and a hardened low pH cement paste (CEM V, based on a recipe provided by ANDRA), using "granitic" water, clay pore water representative for Callovo-Oxfordian clay formations, and saline water/sea water. The characterization methods intended to be used for the evaluation of changes in transport properties and cement microstructure within WP1 comprise porosimetry and permeability techniques (MIP, gas adsorption, oxygen permeametry) and μ-XCT as well as microscopy techniques (SEM, TEM). Secondary cement phases will be characterized by USFD inter alia through NMR and μ-XAS techniques.

The current development and testing of the coupling and extension of the LBM code is carried on a local office workstation. Shared memory computations will be performed to test and validate the code based on simple model setups. Thereafter, scalability tests with increased geometrical and/or chemical complexity will be done with help of the in-house cluster at Jülich (274 cores, 868 GB memory). The final goal is to use High Performance Computing (HPC) resources available at Jülich Supercomputing Centre (JSC).

To analyze up-scaling effects the provided effective diffusion and permeability coefficients from the results of the micro-scale calculations can be reused as input data for continuum based calculations, e.g., using iCP (interface COMSOL-PhreeqC) (Nardi et al., 2014).

Acknowledgements

The research leading to these results has received funding from the European Union's European Atomic Energy Community's (Euratom) Horizon 2020 Programme (NFRP-2014/2015) under grant agreement, 662147 – Cebama.

References

Bentz, D.P. (2005). CEMHYD3D: A three-dimensional cement hydration and microstructure development modeling package. Version 3.0.

Crank, J. (1975). The mathematics of diffusion (Second Edi.). Oxford: Clarendon Press.

FlowKit (2011). Palabos user guide.

Galíndez, J.M. and Molinero, J. (2010). On the relevance of electrochemical diffusion for the modeling of degradation of cementitious materials. Cement and Concrete Composites, 32(5), 351-359.

Giffaut, E., Grivé, M., Blanc, P., Vieillard, P., Colàs, E., Gailhanou, H., Gaboreauc, S., Marty, N., Madé, B., Duro, L. (2014). Andra thermodynamic database for performance assessment: ThermoChimie. Applied Geochemistry, 49, 225-236.

Huber, C., Shafei, B., Parmigiani, A. (2014). A new pore-scale model for linear and non-linear heterogeneous dissolution and precipitation. Geochimica et Cosmochimica Acta, 124, 109-130.

Kang, Q., Chen, L., Valocchi, A.J., Viswanathan, H.S. (2014). Pore-scale study of dissolution-induced changes in permeability and porosity of porous media. Journal of Hydrology, 517, 1049-1055.

Mohamad, A.A. (2011). Lattice Boltzmann Method. London: Springer London.

Nardi, A., Idiart, A., Trinchero, P., de Vries, L.M., Molinero, J. (2014). Interface COMSOL-PHREEQC (iCP), an efficient numerical framework for the solution of coupled multiphysics and geochemistry. Computers & Geosciences, 69, 10-21.

Parkhurst, D.L. and Appelo, C.A.J. (2013). Description of input and examples for PHREEQC Version 3 - A computer program for speciation, batch-reaction, one-dimensional transport, and inverse geochemical calculations. U.S. Geological Survey Techniques and Methods, book 6, chapter A43, 6-43A.

Parkhurst, D.L. and Wissmeier, L. (2015). PhreeqcRM: A reaction module for transport simulators based on the geochemical model PHREEQC. Advances in Water Resources, 83, 176-189.

Patel, R.A., Perko, J., Jacques, D., Schutter, G. De Breugel, V., Ye, G. (2014). Simulating cement microstructural evolution during calcium leaching, Proceedings of the 10th fib International PhD Symposium in Civil Engineering, 173-178.

Raoof, A., Nick, H.M., Hassanizadeh, S.M., Spiers, C.J. (2013). PoreFlow: A complex pore-network model for simulation of reactive transport in variably saturated porous media. Computers and Geosciences, 61, 160-174.

Shafei, B., Huber, C., Parmigiani, A. (2013). Pore-scale simulation of calcite dissolution and precipitation using lattice-Boltzmann method. Mineralogical Magazine, 77(5), 2185.

Succi, S. (2001). The Lattice Boltzmann Equation for fluid dynamics and beyond. Oxford.

Taylor, H.F.W. (1997). Cement chemistry. Thomas Telford.

Yoon, H., Kang, Q., Valocchi, A.J. (2015). Lattice Boltzmann-Based Approaches for Pore-Scale Reactive Transport. Reviews in Mineralogy and Geochemistry, 80 (1), 393-431.

Zhang, M., Xu, K., He, Y., Jivkov, A.P. (2014). Pore-scale modelling of 3D moisture distribution and critical saturation in cementitious materials. Construction and Building Materials, 64, 222-230.

Effect of selection of secondary minerals on H-M-C coupling calculation

Hitoshi Owada[1*], Daisuke Hayashi[1], Atsushi Iizuka[2]

[1] RWMC, Radioactive Waste Management Funding and Research Centre (JP)
[2] Kobe University (JP)

* Corresponding author: owada@rwmc.or.jp

Abstract

Two dimensional (2-D) Hydro-Mechanical-Chemical (H-M-C) coupled calculations with assumptions of various secondary zeolitic minerals were performed to identify the influences of secondary minerals on the change of mechanical and hydraulic properties as the results of long-term alteration of cement-bentonite systems. Analcime, heulandite and clinoptilolite are assumed as the secondary minerals generated by the alkaline alteration of montmorillonite from our previous works and 2nd TRU report in Japan (Fujii et al., 2010; Japan Nuclear Cycle Development Institute and Federation of Electric Power Company, 2007). Predicted rate of the volume change around waste packages after 10 000 and 100 000 years were largely different by the selection of secondary minerals.

Introduction

In the concept of Japanese TRU geological disposal, bentonitic material is anticipated as a mechanical and hydraulic buffer to maintain low water penetration into the disposal vault. Because hydraulic and mechanical properties of bentonitic materials depend on the montmorillonite content and dry density, the dissolution of montmorillonite and precipitation of secondary minerals as a result of the alkaline alteration will directly affect the long term performance of the EBS system. Zeolitic minerals such as analcime, heulandite and clinoptirolite are expected to be secondary phases that will precipitate as a product of cement-bentonite interaction (Fujii et al., 2010; Japan Nuclear Cycle Development Institute and Federation of Electric Power Company, 2007). Given that the type of secondary minerals will differ depending on the environmental conditions, such as temperature, groundwater composition, and pH, it is useful to predict the influence of such uncertainty in the current generic phase of the site selection.

Mechanical properties of bentonite

Dependensce of mechanical properties, such as swelling pressure, swelling capacity and compressive strength, and also hydraulic properties, such as water permeability, on montmorillonite content and dry density of bentonitic materials are predicted from the results of previous studies (Sasakura, 2002; Ishikawa, 1997; Ishibashi, 2011; Takaji, 1999).

Formulation of the alteration using the montmorillonite content

Because hydraulic properties such as water permeability and diffusivity change with not only porosity but also montmorillonite content and those affect to the amount of the secondary minerals which are generated by the reaction between bentonitic minerals and cementitious alkaline solution, change of the total volume of the solid phase must be considered to predict the influence of the dissolution of montmorillonite. Morimoto (2014) had formulated such volume change as shown in Figure 1.

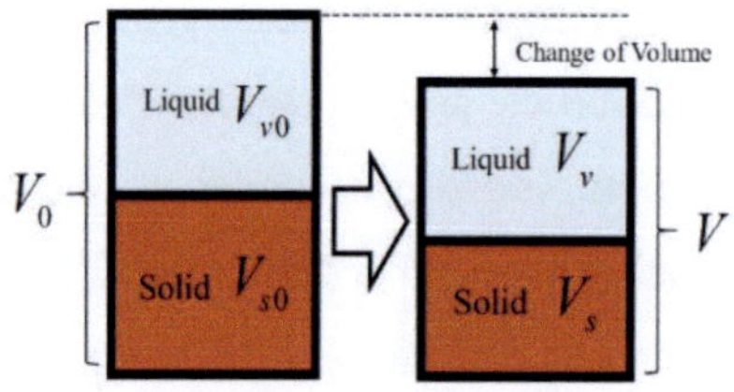

Figure 1: *The Conception diagram of the volume change.*

Input data for the Coupled calculations

Thermodynamic data from JNC-TDB.TRU (JNC, 2005) and our work (Satoh, 2017) are used for the chemical reactions regarding alkaline alteration of bentonitic materials. Data of the density and molar volumes of montmorillonite zeolitic minerals and SiO_2, which precipitates as secondary mineral, are listed in Table 1.

Table 1: *Density and molar volume of minerals.*

Minerals	Density (g/cm^3)	Molar volume (cm^3/mol)
Montmorillonite	2.74	138.07
Analcime	2.74	131.5
Clinoptilolite	2.24	361.01
Heulandite	2.20	210
SiO_2	2.65	22.67

Prediction of the alteration rate

Long term chemical-transport calculation was carried out using the code Phreeqc-TRANS to predict the alteration rate (β) in each element and each time, which is needed to calculate the mechanical properties. Figure 2 shows the dimensions, geometric domains, and boundary conditions of the chemical-mass transport coupled calculation.

Initial conditions after the operational period

Initial conditions, such as the distribution of dry density and volumetric strain in buffer materials, were calculated using the code DACSAR-BA which is a Hydro-Mechanical coupling code for clay materials being developed by Kobe University in Japan. Figure 3 shows the distribution of the volumetric strain before the chemical alteration.

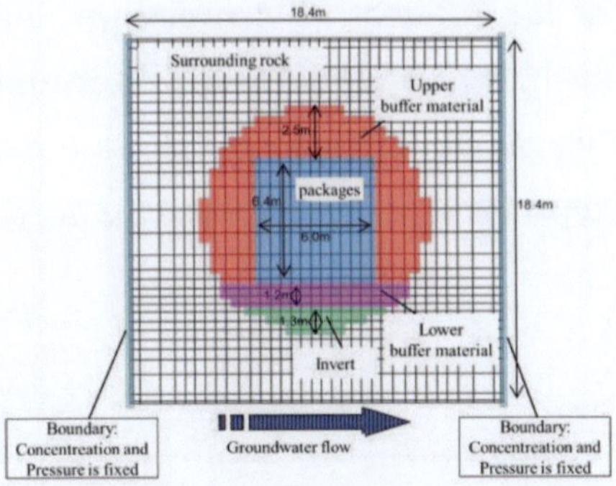

Figure 2: *Geometry of the chemical-mass transport coupled calculation.*

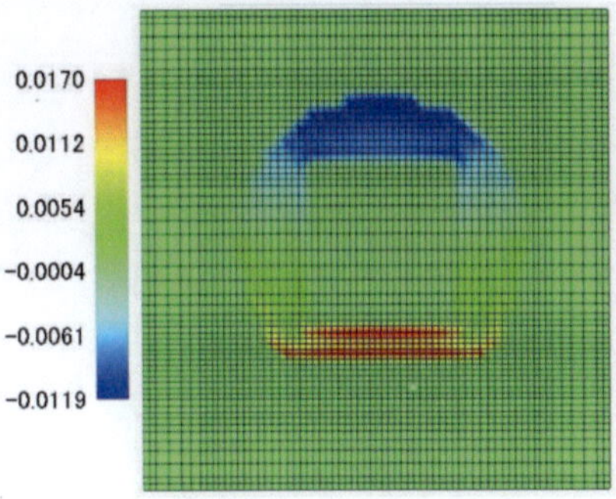

Figure 3: *Distribution of the volumetric strain (-) before the chemical alteration.*

H-M-C coupled calculation using DACSAR-BA

H-M-C coupled calculation is carried out using the code DACSAR-BA. The alteration rate of each position described in the previous section was used for the input to the volume and density calculation for that coupled calculations. To estimate the influence of the different secondary minerals, 4 different calculations were done to consider different scenarios, as shown in Table 2.

Table 2: *Scenarios for the coupled calculations.*

Case	Scenario
Case 1	Montmorillonite changed into analcime
Case 2	Montmorillonite changed into clinoptilolite
Case 3	Montmorillonite changed into Heulandite
Case 4 (references)	Montmorillonite dissolves (no secondary minerals)

Figure 4 shows the distribution of the alteration rate and rate of volume change in each element. Although the distribution of the alteration rate by the chemical calculation was common in all cases, the volume change was different in each of the scenarios considered. The difference in molar volumes of secondary minerals is considered as the dominant cause of the different volume changes. As shown in Figures 5, 6 and 7 the distribution of volumetric stress in the buffer strongly depends on the properties of secondary minerals. It is noted that a change in the volume of the buffer will lead to an increase water permeability and diffusivity. This increase is due to the fact that these properties strongly depend on the dry density of the buffer material.

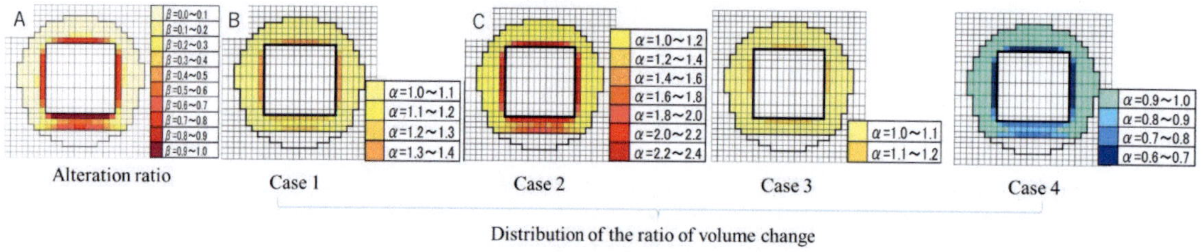

Figure 4: *Distribution of the alteration rate (β) and rate of volume change (α) in each element.*

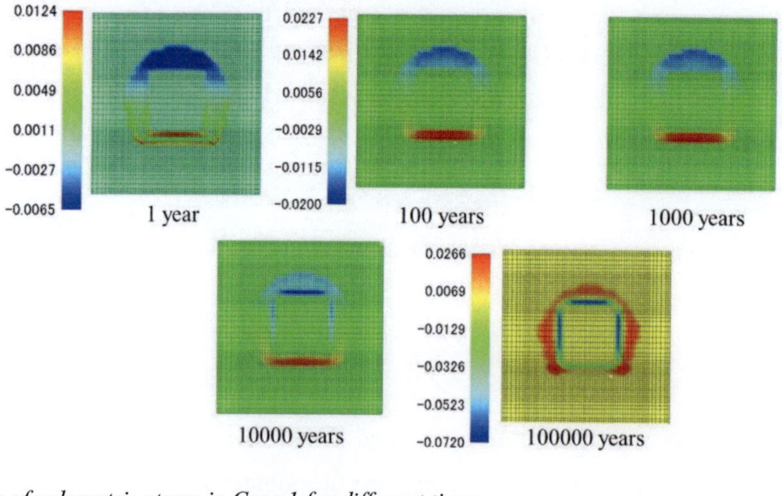

Figure 5: *Distribution of volumetric stress in Case 1 for different times.*

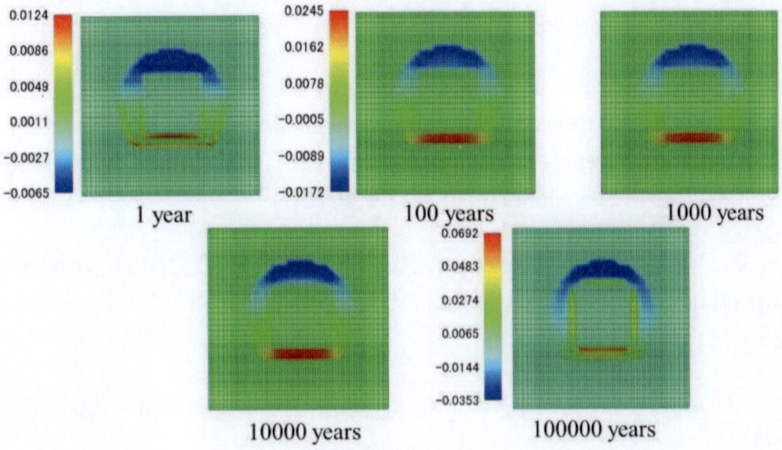

Figure 6: *Distribution of volumetric stress in Case 3 for different times.*

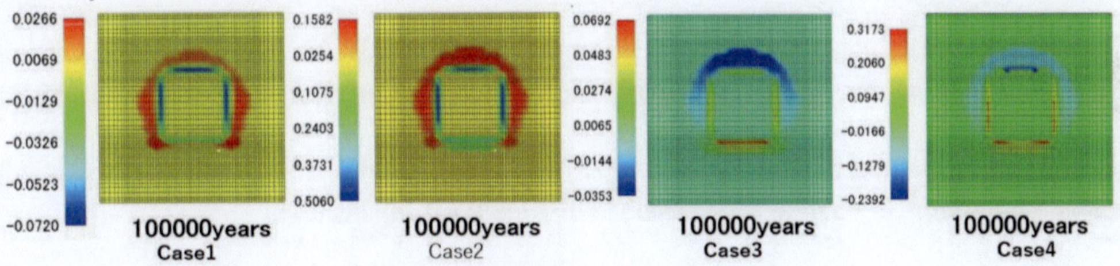

Figure 7: *Distribution of volumetric stress at 100 000 years later of each cases.*

Conclusions and Future work

The mineralogy of the secondary minerals will affect the performance of the clay material. To predict the long term performance of the EBS and near field, including cement-clay interactions, it will be important to identify the secondary minerals and to predict the reaction path.

We will perform a long term alteration test for cement-bentonite coupling that will be useful to determine the secondary minerals in the cement-bentonite systems.

Acknowledgement

This research is a part of "Development of the Treatment and Disposal Techniques for TRU Waste Disposal" (FY2015) under a grant from the Japanese Ministry of Economy, Trade and Industry (METI).

The research leading to these results has received funding from the European Union's Horizon 2020 Research and Training Programme of the European Atomic Energy Community (EURATOM) (H2020-NFRP-2014/2015) under grant agreement n° 662147 (CEBAMA).

References

Fujii, N., Arcilla, C-A., Yamakawa, M., Pascua, C., Namiki, K., Sato, T., Shikazono, N., Alexander, W.-R. (2010). Natural analogue of bentonite reaction under hyperalkaline conditions: overview of ongoing work at the Zambales ophiolite, Philippines. Proceedings of the 13th International Conference on Environmental Remediation and Radioactive Waste Mangement (ICEM2010).

Ishibashi, N., Komine, H., Yasuhara, K., Murakami, S., Mori, T., Ito, H. (2011). Influences of Monmorilonite content on the compaction properties of bentonite. Proceedings of the 46th Conference Japanese Geotechnical Society (in Japanese).

Ishikawa, H., Ishiguro, K., Namikawa, T., Sugano, T. (1997). Compaction Properties of Buffer Material. PNC-TN8410 97-051.

Japan Nuclear Cycle Development Institute and Federation of Electric Power Company (2007). Second Progress Report on Research and Development for TRU Waste Disposal in Japan. JAEA-Review 2007-010, FEPC TRU-TR2-2007-01.

Arthur, R.C., Sasamoto, H., Oda, C., Honda, A., Shibata, M., Yoshida, Y., Yui, M. (2005). Development of Thermodynamic Database for Hyperalkaline Argillaceous Systems. JNC Report, JNC TN8400-2005-010.

Morimoto, K. (2014). Modelling of Mechanical properties of Bentonitic Buffer Materials in Consideration with the Chemical Alteration. Graduation thesis, Kobe University (in Japanese).

Sasakura, K., Kuroyanagi, M., Okamoto, M. (2002). Obtaining the alteration data for the modelling of Bentonite Alteration. JNC Report, JNC TJ8400-2002-025 (in Japanese).

Satoh H. et al. (2017). To be submitted.

Takaji, K. and Suzuki, H. (1999). Static Mechanical Properties of Buffer Material. PNC TN8400 99-041 (In Japanese).

Coupled THCm numerical models of heating and hydration tests to study the interactions of compacted FEBEX bentonite with OPC concrete

Javier Samper [1*], Alba Mon [1], Luis Montenegro[1], Acacia Naves[1], Jesús Fernández[1], Jaime Cuevas[2], Raúl Fernández[2], María Jesús Turrero[3], Elena Torres[3]

[1] Escuela de Ingenieros de Caminos. Universidad de A Coruña (ES)
[2] Facultad de Ciencias. Universidad Autónoma de Madrid (ES)
[3] Centro de Investigaciones Energéticas, Medio Ambientales y Tecnológicas (ES)

* Corresponding author: j.samper@udc.es

Introduction

The main objectives of UDC in the CEBAMA Project include: 1) improving THCm codes and models of the interactions of concrete and bentonite; 2) performing THCm models of the heating and hydration HB tests; 3) performing long-term predictions of the interactions of bentonite, concrete and clay rock; and 4) contributing to the analysis of time scaling issues. Here we present the THCm model of the HB1, HB2, HB3 and HB4 tests (Turrero et al., 2011; Torres et al., 2013). The tests were modelled with INVERSE-FADES-CORE. The first version of INVERSE-FADES-CORE was developed within FEBEX I and was later improved within the FEBEX II, NFPRO, and PEBS Projects (Zheng et al., 2010; Samper et al., 2013).

Tests and model description

Several sets of tests were performed to study the interactions of concrete and bentonite (Turrero et al., 2011; Torres et al., 2013). The tests aimed at studying the interactions of concrete and bentonite pore water at the conditions prevailing in the EBS during the early hydration stage. This set of tests are denoted as HB and were performed on medium-size cells containing a 7.15 cm thick bentonite sample in contact with a 3 cm thick piece of concrete. The concrete was sulphate-resistant ordinary Portland cement (CEM I-SR) following the recipe by CSIC-IETcc CEM-SR: 400 kg of cement, 911 kg of sand (0 - 5 mm), and 946 kg of aggregates (6 - 16 mm) with a w/c ratio of 0.45. Each cell was hydrated at a constant pressure at the top of the cell through the concrete while the temperature was maintained constant at 100ºC at the bottom of the cell (Figure 1). The HB cells provide data on the concrete and bentonite interface at several times: they were dismantled after 6, 12, 18, 54, and 80 months for the HB1 to the HB5 tests and 104 months for the HB6 test, which will be dismantled within the course of CEBAMA Project. The full set of the experimental results of the HB5 tests are not yet available for modelling.

HB1 to HB4 tests were modelled with a 1D grid (see Figure 1). Bentonite and concrete parameters were taken from the previous model of the HB4 cell in Samper et al. (2013). The initial porosity of the bentonite is 0.4 and the initial water content is 13.3%, which corresponds to a saturation of 57% and a suction of $1.27 \cdot 10^8$ Pa. The concrete has a porosity of 0.125 and a gravimetric water content of 2.2%. The initial temperature is 22ºC along the cell. Similar to previous THMC models of the FEBEX bentonite (Zheng et al., 2010), the initial total stress is assumed to be uniform, isotropic and equal to 250 kPa. There were experimental problems to maintain a constant water injection pressure during the experiment. Given the lack of reliable water injection pressure data, the injection liquid pressure was estimated from measured cumulative inflow data. Its value is equal to 100 kPa.

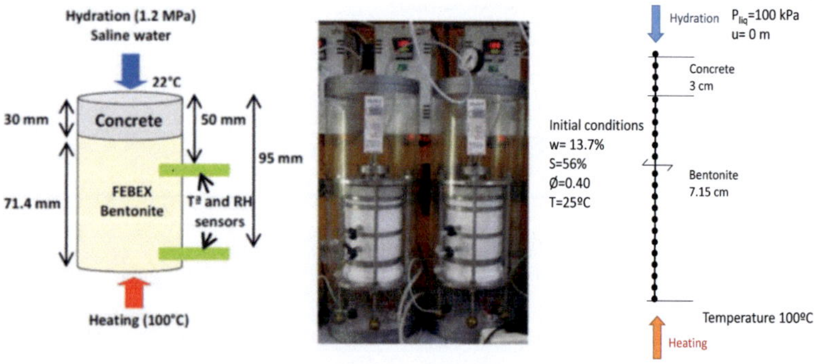

Figure 1: *Setup of the concrete-bentonite HB tests* ***(left)*** *on medium-size cells and finite element mesh and boundary conditions for the numerical model* ***(right)****.*

The cells were hydrated with a synthetic Spanish Reference Clay Porewater (Turrero et al., 2011; Torres et al., 2013). The initial composition of the OPC concrete porewater was derived from speciation runs performed with EQ3/6 (Wolery, 1992) by assuming that the concentration of dissolved Ca^{2+} is controlled by local chemical equilibrium with respect to portlandite, HCO_3^- is at equilibrium with respect to calcite, Mg^{2+} with brucite, Al^{3+} with ettringite and SiO_2(aq) with C1.8SH. The initial mineral volume fractions in the concrete are: 7.4% for portlandite, 2.2% for ettringite, 14.6% for C1.8SH, 1% for brucite, 0.1% for calcite and 62.2% for quartz. Quartz dissolution is small and therefore this mineral is almost a nonreactive phase. The initial pore water composition of the FEBEX bentonite was taken from Fernández et al. (2004). The initial mineral volume fractions in the bentonite are: 0.36% for calcite, 0.08% for gypsum and 1.192% for cristobalite. The smectite was assumed to be not reactive.The model allows the precipitation of the following secondary minerals: sepiolite, C0.8SH, anorthite and anhydrite. The dissolution/precipitation of portlandite, ettringite, C1.8SH, C0.8SH, quartz and cristobalite is simulated under kinetic control (Fernández et al., 2009).

Model results

THM results

Computed water content and porosity results reproduce the general trend of the measured data at the end of the HB1, 2, 3, and 4 tests (see Figure 2). The concrete is fully saturated at 7 days and then, the bentonite hydrates through the concrete. Water content increases near the concrete and decreases close to the heater due to the evaporation. The porosity increases in bentonite interface due to bentonite swelling. The results of the HB1 to HB4 tests are similar because the water intake is fast in the first months and later slows down after 10 months.

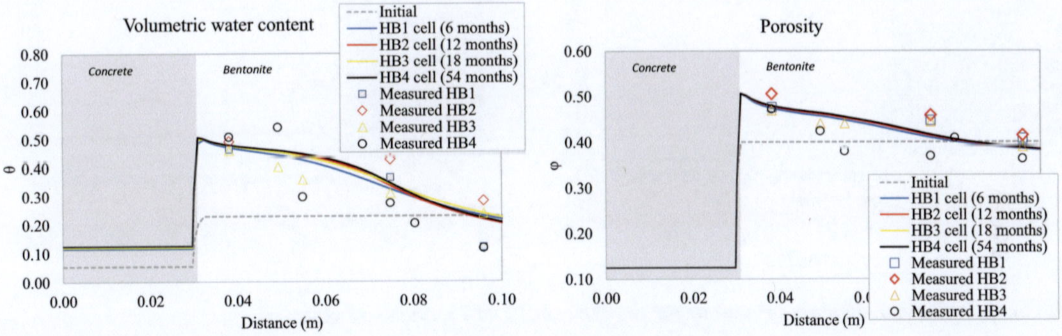

Figure 2: *Computed (lines) volumetric water content* ***(left)*** *and porosity* ***(right)*** *and measured data (symbols) for the HB1, HB2, HB3, and HB4 cells.*

Chemical results

Figure 3 shows the computed mineral volume fractions and pH for the HB1 to HB4 tests. Table 1 shows the summary of the main experimental observations on mineral phases at the end of the HB1, HB2, HB3, and HB4 tests and computed results.

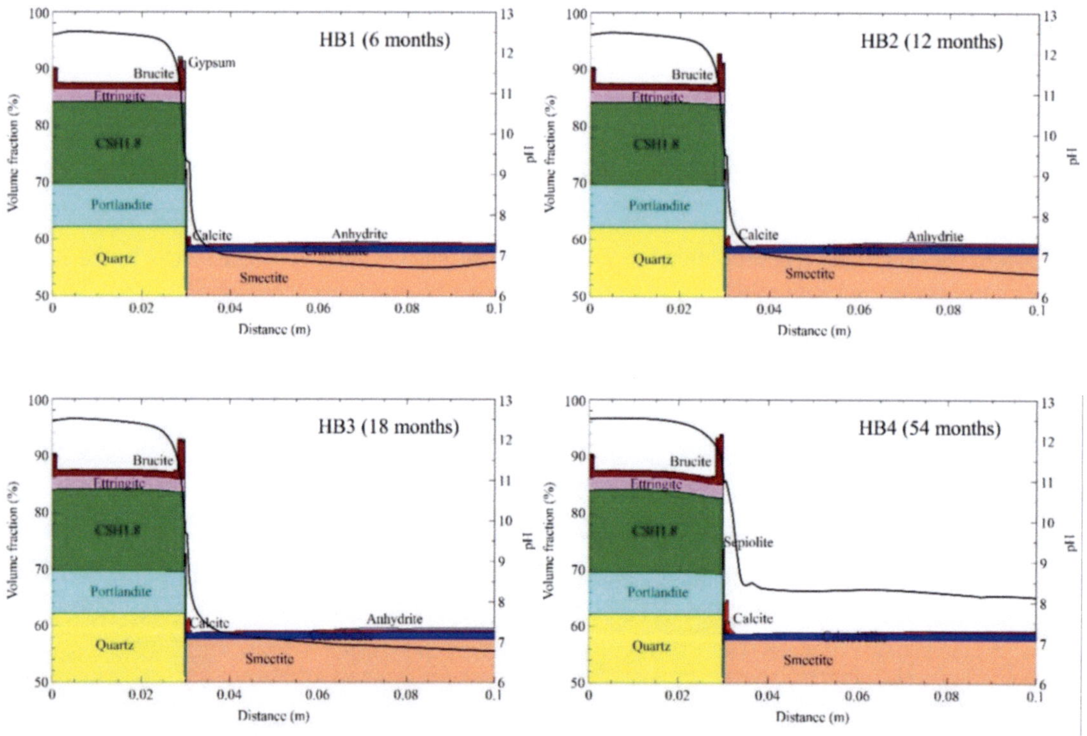

Figure 3: *Computed mineral volume fractions for the HB1, HB2, HB3, and HB4 cells.*

These observations are compared qualitatively to the computed values of dissolved/precipitated mineral phases. For the most part, the numerical model captures the main trends of the experimental mineralogical observations. However, there are some discrepancies, especially for ettringite and C-S-H precipitation. For these phases, the numerical model predicts much less precipitation than the observed values. Model discrepancies could be caused by uncertainties in: 1) CSH, MSH and CASH phases; 2) Kinetic parameters (rate law and specific surface); and 3) The relevant secondary clay minerals. The porosity decreases in the bentonite/concrete interface and at the hydration surface mainly due to the precipitation of brucite and calcite. The front of the pH diffuses into the bentonite. The penetration of the pH front into the bentonite is 1 cm after 54 months. The diffusion of the alkaline plume is retarded by the mineral precipitation at the bentonite-concrete interface.

Conclusions and future work

The results of the THCm model of the HB1, HB2, HB3, and HB4 tests have been presented. The numerical model results reproduce the general trends of the measured water content, porosity, and temperature and the observed patterns of minerals. More adequate representation of C-S-H and M-S-H phases, smectite dissolution, secondary clay minerals, and feedback on porosity changes will be taken into account to improve the model results and reproduce the laboratory observations.

Acknowledgements

The research leading to these results has received funding from the European Union's Horizon 2020 Research and Training Programme of the European Atomic Energy Community (EURATOM) (H2020-NFRP-2014/2015) under grant agreement n° 662147 (CEBAMA).

References

Fernández, A.M., Bayens, B., Bradbury, M., Rivas, P. (2004). Analysis of pore water chemical composition of a Spanish compacted bentonite used in an engineered barrier. Physics and Chemistry of the Earth, 29, 105-118.

Fernández, R., Cuevas, J., Mäder, U.K. (2009). Modelling concrete interaction with a bentonite barrier. European Journal of Mineralogy, 21, 177-191.

Samper, J., Mon, A., Montenegro, L., Pisani, B. Naves, A. (2013). Report on testing multiple-continua THCm models with laboratory and large-scale tests. Deliverable 3.4-1 of the PEBS Project.

Torres, E., Turrero, M.J., Escribano, A., Martín, P.L. (2013). Geochemical interactions at the concrete-bentontie interface of column experiments. Deliverable 2.3-6-1 of PEBS Project.

Turrero, M.J., Villar, M.V., Torres, E., Escribano, A., Cuevas, J., Fernández, R., Ruiz, A.I., Vigil de la Villa, R., del Soto, I. (2011). Laboratory tests at the interfaces, Final results of the dismantling of the tests FB3 and HB4. Deliverable 2.3-3-1 of PEBS Project.

Wollery, T.J. (1992). EQ3/3. A software package for geochemical modeling of aqueous system: package overview and installation guide version 7.0. UCRL-MA-110662-PT-I, Lawrence Livermore National Laboratory, Livermore, California.

Zheng, L., Samper, J., Montenegro, L., Fernández, A.M. (2010). A coupled model of heating and hydration laboratory experiment in unsaturated compacted FEBEX bentonite. Journal of hydrology, 386, 80-94.

Table 1: *Qualitative comparison of the laboratory observations and model results for the mineral precipitation/dissolution.*

	HB1		HB2		HB3		HB4	
	Laboratory observations	Model results	Laboratory observations	Model results	Laboratory observations	Model results	Laboratory observations	Model results
Concrete (Hydration)	➢ **Portlandite** dissolution	√	➢ **Portlandite** dissolution	√	➢ **Portlandite** dissolution	√		
	➢ **Calcite** precipitation	√	➢ **Calcite** precipitation	√	➢ **Calcite** precipitation	√	➢ **Calcite** precipitation	√
	➢ **Brucite** precipitation	√	➢ **Brucite** precipitation	√	➢ **Brucite** precipitation	√	➢ **Brucite** precipitation	√
Concrete (30 mm)	➢ **Portlandite** precipitation	√	➢ **Portlandite** precipitation	√	➢ **Portlandite** precipitation	√	➢ **Portlandite** dissolution	√
		√	➢ **Calcite** precipitation in concrete	√	➢ **Calcite** precipitation in concrete	√	➢ **Calcite** precipitation in concrete	√
	➢ **CSH gels** precipitation	~	➢ **CSH gels** precipitation	~	➢ **CSH gels** precipitation	~	➢ **CSH gels** precipitation	~
	➢ **Quartz** precipitation	*N*	➢ **Quartz** precipitation	*N*	➢ **Quartz** precipitation	*N*		
	➢ **MSH** precipitation	√	➢ **MSH** precipitation	√	➢ **MSH** precipitation	√		
							➢ **Ettringite** precipitation	*N*
					➢ **Gypsum** and **thaumasite** precipitation	√		
	➢ **Zeolites** precipitation	Not considered	➢ **Halite** and **K-feldspar** precipitation	Not considered	➢ **Vaterite** and **aragonite** precipitation	Not considered		
Concrete/ bentonite interface (some mm)	➢ **Brucite** precipitation	√	➢ **Brucite** precipitation	√	➢ **Brucite** precipitation	√	➢ **MSH** precipitation	√
	➢ **Portlandite** precipitation	*N*	➢ **Portlandite** precipitation	*N*	➢ **Portlandite** precipitation	*N*		
	➢ **Calcite** precipitation	√	➢ **Calcite** precipitation	√	➢ **Calcite** precipitation	√	➢ **Calcite** precipitation	√
	➢ **CSH gels (tobermorite and jennite type)** precipitation	*N*	➢ **CSH gels (tobermorite and jennite type)** precipitation	*N*	➢ **CSH gels (tobermorite** and **jennite type)** precipitation	*N*	➢ **CSH gels (0.8)** precipitation	*N*
			➢ **Ettringite** precipitation	*N*	➢ **Ettringite** precipitation	*N*	➢ **No Ettringite** precipitation	√
			➢ **Gypsum** precipitation	√	➢ **Gypsum** precipitation	√		
					➢ **Quartz dissolution**	√		
	➢ **Smectite** alteration not observed	Not considered	➢ **Smectite alteration** (brucite saponite-smectite mixed phase")	Not considered	➢ **Smectite alteration** (brucite saponite-smectite mixed phase")	Not considered		
					➢ **Saponite** formation	Not considered		
Bentonite	Unaltered bentonite mineralogy	~	Unaltered bentonite mineralogy	~	Unaltered bentonite mineralogy	~	➢ **Calcite** dissolution	√
							➢ **Gypsum** and **chloride** precipitation	√

Methods of evaluation of through-diffusion experiments on sandwich bentonite-cement layers in a simple experimental set-up

Dušan Vopálka[*1], Aleš Vetešník[1], Eva Hofmanová[1,2]

[1] Department of Nuclear Chemistry, Czech Technical University in Prague (CZ)
[2] Fuel Cycle Chemistry Department, ÚJV Řež a.s (CZ)

* Corresponding author: dusan.vopalka@fjfi.cvut.cz

Abstract

In relation with the Czech program of radioactive waste disposal, the methods for evaluation of diffusion experiments of critical radionuclides in barrier materials have been improved. The module for the modelling and evaluation of through-diffusion experiments implemented in the GoldSim modelling environment enables now to obtain characteristic data of porous materials by fitting of experimental data of different nature. Namely, the concentration changes in time in two working containers and profiles of total concentration of studied species in the porous layers (e.g., compacted bentonite and/or hydrated cement paste). A simple method for evaluation of diffusion experiments at stationary state was developed and verified which can also take into account the concentration changes in the working reservoirs during the diffusion experiment and the sandwich structure of body through which diffusion is studied.

Introduction

The knowledge of macroscopic diffusion properties of barrier materials is necessary for the performance assessment of the final disposal of radioactive waste. Therefore it is advantageous to have at hand methods for the determination of macroscopic diffusion coefficients, namely effective diffusion coefficient D_e and apparent diffusion coefficient D_a. These methods can then be used for the evaluation of real diffusion experiments and results of simulations that take into account both the microscopic structure of the studied materials and sophisticated multicomponent descriptions of interaction of radionuclides of interest with the solid phase.

The first goal of the modeling in CTU-DNC (Department of Nuclear Chemistry) within CEBAMA project, in the context of previous laboratory work, is to improve the evaluation methods of diffusion experiments that are planned to be performed in cooperation with Czech laboratories (UJV and CTU-CEG). Planned works include there long term laboratory and in-situ experiments in the Underground laboratory Josef and subsequent laboratory analyses of cementitious and bentonite samples from these experiments. The main aim is the study of changes in bentonite and cement properties, including characteristics of diffusion transport, due to the mutual interaction under in-situ and laboratory conditions.

The diffusion experiments planned for the study of bentonite and/or cementitious materials changes caused by the interaction of these materials will be initially through-diffusion tests with non-interacting species using an arrangement with limited volume of working reservoirs. Concentrations in these reservoirs will be monitored, but not kept constant, during the experiment. This type of experiment can be operated in a simple way, but its evaluation needs a modification of the broadly used *time-lag* method of evaluation of through-diffusion experiments.

Development of models

Model of through diffusion experiments implemented in the GoldSim environment

The commonly used methods of evaluation of diffusion experiments are based on analytical solutions of the diffusion equation for idealized initial and boundary conditions (e.g., Shackelford and Moore, 2012). In order to evaluate the experiments performed with diffusion cells with filters separating bentonite layer that use small volumes of working reservoirs, numerical methods are needed, as the boundary conditions, for which analytical solutions were derived, are not fulfilled. For the evaluation of such experiments in blocks of compacted bentonite we developed a tool using the software GoldSim (GoldSim Technology Group, 2002). This tool enables to model diffusion through the sandwich layers consisted of layers of different nature, e.g., separating filters and the studied layer in case of compacted bentonite (Vopalka et al., 2006). This model was thoroughly tested in numerous studies that showed for simple initial and boundary conditions good agreement between modelled solutions of the 1D diffusion equation and corresponding analytical solutions. The model also proved to be consistent with theory in case of retardation due to the sorption described by linear isotherm (K_d model) and is able to model and evaluate numerous types of diffusion experiments. These include the above mentioned experiments with small volumes of working containers. The characteristic diffusion coefficients, D_e and D_a, could be obtained with the use of the model not only for experiments at stationary state, characterized by a constant mass flow going into and out of the studied layer, but also for experiments terminated sooner – i.e., in transient, or non-stationary, state. This feature enables to shorten the duration of diffusion experiments. Recently, we extended this model with the aim to prepare methods for evaluation of diffusion experiments within CEBAMA project, in two respects: (i) the non-linear sorption isotherm (Freundlich and Langmuir) can be now used for the description of interaction of studied species with the solid phase; and (ii) the Box's method (Box, 1965) implemented in GoldSim software is used for fitting the experimental results with the model – i.e., for determination of model parameters. The first innovation enables to study for one species the influence of the shape of the interaction isotherm on the retardation of this species in the diffusion transport in a homogeneous porous layer. The sorption isotherm can be determined experimentally or obtained as a result of a multicomponent equilibrium model using a chemical speciation code. The second innovation enables to determine the unknown parameters of the studied system (diffusion coefficients, parameters of sorption isotherm, porosity) more rigorously. Three sets of data could be taken into consideration: concentration changes in both input and output containers and concentration profile at the time of termination of the diffusion experiment. The application of the χ^2 test to the comparison of experimental and modelled results enables to assess the goodness-of-fit and helps to determine which set of fitted parameters is optimal, e.g., by choosing between linear and non-linear interaction model.

A simple method for the evaluation of diffusion experiments in the stationary state

The numerous simulation studies with the developed model of the through-diffusion experiment helped us to develop a simple tool for the determination of both diffusion coefficients for structured layers. This tool, that can be called *method-of-lines*, is intended for the determination of diffusion coefficients in cases where all individual parallel layers in the sandwich structure are homogeneous and fully saturated by water and the stationary state is attained. This method is only applicable for materials in which sorption can be described by the linear isotherm. The tool is based on the application of Fick's first law and reflects the knowledge of mass transfer resistance r (equal to the reciprocal value of the equivalent flux $Q_{eq} = D_e \cdot S/L$, where S and L are area of the cross section and width of the block in which diffusion is studied, respectively, see e.g., Neretnieks et al., 2010) that is expressed, for two layers of porous materials characterized by respective effective diffusion coefficients and widths, as

$$r_{x+y} = r_x + r_y \quad \Rightarrow \quad \frac{x+y}{D_{e\_tot}} = \frac{x}{D_{e\_x}} + \frac{y}{D_{e\_y}} \qquad \textit{Eq. 1}$$

The method developed is divided into following steps:

- determination of the mean concentration in the liquid phase in the central layer after the termination of the experiment using concentration in both working reservoirs in this time and the assumption about the same concentration drop in filters, which is relevant in the stationary state,
- determination of the mean total concentration in the central layer from the measured concentration profile,
- calculation of capacity factor α (= $\varepsilon + \rho_d K_d$, where ε represents porosity, ρ_d dry density of bulk material and K_d distribution coefficient) as the ratio of concentration values obtained in previous two steps,

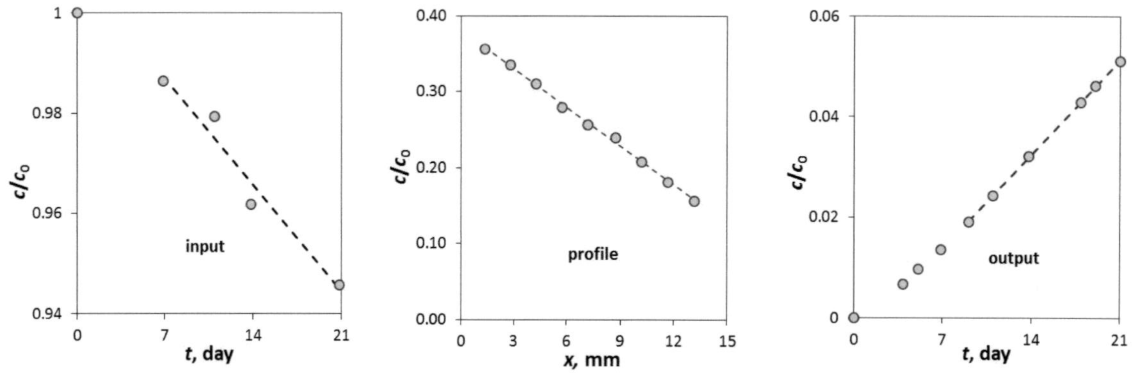

Figure 1: *Illustration of the application of the method-of-lines for the evaluation of a real through diffusion experiment (diffusion of HTO through compacted bentonite) performed in the apparatus in which concentrations in working reservoirs are not kept constant.*

- if ε and ρ_d values in material in the central layer are known, the K_d value characterizing the sorption in the studied material can be evaluated from the determined value of capacity factor α,
- the mass flow through the sandwich layer J is obtained using known values of volumes of the working containers and the area of the cross section of the studied block and the determined tangents of concentration changes in both working reservoirs for time of termination of diffusion experiment,
- as the mass flow in the stationary state through all the system of parallel layers is the same for all layers, both effective (D_e) and apparent (D_a) diffusion coefficients for the central layer are determined using the Fick's first law,
- as the determination of uncertainty of parameters of three lines that are used for the evaluation of experiment is simple, also the uncertainty of D_e and D_a values could be assessed.

The application of this method for the evaluation of diffusion of HTO in the sandwich layer, consisting of the block of compacted bentonite and two separating filters, is shown in Figure 1.

This evaluation of diffusion experiments was implemented using Excel. We recommend to use Equation 1 for the determination of D_e corresponding to the layer of hydrated cement for the description of diffusion through cement-bentonite-filter sandwich layers in experiments in which the concentration profile in the layer of hydrated cement is not determined.

Conclusions and Future work

Two methods have been developed for the determination of diffusion coefficients that are able to take into account the concentration changes in working containers during the through-diffusion experiments and to determine the values of characteristic diffusion coefficients (D_e and D_a) for all individual parallel layers in the sandwich arrangement. These methods will help to evaluate the diffusion experiments that are planned to be performed in cooperation with UJV and CTU-CEG laboratories for the characterization of changes of bentonite and hydrated cement paste after their contact. Both presented methods could be also applied to the determination of macroscopic diffusion coefficients from results of simulations of diffusive transport in which the microscopic characterization of porous materials and advanced approaches to the description of interaction of studied species with the solid phase will be used.

Acknowledgement

The research leading to these results has received funding from the European Union's Horizon 2020 Research and Training Programme of the European Atomic Energy Community (EURATOM) (H2020-NFRP-2014/2015) under grant agreement n° 662147 (CEBAMA). This contribution is partially a result of Radioactive Waste Repository Authority project "Research support for Safety Evaluation of Deep Geological Repository".

References

Box, M.J. (1965). A new method of constrained optimization and a comparison with other methods. Computer Journal, 8, 42-52.

GoldSim Technology Group (2002). GoldSim Contaminant Transport Module, Manual, Version 1.30.

Neretnieks, I., Liu, L., Moreno, L. (2010). Mass transfer between waste canister and water seeping in rock fractures. Revisiting the Q-equivalent model. SKB Technical Report, TR-10-42.

Shackelford, Ch.D. and Moore, S.M. (2012). Fickian diffusion of radionuclides for engineered containment barriers. Engineering Geology, 152, 133-147.

Vopálka, D., Filipská, H., Vokál, A. (2006). Some methodological modifications of determination of diffusion coefficients in compacted bentonite. MRS Proceedings, 932, 983-990.

Multiscale modeling of concrete mechanical behaviour with chemical degradation

Hamid Ghorbanbeigi[1], Wanqing Shen[1], Jianfu Shao[1*]

[1] University of Lille I, Laboratory of Mechanics of Lille (FR)

* Corresponding author: jian-fu.shao@polytech-lille.fr

Abstract

Cement-based materials are chemically active media and are in constant reaction with their nearby environment. The chemical reactions take place at the core of these materials which is the cement matrix and alter it at different scales. Micro-macro models can take into account these microscopic changes and relate it to the macroscopic behaviour. This study is devoted to the micromechanical modelling of concrete whose matrix is reinforced by small and big inclusions at different scales. At mesoscopic scale, the material is considered as a porous matrix reinforced by big elastic inclusions which globally represent the presence of sand and large aggregates. The porous matrix is constituted of a solid phase and randomly distributed pores at microscopic scale. Due to the chemical alteration of the material, small elastic inclusions are generated by the presence of calcite and are embedded in the solid phase, which is described by a pressure sensitive elastic-plastic model. By adopting the modified secant method, a three-step nonlinear homogenization technique has been used to establish an explicit macroscopic criterion to describe the overall responses of the studied material with a non-associated plastic flow rule and a plastic hardening law. The proposed micro-macro model can be applied to various cement-based materials such as concrete, mortar and cement paste for different kinds of degradation for instance carbonation and calcium leaching. Some numerical simulations are compared with sound and degraded samples under uniaxial and triaxial compression tests in order to show the capacity of this model to capture the main features of the mechanical behaviour of the studied materials.

Introduction

In many engineering applications such as nuclear power plants, disposal and storage of radioactive wastes, drilling of oil wells, and sequestration of residual gases in depleted reservoirs, cement-based materials undergo significant mechanical and chemical degradations throughout their lifetime. Carbonation and calcium leaching are two main chemical processes occurring in structures in contact with carbon dioxide and Ca-poor water. These two chemical processes cause significant changes in the microstructure and mechanical behaviour of cement-based materials. In most cases, for predicting the behaviour of such materials through time, phenomenological elastoplastic and damage models were used. The formulation of these models is essentially based on the standard framework of thermodynamics and experimental evidences. However, it is admitted that the mechanical behaviour of cement-based materials is inherently related to the chemical composition and mechanical properties of its constituents. Phenomenological models can not properly take into account such relationships between microstructure and macroscopic behaviour. The main role of micro-macro models is to overcome this problem and link the change occurring in the microstructure to the macroscopic behaviour. This is accomplished when the model explicitly takes into account such physical changes at smaller scales.

Micromechanical based approach for Concrete Carbonation

Microstructure of carbonated concrete

At the macroscopic scale (some centimetres), concrete is a composite material mainly constituted by aggregates (sand and gravel) and cement paste. The real microstructure is generally complex. In view of the modelling, some simplifications are adopted.

As it can be seen in Figure 1, due to the effect of carbonation, at a very small scale (nm-µm) calcite grains are formed embedded into the solid phase. At a slightly bigger scale (µm-mm) pores can be distinguished from the reinforced solid phase. At a larger scale (mm-cm) aggregates are distributed into the porous matrix. This framework mentions that aggregates are systematically greater than pores, and pores greater that calcite grains. It should be mentioned that, in view of obtaining an analytical expression for the macroscopic criterion of concrete considering it as a composite made out from materials at different scales, pores and inclusions are assumed to be spherical, elastic and randomly distributed.

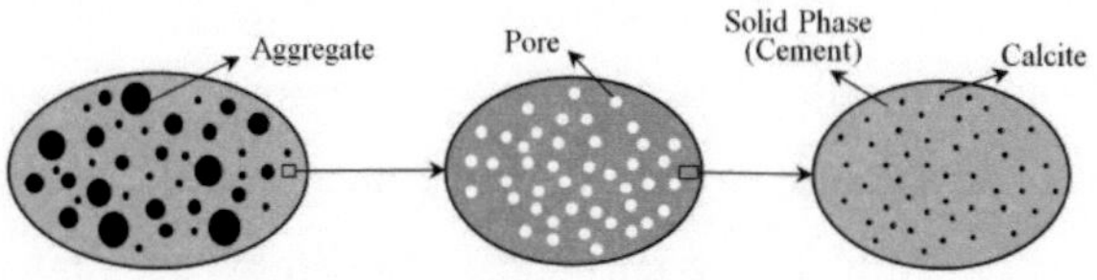

Figure 1: *Representative volume element (RVE) of carbonated concrete.*

Non-associated micro-macro model for concrete carbonation

The macroscopic criterion of carbonated concrete is formulated from three homogenization steps using the modified secant method (Suquet, 1995). The first homogenization step is devoted to describe the effect of calcite grains generated by carbonation in the solid phase. In the second step, the influence of pores in the porous cement paste will be considered. At the third step, the aggregates in the concrete composite are taken into account. In order to avoid too much homogenization steps, the latter step includes all types of aggregates namely sand and gravel. For the case where different aggregates size are used, the impact on the elastic properties can be minimized by assuming an overall properties for aggregates based on their respective elastic properties and volume fractions. This would be a very logical and close approximation and facilitates a lot of complicated analytical calculations saved from further homogenizations. The explicit expression of the macroscopic plastic criterion can be expressed as Equation 1.

$$F_c = \frac{\dfrac{1+\frac{2}{3}f}{\tilde{T}^2} + \dfrac{2}{3}\rho_a\left(\dfrac{3f}{2\tilde{T}^2}-1\right)}{\dfrac{4\tilde{T}^2-12f-9}{6\tilde{T}^2-13f-6}\rho_a+1}\Sigma_d^2 + \left(\frac{3f}{2\tilde{T}^2}-1\right)\Sigma_m^2 + 2(1-f)h\Sigma_m - \left(\frac{3f\rho_a}{3+2f}+1\right)(1-f)^2 h^2 = 0 \qquad \textbf{\textit{Eq. 1}}$$

Inclusions are considered to be elastic however the solid phase is supposed to obey a Drucker-Prager type plastic criterion where the frictional coefficient T and the hydrostatic tensile strength h are included in it. $\tilde{T}$ is the frictional coefficient of the reinforced matrix which is impacted by the presence of small inclusions. Σ_d and Σ_m are the equivalent and mean stresses. ρ_a, ρ_c and f are respectively the volume fractions of aggregates, calcite and pores.

$$\tilde{T} = T\sqrt{1+\frac{3}{2}\rho_c}\sqrt{1-\frac{2}{3}\rho_c t^2}\Big/\left(1-\frac{2}{3}\rho_c tT\right) \qquad \textbf{\textit{Eq. 2}}$$

Inspired from (Maghous et al. 2009) and (Shen et al. 2012), a non-associate plastic flow rule is proposed here:

$$G_c = \frac{\dfrac{1+\frac{2}{3}f}{\tilde{T}\tilde{t}} + \dfrac{2}{3}\rho_a\left(\dfrac{3f}{2\tilde{T}\tilde{t}}-1\right)}{\dfrac{4\tilde{T}\tilde{t}-12f-9}{6\tilde{T}\tilde{t}-13f-6}\rho_a+1}\Sigma_d^2 + \left(\frac{3f}{2\tilde{T}\tilde{t}}-1\right)\Sigma_m^2 + 2(1-f)h\Sigma_m - \left(\frac{3f\rho_a}{3+2f}+1\right)(1-f)^2 h^2 \qquad \textbf{\textit{Eq. 3}}$$

Where $\tilde{t}$ is the dilation coefficient of the reinforced matrix impacted by the presence of small inclusions and t is the dilation coefficient of the solid phase.

$$\tilde{t} = t\sqrt{\left(1+\frac{3}{2}\rho_c\right)\Big/\left(1-\frac{2}{3}\rho_c tT\right)} \qquad \textbf{\textit{Eq. 4}}$$

Experimental data show that concrete exhibits significant plastic hardening. Therefore, in the present study, the plastic hardening of the solid matrix is taken into account via the evolution of the frictional coefficient T as a function of the equivalent plastic strain ε^p. Denoting T_0 the initial threshold and T_m the asymptotic value of the frictional coefficient, the following exponential form is adopted for the hardening law:

$$\hat{T} = T_m - (T_m - T_0)e^{-b_1\varepsilon^p} \qquad \textbf{\textit{Eq. 5}}$$

As the rate of volumetric dilatancy generally varies with plastic deformation history, it is assumed that t is a function of the plastic hardening variable:

$$\hat{t} = t_m - (t_m - t_0)e^{-b_2\varepsilon^p} \qquad \textbf{\textit{Eq. 6}}$$

The plastic flow rule is given by:

$$\boldsymbol{D}^p = \dot{\lambda}\frac{\partial G}{\partial \boldsymbol{\Sigma}}(\boldsymbol{\Sigma}, \rho_c, f, \rho_a, \hat{T}, \hat{t}) \qquad \textit{Eq.7}$$

The equivalent plastic strain of the solid matrix is given by:

$$\dot{\varepsilon}^p = \frac{\Sigma : D^p}{(1-f)(1-\rho_a)(1-\rho_c)\left[\hat{T}h + (\hat{t}-\hat{T})\frac{\Sigma_m}{(1-f)(1-\rho_a)(1-\rho_c)}\right]} \qquad \textit{Eq. 8}$$

Due to the contracting dilating behavior of the solid phase the volume fraction of constituent change since the overall volume is in constant evolution.

$$\dot{f} = \frac{1-f}{1-\rho_a} tr\boldsymbol{D}^p - (1-f)(1-\rho_c)\hat{t}\dot{\varepsilon}^p \; ; \; \dot{\rho}_c = -\rho_c(1-\rho_c)\hat{t}\dot{\varepsilon}^p \; ; \; \dot{\rho}_a = -\rho_a \, tr\boldsymbol{D}^p \qquad \textit{Eq. 9}$$

It is worth mentioning that the presented model final peak stress can change. Indeed if the material has a compressible behaviour the peak stress would slightly get higher throughout the test. However if the material is dilating it could result in a decrease of stress. Therefore, the non-associated model provides a larger ability to describe complex volumetric strain evolutions than the associated one.

Application to cement-based materials

Case of concrete Carbonation (ρ_a, ρ_c, f)

The effect of carbonation on concrete is an increase in calcite grains volume fraction (ρ_c) and a decrease in porosity (f). Figure 2 shows the comparison between the numerical prediction and experimental data with different degrees of carbonation.

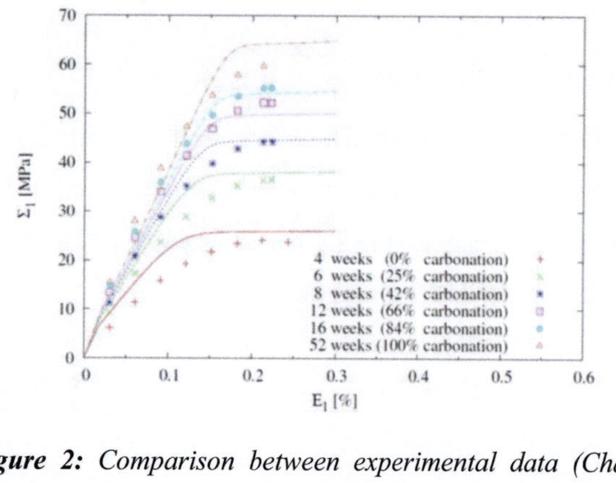

	Cement	Calcite	Aggregate
Elastic Parameters	$E_s = 27$ GPa	$E_c = 95$ GPa	$E_a = 75$ GPa
	$\nu_s = 0.27$	$\nu_c = 0.3$	$\nu_a = 0.2$
Plastic Parameters	$T_0 = 0.3$		
	$T_m = 0.91$		
	$b_1 = 750$		
	$h = 15$ MPa		

Carbonation	0%	25%	42%	66%	84%	100%
ρ_c	0	0.129	0.188	0.228	0.234	0.239
f	0.385	0.316	0.291	0.269	0.252	0.234
ρ_a	0.67	0.67	0.67	0.67	0.67	0.67

***Figure 2:** Comparison between experimental data (Chang and Chen, 2005) and numerical simulations on uniaxial compression tests.*

Case of concrete leaching (ρ_a, f, $\rho_c = 0$)

In the second application, the effect of leaching on concrete is considered which increases porosity by leaching the calcium of cement paste hydrates. Considering this effect of lixiviation, the non-associated model obtained in the previous section can be adopted by eliminating small inclusions of calcite in the solid phase ($\rho_C = 0$). In this case, the criterion obtained coincides exactly to the criterion obtained by (Shen et al., 2012) for a porous medium reinforced by big rigid inclusions. Figure 3 shows numerical simulations and for a concrete from sound to complete degraded state.

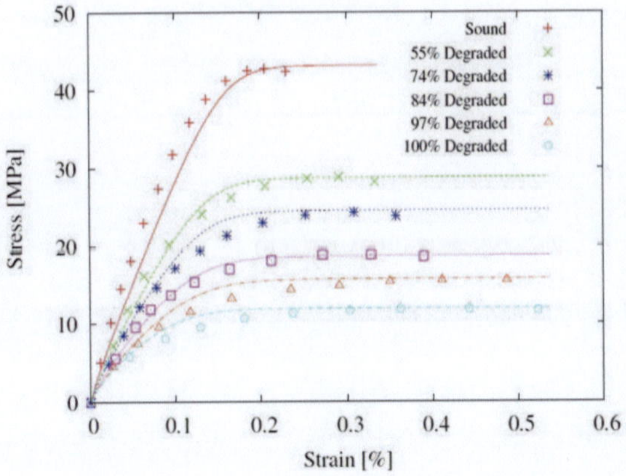

Parameters	Sound	Degraded [%]				
		55	74	84	97	100
Elastic	E_s=40000 MPa; ν_s=0.3; E_a=75000 MPa; ν_a=0.2					
Plastic	T_m=1	T_m=1	T_m=1	T_m=1	T_m=1	T_m=0.9
	T_0=0.01 ; b_1=2250 ; h=20 MPa					
Porosity [%]	39.3	53	58	63	67.5	68.7
Aggregates [%]	66	66	66	66	66	66

Figure 3: *Numerical simulations on sound and degraded concrete. Data from Nguyen (2005).*

Case of cement paste carbonation (ρ_c, f, $\rho_a = 0$)

In the case of cement carbonation, the RVE can be presented as a porous medium having a reinforced matrix. The proposed model can be adopted by setting $\rho_a = 0$. Cement paste microstructure evolves during carbonation. There is a decrease in porosity as well as an increase in calcite volume fraction which is taken into consideration in the criterion. Figure 4 shows an example of the numerical prediction of the model. E_{11} is the axial macroscopic strain, E_{33} the lateral strain, and Σ_{11} the axial stress.

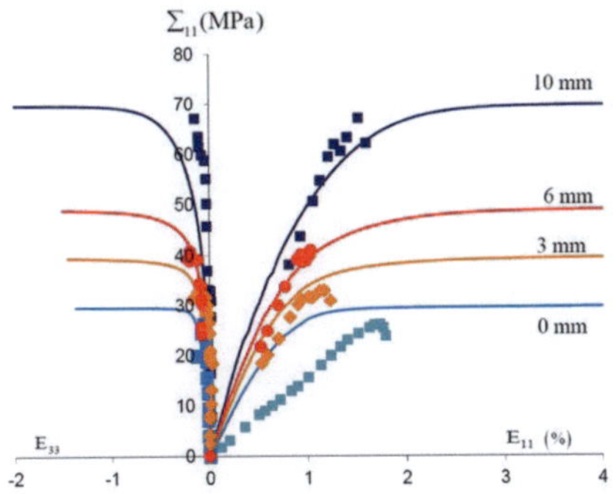

Parameters	Solid phase	Calcite
Elastic parameters	E_s=20GPa, ν_s=0.2	E_i=95GPa, ν_i=0.3
Plastic parameters	T_0=0.01 T_m=1.21 b_1=500 t_0=-1 t_m=0.15 b_2=120 h = 20	
porosity and calcite		
sound samples	f=0.3516	ρ=0
3mm carbonated samples	f=0.3	ρ=0.12
6mm carbonated samples	f=0.286	ρ=0.25
10mm carbonated samples	f=0.286	ρ=0.50

Figure 4: *Comparison between model predictions and data of cement on different states of carbonation. Data from Takla et al. (2011).*

Case of cement paste leaching ($f, \rho_a = 0, \rho_c = 0$)

The RVE of the cement paste leaching can be seen as a porous medium made of a solid phase (matrix) and pores. As mentioned above, leaching induces an increase in porosity and decreases the mechanical properties. In the proposed constitutive model, there are no aggregates and calcite grains anymore ($\rho_a = 0, \rho_c = 0$). Figure 5 shows an example of the numerical prediction of the model.

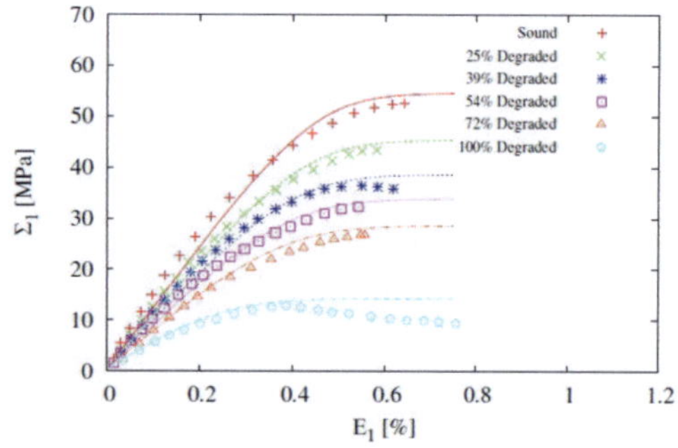

Parameters	Sound	Degraded [%]				
		25	39	54	72	100
Elastic	E_s=40000 MPa; ν_s=0.3					
Plastic	T_m^0=1.76	T_m^{25}=1.7	T_m^{39}=1.64	T_m^{54}=1.61	T_m^{72}=1.55	T_m^{100}=1.2
	T_0=0.01 ; t_0=-0.5 ; t_m=0.3 ; b_1=2250 ; b_2=5000 ; h=20 MPa					
Porosity [%]	41.9	46.1	49.1	52.3	56.1	62.1

Figure 5: *Uniaxial compression tests of cement paste from sound state completely leached samples state. Data from Carde et al. (1996).*

Conclusions and Future work

We have proposed a micromechanics-based elastic–plastic model for carbonated concrete. This model is obtained from a three-step nonlinear homogenization procedure by assuming a non-associated flow rule for the plastic deformation in the cement matrix. Compared with classical phenomenological models, the micromechanics-based model can explicitly take into account the effects of calcite grains generated during the carbonation process, the porosity of cement paste as well as the large aggregates. This model is then applied to consider the effects of leaching for both concrete and cement paste. Further works could include an extension of the proposed model to time dependent behaviours.

Acknowledgements

The research leading to these results has received funding from the European Union's Horizon 2020 Reasearch and Training Programme of the European Atomic Energy Community (EURATOM) (H2020-NFRP-2014/2015) under grant agreement n° 662147 (CEBAMA).

References

Chang, C.-F. and Chen, J.-W. (2006). The experimental investigation of concrete carbonation depth. Cement and Concrete Research, 36(9), 1760-1767.

Carde, C., François, R., Torrenti, J.-M. (1996). Leaching of both calcium hydroxide and C-S-H from cement paste: Modeling the mechanical behavior. Cement and Concrete Research, 26(8), 1257-1268.

Maghous, S., Dormieux, L., Barthélémy, J. F. (2009). Micromechanical approach to the strength properties of frictional geomaterials. European Journal of Mechanics - A/Solids, 28(1), 179-188.

Nguyen, V.-H. (2005). Couplage dégradation chimique - comportement en compression du béton. PhD Thesis, Ecole des Ponts ParisTech.

Shen, W.Q., Kondo, D., Dormieux, L., Shao, J.F. (2013). A closed-form three scale model for ductile rocks with a plastically compressible porous matrix. Mechanics of Materials, 59, 73-86.

Suquet, P. (1995). Overall properties of nonlinear composites: a modified secant moduli theory and its link with Ponte Castañeda's nonlinear variational procedure. Comptes Rendus de l'Académie Des Sciences. Série II, Mécanique, Physique, Chimie, Astronomie, 320(11) 563-571.

Takla, I., Burlion, N., Shao, J., Saint-Marc, J., Garnier, A. (2011). Effects of the Storage of CO_2 on Multiaxial Mechanical and Hydraulic Behaviors of Oil-Well Cement. Journal of Materials in Civil Engineering, 23(6), 741-746.

POSTERS

TABLE OF POSTERS

Experiments on Interface Processes at the Cement/Callovo-Oxfordian Claystone Interface and the Impact on Physical Properties; Progress

Robert Cuss[1*], Andrew Wiseall[1], Jon Harrington[1], Jean Talandier[2] and Xavier Bourbon[2]
[1] British Geological Survey, Keyworth, Nottingham, NG12 5GG, United Kingdom (*rjcu@bgs.ac.uk)
[2] Andra, 1-7, rue Jean Monnet, 92298 Chatenay-Malabry, France

1. Introduction

The weakest part of any gallery or deposition hole seal is likely to be at the interface between the sealing components and the host rock. A single seal completion may comprise a number of elements reflecting different design criteria in order to address specific engineering challenges associated with changes in geochemistry and stress. The interaction of these components with the host rock, their evolution in terms of strength/bonding, cation exchange behaviour and interfacial permeability, and the sensitivity of these properties to an evolving geochemical and physical environment will be key factors in determining the long-term seal performance. In the French repository concept, low-alkali cement is proposed as a mechanical support for bentonite seals as they slowly hydrate to isolate sections of the repository system. The interaction between the cement, bentonite and host rock must be understood in order to inform performance assessment. As such, a matrix of bespoke tests using state-of-the-art experimental systems capable of simulating key repository components will be performed.

2. Experimental apparatus

BGS have three bespoke shear rigs. The Direct Shear Rig (DSR) has been used to investigate the flow characteristics of rock (Opalinus Clay, Callovo-Oxfordian Claystone). The apparatus has the following specifications:

- Normal/vertical load up to 20 MPa
- Shear as slow as 1 mm per 3 month
- Vertical displacement accurate to <60nm
- Sample dimension 60 × 60 mm
- Pore pressure 0.5 - 25 MPa
- Central injection point ensuring pore fluid is delivered directly to the fracture

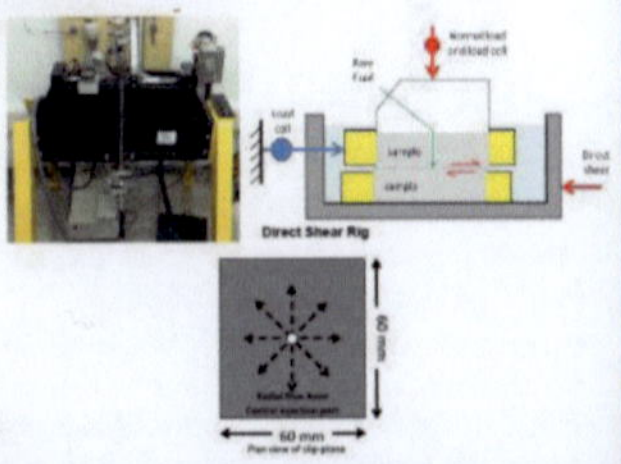

3. Test programme

			COx	Concrete		Test
Cebama_01	Shear test	COx/Conc	Repository	T_L	Rough	Yr0
Cebama_02		COx/Conc	Repository	T_L	Rough	Yr1
Cebama_03		COx/Conc	Repository	T_L	Rough	Yr2
Cebama_04		COx/Conc	Repository	T_L	Rough	Yr3
Cebama_05		COx/Conc	Repository	T_L	Planed	Yr0
Cebama_06		COx/Conc	Repository	T_L	Planed	Yr1
Cebama_07		COx/Conc	Repository	T_L	Planed	Yr2
Cebama_08		COx/Conc	Repository	T_L	Planed	Yr3
Cebama_09		COx/Conc	High Carb	T_L	Rough	Yr0
Cebama_10		COx/Conc	High Carb	T_L	Rough	Yr1
Cebama_11		COx/Conc	High Carb	T_L	Rough	Yr2
Cebama_12		COx/Conc	High Carb	T_L	Rough	Yr3
Cebama_13		COx/Conc	High Carb	T_L	Planed	Yr0
Cebama_14		COx/Conc	High Carb	T_L	Planed	Yr1
Cebama_15		COx/Conc	High Carb	T_L	Planed	Yr2
Cebama_16		COx/Conc	High Carb	T_L	Planed	Yr3
Cebama_17		Conc/Conc	/	T_L / T_L	Rough	Yr1
Cebama_18		Conc/Conc	/	T_L / T_L	Rough	Yr3
Cebama_19		Conc/Conc	/	T_L / T_L	Planed	Yr1
Cebama_20		Conc/Conc	/	T_L / T_L	Planed	Yr3
Cebama_21		COx/Conc	Repository	T_L	Rough	Yr1
Cebama_22		COx/Conc	Repository	T_L	Rough	Yr3
Cebama_23		COx/Conc	High Carb	T_L	Rough	Yr1
Cebama_24		COx/Conc	High Carb	T_L	Rough	Yr3
Cebama_25	Uniaxial test	Conc	/	T_L		Month 3
Cebama_26		Conc	/	T_L		Month 6
Cebama_27		Conc	/	T_L		Month 12
Cebama_28		Conc	/	T_L		Month 20
Cebama_29		Conc	/	T_L		Month 28
Cebama_30		Conc	/	T_L		Month 36

4. Starting material and casting

Callovo-Oxfordian claystone (COx, right) is used from the Meuse/Haute-Marne underground research laboratory at Bure in France.

Calcite and quartz represent 40 – 55% of the rock. Clay represents 20 – 55%, with secondary minerals forming less than 5 %. Clay minerals include illite and illite-smectite with subordinate kaolinite and chlorite.

		T_L (kg)		Supplier
Binder	Super Plasticizer	5.7	Fluid Optima 175	CHRYSO
	Cement	76.0	CEM I 52.5 PM EC CP2	Lafarge in Le Teil
	Silica fume	123.5	Ref DM95	Condensil
	Blast furnace slag	180.5		ECOCEM
	Sand	855	0/4 mm	
	Gravel	949	Gravels 4/12 mm	
	Effective water E_{ff} E_{ff}/L G/S	152 0.4 1.1		

TL variant concrete is used, recipe above

Care is taken to ensure full saturation of the concrete and COx (right)

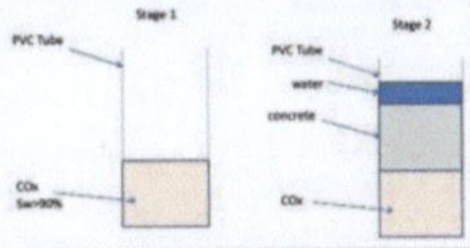

5. Test procedure I: Fracture formation

Samples of COx will be placed in the apparatus and re-hydrated to full saturation under a normal stress of 10 MPa.

The samples will be sheared for 5 days at a rate of 1 mm per day. Data recorded will be pore-fluid flow, shear displacement, shear stress, normal stress, and vertical displacement. This will give; initial flow rate, shear yield stress, shear modulus (G), shear peak strength, shear residual strength, dilation/contraction during shear, and change in flow during shear.

Fracture surface will be scanned using a NextEngine 3D laser scanner. This will determine: roughness average, root mean square roughness, peak-to-valley height, kurtosis, skewness, texture direction, texture aspect ratio, and texture direction index.

6. Test procedure II: Shear testing of Cox/Conc interface

A 4 mm-diameter bore will be drilled through the COx to the depth of the interface between the COx and concrete.

Normal load of 8 MPa will be used with a target pore pressure of 1 MPa will initially be used, dependent on the overall permeability of the COx/concrete interface.

The sample will be sheared 5 mm at a rate of 0.25 mm/day.

Flow along the interface and the shear properties will be determined.

The failure surfaces will be laser scanned to determine fracture roughness parameters.

Fluorescent scanning will identify flow paths of the injection water.

SEM analysis will determine chemical aging of the concrete, geochemical reactions between the COx and cement and identify flow paths.

7. Test procedure III: Uniaxial testing

The uniaxial tests on concrete will be performed using an MTS apparatus and will determine;

- Young's modulus (E),
- Poisson's ratio (v),
- bulk modulus of compressibility (K),
- yield stress (σ_y),
- unconfined compressive strength (q_u).

Repeat testing at month 3,6,12,20,28, and 36 after casting will show how these parameters vary as the concrete ages

8. Equipment modifications

The Cebama test programme requires a total of 14 cylindrical samples to be manufactured

Experience has shown that production of cubic samples is problematic (see right)

Necessity to modify the DSR to work on cylindrical samples

Failure rate of cubic samples production is around 50 %, compared with less than 10 % for cubic samples

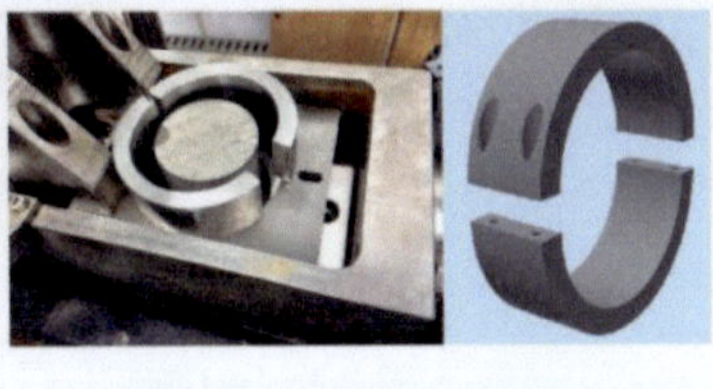

Effort in Year 1 has concentrated on equipment modification to allow 24 tests to be conducted using cylindrical samples

New configuration is shown left

9. Future activities

- Testing is due to start in June 2016
- Sample preparation and concrete casting will take the remainder of 2016
- Testing of COx/Conc interface will begin early in 2017
- Project has been delayed and final suite of tests may have to be conducted after the conclusion of the Cebama project.
- Uniaxial and swelling potential tests will be conducted during 2016

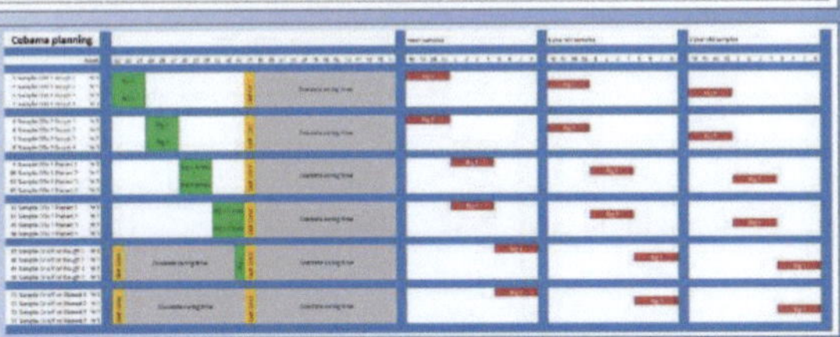

British Geological Survey
NATURAL ENVIRONMENT RESEARCH COUNCIL

The research leading to these results has received funding from the European Union's Horizon 2020 Research and Training Programme of the European Atomic Energy Community (EURATOM) (H2020-NFRP-2014/2015) under grant agreement n° 662147 (CEBAMA).

Contact:

Dr Robert Cuss
Fracture Physics Laboratory
British Geological Survey, Keyworth, Nottingham, NG12 5GG, UK
Tel: +44 (0)115 936 3486 Email: rjcu@bgs.ac.uk

OBSERVATION OF GROWTH OF THE ALTERED ZONE REGARDING CEMENT-BENTONITE INTERACTION BY USING CA-XAFS ANALYSIS

Hitoshi Owada*, Noki Fujii and Daisuke Hayashi

Radioactive Waste Management Funding and Research Center (Japan RWMC)
* Corresponding author : owada@rwmc.or.jp

Overall objectives:

- To obtain the realistic data for the expansion of altered zone around the contact between cementitious and bentonitic materials.
- To obtain the fundamental data of change of mechanical properties arose from the chemical alteration.

Because mechanical and Hydraulic characteristics of bentonitic materials are modeled as functions of effective montmorillonite density, i.e. content of montmorillonite and dry-density of that material, dissolution and/or precipitation of montmorillonite and secondary minerals are important data.

System to be studied:

- OPC and OPC/Fly ash mixed cement (pH=12.5 or around 11.5)
- Bentonitic materials (Kunigel V1 or MX80)

Description of Work to be Performed

- Ca-XAFS observation of cement bentonite contact zone.
  - This experiment will be performed till FY2017 and 1 or 2 specimen per year because of the limitation of the machine time of synchrotron-XAFS at High Energy Accelerator Research Organization (72 to 96hrs/year).
  - OPC-Kunigel interface boring core of the formally GMT in Grimsel and 11 years immersed (at RT) OPC or Fly ash mixed cement - bentonite (Kunigel or MX80) coupled specimen that is similar to the diffusion test.

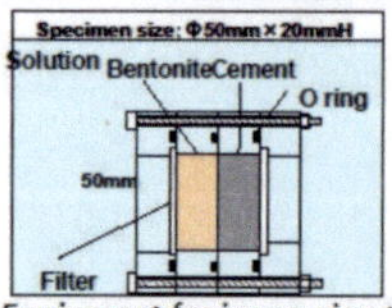

Equipment for immersion test

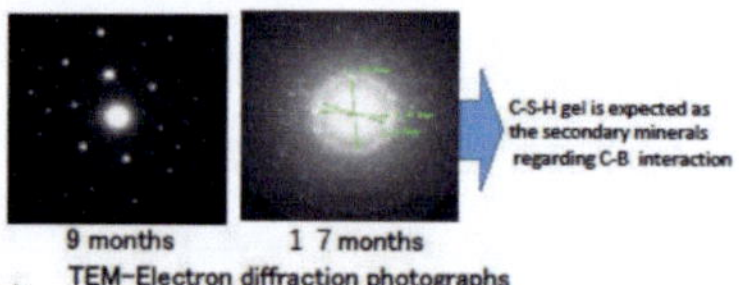

TEM–Electron diffraction photographs of bentonite near the boundary.

To investigate C-S-H precipitation at the boundary of the cement and bentonite materials, Ca *K*-edge (4.04 keV) XAFS measurements were carried out at beamline BL9A with a Si(111) double crystal monochtometer at the KEK Photon Factory (Ibaraki, Japan).

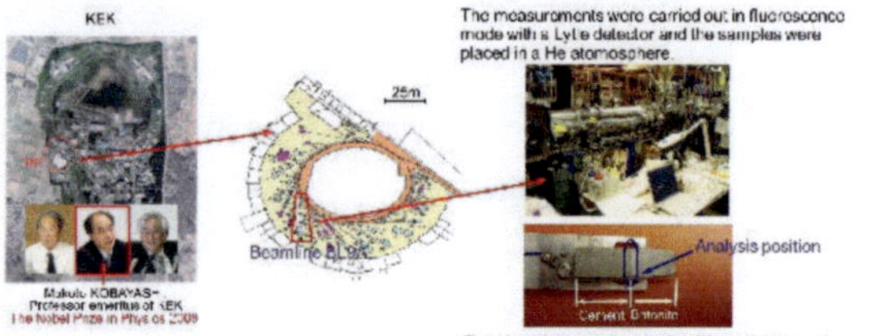

The measurements were carried out in fluorescence mode with a Lytle detector and the samples were placed in a He atomosphere.

The alteration samples obtained through immersion experiments were cut into small pieces (1x4x0.6 cm³) and used directly for XAFS measurements.

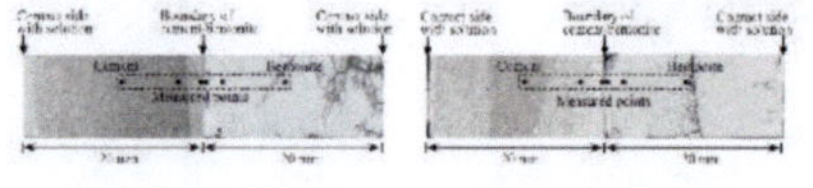

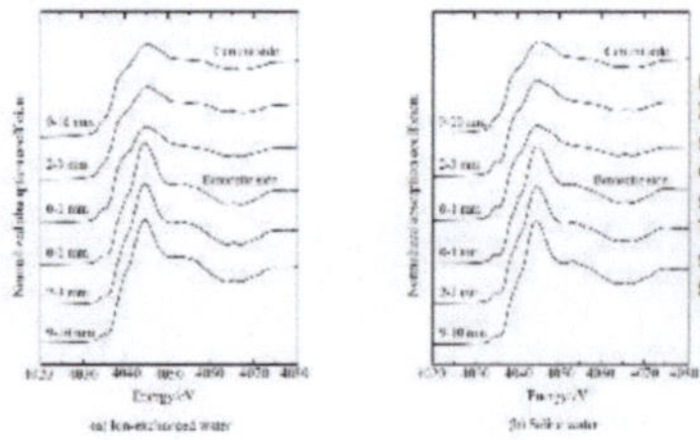

In this work, pattern-fitting analysis of the XANES spectra was applied to the alteration samples, assuming that the unknown spectra could be reproduced by linear combination spectra analysis of the reference spectra. The pattern-fitting analysis of the XANES spectra were performed using REX2000 software.

Figure . Ca *K*-edge XANES spectra of alteration samples immersed in ion-exchanged water and saline water.

Kunigel-V1 was measured as the reference material for XAFS analysis. Also, several compounds were measured as reference materials on the basis of Ca compounds in cement and the likely chemical form from cement and bentonite interaction.

Table List of reference minerals and materials

Kunigel-V1 (Na-type bentonite)
Ca compounds in cement: Quicklime (CaO), Calcite ($CaCO_3$), Portlandite ($Ca(OH)_2$ hydrogarnet ($3CaO{\cdot}Al_2O_3{\cdot}6H_2O$) ettringite ($3CaO{\cdot}Al_2O_3{\cdot}3CaSO_4{\cdot}32H_2O$) C-S-H†
Ca compounds that are considered to cause cement-bentonite interaction: Mono-sulfate ($3CaO{\cdot}Al_2O_3{\cdot}3CaSO_4{\cdot}12H_2O$) mono-carbonate ($3CaO{\cdot}Al_2O_3{\cdot}3CaSO_3{\cdot}12H_2O$) Friedel's salt ($3CaO{\cdot}Al_2O_3{\cdot}3CaCl_2{\cdot}10H_2O$) C-S-H†
Ca-type bentonite (Ca-exchanged montmorillonite using a $CaCl_2$ solution)

†Four kinds of C-S-H are synthesized with different Ca/Si mole ratios (Ca/Si=0.6, 0.83, 1.0, 1.4)

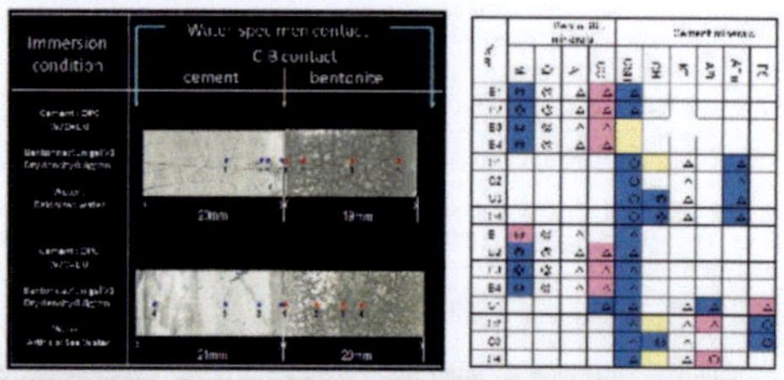

M : Montmorilonite Q : Quarts A : feldspar CC : $CaCO_3$ CSH : C-S-H gel CH : $Ca(OH)_2$
KT : Katoite AF t : Ettringite AFm : Mono Sulfate FS : Freedel's salt
◎ : Very high intensity ○ : High Intensity △ : Weak Blank : very weak or not Identified
■ Identified by XRD and XAFS ■ Identified by XRD □ Identified by XAFS analysis

○Dominant Secondary mineral in Bentonite is C-S-H gel
○Secondary minerals in cement are AFm for Deionized water and AFt and Freedel's salt for saline water.

Ca-XAFS analysis of 12years old specimen from GMT Sand box

Cement : OPC Bentonite : Kunigel V1:Silica sand=20:80
Over 90%saturation by water for 12years.
Temperature : about 15drgree C

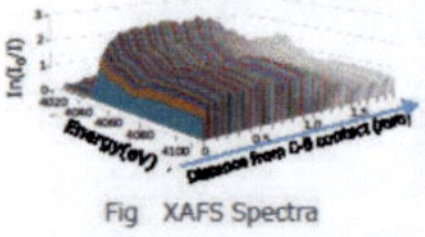

Fig XAFS Spectra

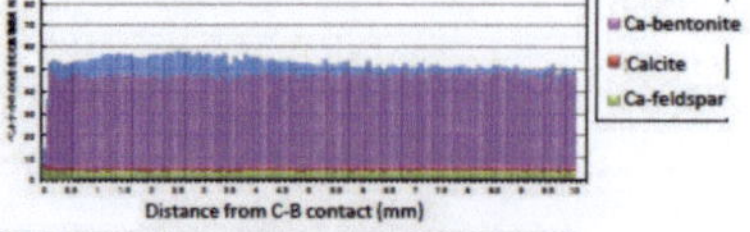

Fig Results of the quantification of Ca-containing minerals by pattern fitting method

Depth of the altered zone in which C-S-H precipitation was observed was only 5 to 6 mm from the C-B contact.

This research is a part of "Development of the Treatment and Disposal Techniques for TRU Waste Disposal" (FY2013-2015)" under a grant from the Japanese Ministry of Economy, Trade and Industry (METI).
The research leading to these results has received from the European Union's European Atomic Energy Community's (Euratom) Horizon2020 Programme (NFRO-2014/2015) under grant agreement, 662147-Cebama.

Interaction between cement and Czech bentonite under temperature load and in in-situ conditions: an overview of experimental program

P. Večerník[1], L. Hausmannová[2], R. Červinka[1], R. Vašíček[2], M. Roll[2], J. Hloušek [2], V. Havlová[1]
[1] ÚJV Řež, a. s. [2]Centre of Experimental Geotechnics, Czech Technical University in Prague

Introduction

The objective of the Czech partners (ÚJV Řež, a. s. and CTU) is to contribute to experimental study on interaction between cement based materials and bentonite. Planned works include long term laboratory and in-situ experiments in the Underground laboratory Josef and subsequent laboratory analyses of cementitious and bentonite samples from these experiments. The effects of cement / bentonite interaction under defined temperature load will be studied. The samples from the in-situ experiment will be used for diffusion radioactive tracer experiment. One of the most important material property, influencing safety function of engineered barrier function, is diffusivity. Both total and available diffusivity can be easily quantified using radioactive tracer (^{3}H, ^{36}Cl). Changes in physical, chemical and geotechnical properties will be also studied on original unaffected and altered/interacted materials.

Main Aims

- Study of changes of bentonite and cement properties due to mutual interaction under in-situ and laboratory conditions (diffusivity, porosity, hydraulic conductivity, swelling pressure, fluid composition, volume density, uniaxial compressive strength)
- Continuation of research works done within the Czech project *MPO TIP FR-TI1/362* (2009-2013)

Materials

- Czech bentonite „Bentonit 75" from 2010 (denoted as B75, Keramost, a.s.)
  - suspension (S:L ratio 1:5)
- Cement material CEM II A-S 42.5R (denoted as CEM II, Lafarge Cement, a.s.)
  - cylindrical samples 50mm in diameter and height of 8.2mm
- Low pH binder from VTT (denoted as LPC)
  - cylindrical samples 50mm in diameter and height of 8.2mm
- Groundwater from Underground laboratory Josef (denoted as GW Josef)

Composition of GW Josef and equilibrated GW Josef with bentonite B75

Species / concentration [mg/l]	GW Josef	Equilibrated GW Josef with bentonite B75 (squeezed porewater)
pH	7.7	8.8
Na^+	18.3	349.0
K^+	2.0	13.0
Ca^{2+}	76.1	5.0
Mg^{2+}	15.4	3.9
Cl^-	14.7	21.4
F^-	0.7	1.2
SO_4^{2-}	106.2	159.5
HCO_3^-	199.6	604.1
NO_3^-	0.1	9.7

Experimental Program

Interaction and its time evolution of the Czech bentonite, cement CEM II, low pH binder and GW Josef is studied. Summary of ageing procedures under in-situ conditions (cartridges and pressure vessels at 8-10°C) and in laboratory (pressure vessels at 95°C) is shown in the table below.

Experimental program - new laboratory procedures and aged cartridges

	Ageing procedures	Bentonite B75	GW Josef	Cement CEM II 42.5R	Low pH cement	Temperature [95°C]
New procedures: ageing 9/18/27 months	suspension + CEM II (LPC) + 95°C	x	x	x	x	x
	suspension + CEM II (LPC)	x	x	x	x	
	water + CEM II (LPC?) + 95°C		x	x	?	x
	water + CEM II (LPC?)		x	x	?	
	bentonite suspension	x	x			
	CEM II (LPC)			x	x	
Aged samples: 60 months	cartridge (bentonite)	x	x			
	cartridge (bentonite + CEM II)	x	x	x		

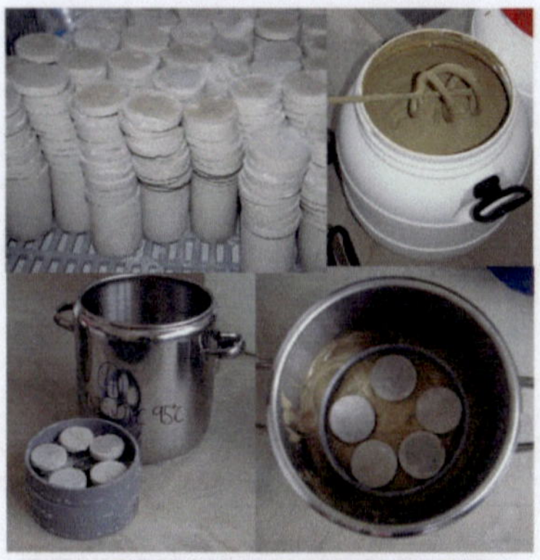

Laboratory procedures: cementitious thin plates and bentonite suspension

Laboratory procedures: material will be analysed in time period of 9, 18 and 27 months to see dynamics of the process and changes of key parameters. In 2 weeks intervals Ca^{2+} and Mg^{2+} concentration, TDS and pH is measured in the cement x water system.

- lab procedures with OPC started on 1st March 2016
- lab procedures with Low pH binder will start later in 2016

Cartridges: filled with compacted bentonite and cement paste (ongoing in-situ test from 2010 in Underground laboratory Josef) will be dismantled, analysed this year (more then 5 years old) and results compared with previous data.

Cartridges: bentonite and cementitious samples preparation for placing into cartridge

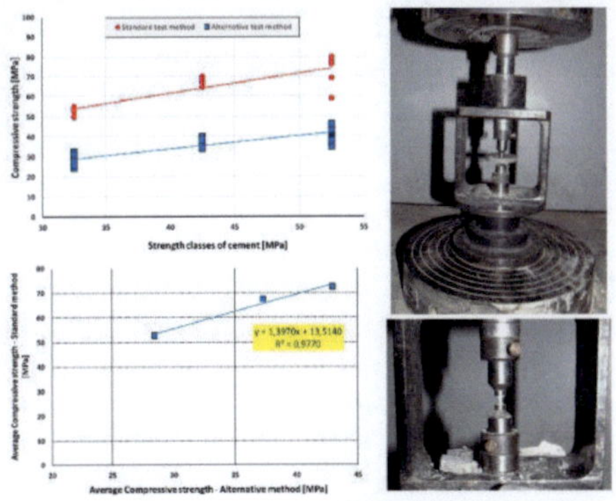

Alternative method for evaluation of compressive strength of thin samples – results (comparison to standard method), frame with sample before and after testing

Alternative Testing Method

It is supposed that in the given limited interaction time only surface part of samples will be affected. Standard procedures for testing of strength properties require samples with prescribed shape and low Surface Area/Volume ratio. That means that effect of interaction would be very difficult to study and alternative sample shape and methods have to be developed and used.

In the first half of the year CTU verified the using of alternative method for evaluation of uniaxial compressive strength of thin samples (experimentally used for rock). The results were statistically evaluated and method was found applicable for cement paste samples. The results and their comparison to standard test method is shown in graphs .

Acknowledgement

The research leading to these results has received funding from the European Union's European Atomic Energy Community's (Euratom) Horizon 2020 Programme (NFRP-2014/2015) under grant agreement, 662147 – Cebama.

The University Of Sheffield. NucleUS Immobilisation Science Laboratory midas Radioactive Waste Management HORIZON 2020 Cebama

Characterisation of UK and French geological disposal cement materials prior to groundwater leaching

Rita. G. W. Vasconcelos*[1], Neil C. Hyatt[1], John. L. Provis[1] and Claire. L. Corkhill[1]

NucleUS Immobilisation Science Laboratory, Department of Materials Science and Engineering, University of Sheffield, Sheffield, United Kingdom
*presenting author: rgwvasconcelos1@sheffield.ac.uk

1. Introduction

Nirex Reference Vault Backfill (NRVB) is a high pH cementitious material considered for use as a backfill material in one of the conceptual scenarios for the UK geological disposal of intermediate level waste (Fig. 1). The aim of this project is to understand how the physico-chemical properties of NRVB will be affected by interactions with groundwater, in comparison with low pH cements proposed for use in geological disposal by France and Finland. We present the results of a full characterisation of NRVB at 28 days of curing and for a low pH, silica fume, blast furnace slag, Portland cement blend from the French waste disposal programme.

(1)

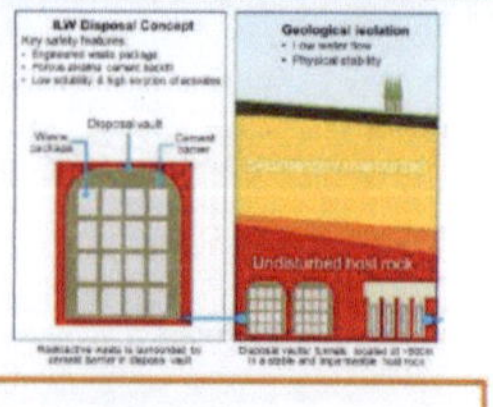

2. High pH backfill cement (NRVB)

Samples of NRVB were prepared according to Table 1 [Francis et al., 1997] and mixed by using Kenwood benchtop mixer for 5 minutes. To determine the workability of NRVB, a mini-slump test was performed, which gave a diameter of 56.5 ± 0.8 mm.

Table 1. NRVB composition

CEM I 52 5N	450 kg m^{-3}
Calcium hydroxide	170 kg m^{-3}
Calcium carbonate	495 kg m^{-3}
Water	615 kg m^{-3}

(2)

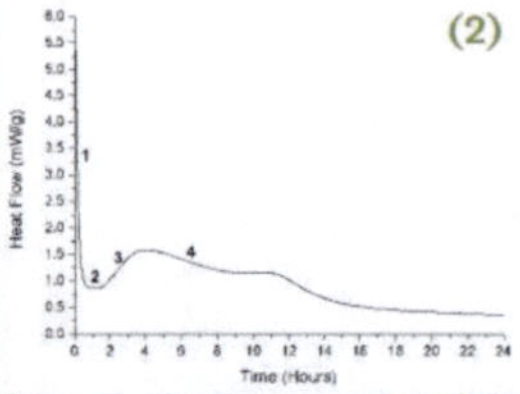

Isothermal calorimetry was performed at 25 °C within minutes of mixing the paste (Fig. 2). Several thermal features were identified: (1) dissolution and C_3A reaction; (2) dormant period; (3) C-S-H formation; and (4) sulphate depletion.

The cement paste was cured at 20°C at relative humidity. After 28 days, the density was determined to be 2.2 ± 0.001 g/cm^3 and the compressive strength was 8.2 ± 0.2 MPa.

(3)

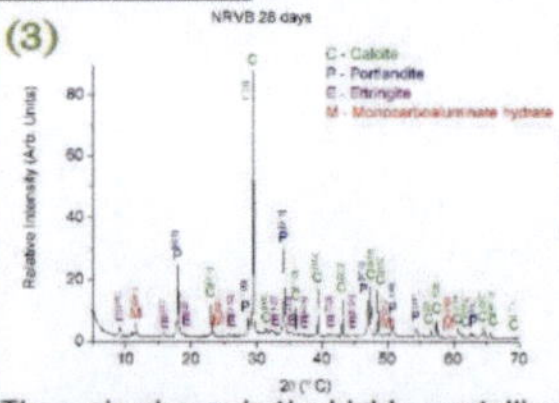

(4)

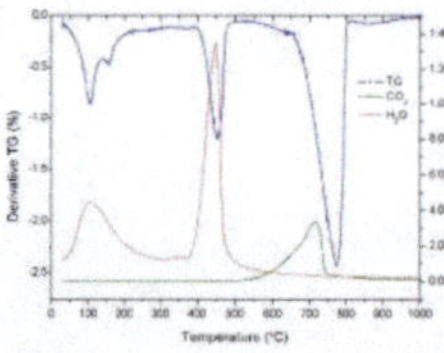

The main phases in the highly crystalline hydrate product assemblage were determined by XRD (Fig. 3) as calcite ($CaCO_3$), portlandite [$Ca(OH)_2$], ettringite (AFt phase) and monocarboaluminate hydrate (AFm phase). TG-MS (Fig. 4) confirmed these phases; two peaks between 100 and 300 °C correspond to ettringite and monocarboaluminate hydrate, while peaks between 400 to 500 °C, and 650 to 800 °C are related to Portlandite and calcite, respectively.

(5)

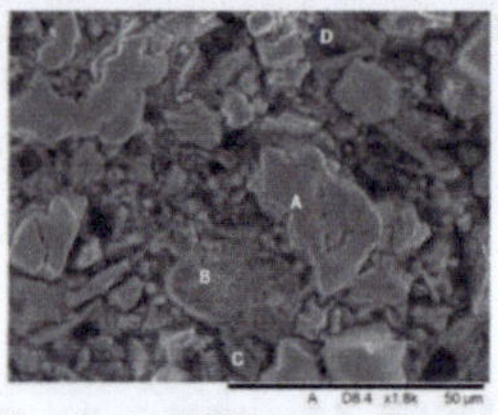

(6)

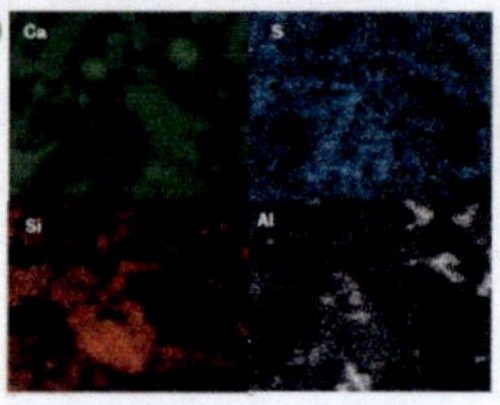

SEM imaging and EDX analysis (Figs. 5 & 6) confirmed the presence of the same phases identified by XRD and TG-MS: (A) Portlandite; (B) C-S-H; (C) ettringite; and (D) monocarboaluminate hydrate.

Further characterisation is currently underway to determine the porosity and pore size distribution at 28 days of curing (gas adsorption, MIP and XCT techniques). The cement hydrate phase assemblage and microstructure will also be evaluated at 90 days.

3. Low pH geological disposal cement blend (SF / BFS / PC)

Samples were prepared according the Andra recipe (Table 2) and mixed in the same way as NRVB. The workability was determined through mini-slump analysis to be 67.3 ± 1.9 mm; this was higher then NRVB due to the addition of 1.5 wt% superplasticiser.

Table 2. SF/BFS/PC composition

CEM I 52 5N	20 g
Silica fume	32.5 g
Blast furnace slag	47.5 g
Superplasticiser	1.5 g
Water	40.6 g

(7)

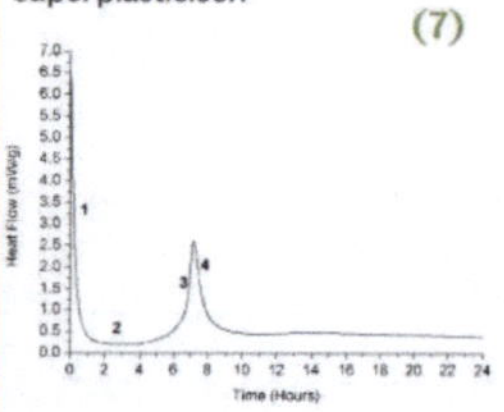

A narrow peak was observed between 6 and 8 hrs by isothermal calorimetry (at 25 °C, Fig. 7), which we attribute to a low sulfate content. Future formulations will include 1 wt% gypsum [$Ca(SO_4)$] to extend the kinetics of the hydration reaction.

After 28d of curing at 20 °C, the density observed 2.2 ± 0.002 g/cm^3 was similar to NRVB. On the contrary, the compressive strength value was significantly higher, at 55.2 ± 0.6 MPa.

(8)

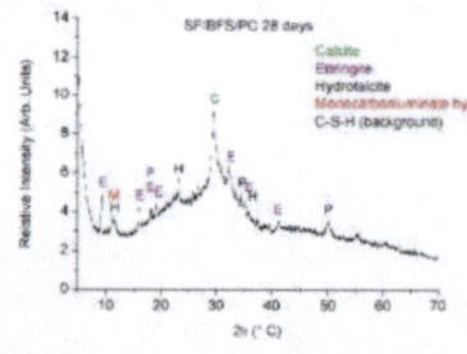

(9)

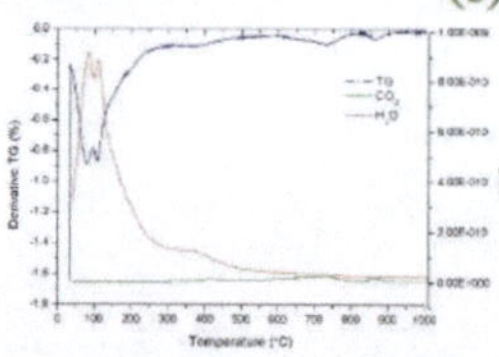

The broad area of diffuse scattering in the XRD analysis (Fig. 8) highlights the amorphous nature of this cement material. The main phases identified were calcite, ettringite and hydrotalcite ($Mg_6Al_2CO_3(OH)_{16}\cdot 4(H_2O)$). Some monocarboaluminate was also identified. TG-MS analysis (Fig. 9) revealed peaks characteristic of ettringite and AFm phase (monocarboaluminate) between 50 and 100 °C. The peak located between 300 - 400 °C corresponds to the presence of hydrotalcite.

(10)

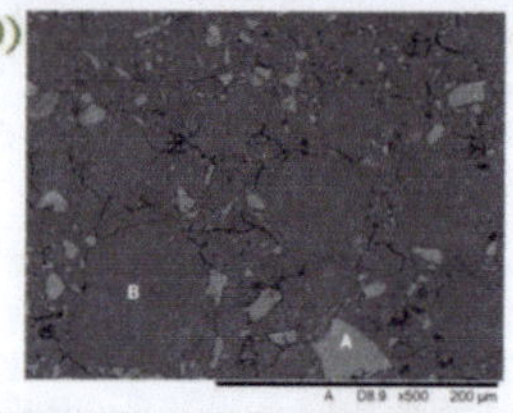

(11)

SEM imaging and EDX analysis (Figs. 10 & 11) reveal the presence of particles of: (A) unreacted blast furnace slag and (B) large particles of silica fume.

As future work, some improvements will be made to this formulation, for example, dispersion of silica fume by using ultrasonic bath and the addition of gypsum.

4. Future work:

Samples of NRVB and low pH cement cured for 28d will be leached in three different types of groundwater (Fig. 12), with regular sampling intervals between 1 and 52 weeks. Characterisation of microstructure and the hydrate phase assemblage as a function of time will be performed laboratory methods and high-resolution x-ray synchrotron techniques.

(12)

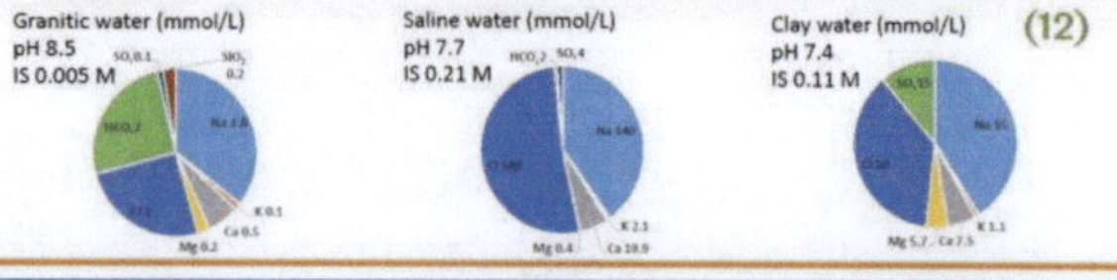

@ISL_Sheffield
@cementsatshef

References
A.J. Francis et al (1997), NSARP Report no: S/97/014

Acknowledgements: The research leading to these results has received funding from the European Union's European Atomic Energy Community's (Euratom) Horizon 2020 Programme (NFRP-2014/2015) under grant agreement, 662147 – Cebama. Steve Williams (RWM) is thanked for his comments and input to this work

Kinetic parameters for assessing the pH evolution of low-pH concretes

Tapio Vehmas • Markku Leivo • Erika Holt
VTT Technical Research Centre of Finland Ltd

Introduction / background

Lifetime of long-term nuclear waste repositories (NWR) is extremely long. Experimental research is incapable to study the long-term behaviour and interactions of materials in the NWR environment. Modelling is needed to understand and forecast the material behaviour and interactions throughout the lifetime of NWR. Thermodynamic modelling and reactive transport modelling has been a widely utilized tools to understand the long-term behaviour. Kinetic factors must be included in the models to predict the time dependent behaviour. In low-pH Portland cement-based materials, the time dependency is very important aspect because the safety of the multi-barrier system depends on the pH –plume, caused by degradation of cementitious materials. The pH of cement-based materials is primarily controlled by the propagation of pozzolanic reaction which first transforms portlandite into calcium-silicate-hydrates (C-S-H) and later on transforms C-S-H into low-calcium C-S-H.

Materials and methods

Implementation of kinetics parameters into modelling is not a straightforward task. Although the parameters are easily integrated into the modelling, their physical relevance is questionable. A good kinetic parameter can distinguish the various processes that has effect on kinetics. The kinetics can depend on the physical factors such as surface area and volume of the pozzolanic materials, but also the rate of the chemical reaction. Arrhenius equation is widely used equation to describe the kinetics of a chemical reaction. Arrhenius equation (Equation 1) is able to distinguish the rate of the chemical reaction and the physical factors. Chemical reaction is characterized with activation energy ΔE. Physical parameters are characterized with frequency factor A. In this study, the applicability of Arrhenius equation to estimate kinetics of pozzolanic reaction was studied.

Pozzolanic reaction of granulated silica fume and crystalline calcium hydroxide was measured with isothermal calorimetry in 15, 25 and 40°C temperatures. The amount of calcium hydroxide and ion exchanged water was 2,0g in each sample. The amount of silica fume varied from 0,5g to 3,0g.

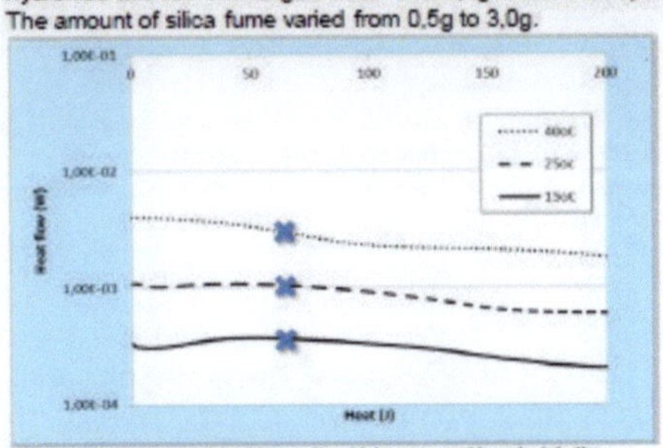

Figure 1. Isothermal calometric results which were used to calculate the parameters of Arrhenius equation. Crosses demonstrates a single point were the activation energy (ΔE) calculation was performed.

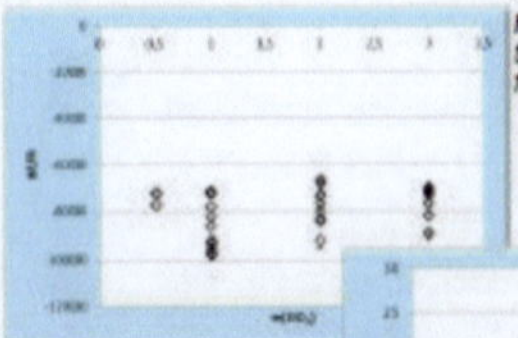

Figure 2. Activation energy and the amount of granulated silica fume in the samples.

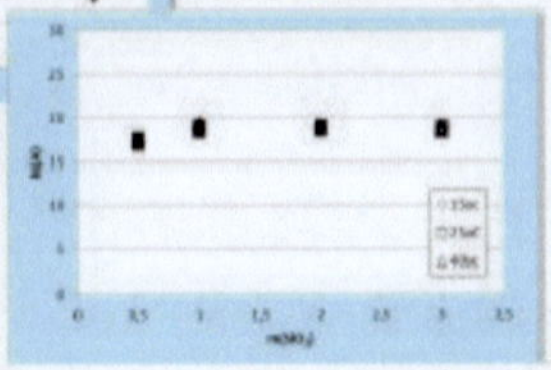

Figure 3. Frequency factor and the amount of granulated silica fume in the samples.

Results

Heat flow rates from the various reaction degrees was used to calculate the activation energy. The ΔE/R value was constant -8000±900, and the value was irrespective to the amount of silica in the sample (Figure 2.). Frequency factor was calculated from each individual heat flow values. When silica mass was used as a factor in the fitting, the frequency factor values become irrespective to the sample composition (Figure 3). The Arrhenius equation (equation 1) seems to be powerful tool to implement the pozzolanic reaction into the thermodynamic and reactive transport modelling.

$$k = Ae^{\Delta E/RT} \quad (Eq.1)$$

Future work

Although the kinetic parameters for granulated silica fume and crystalline calcium hydroxide was successfully determined, a lot of work is still to be done. Activation energy of C-S-H transformations must be defined and the physical parameters for various pozzolanic materials must be defined. Only after that the parameters are solid enough to be used in the modelling. This work is being utlized by Posiva Oy.

Acknowledgements

The research has received funding from The European Union's European Atomic Energy Community's (Euratom) Horizon 2020 programme (NFRP-2014/2015) under grant agreement 662147 –Cebama (www.cebama.eu).

Conclusions

Activation energy of pozzolanic reaction of crystalline calcium hydroxide and granulated silica fume was successfully determined (Equation 2). Where γ=correlation coeffient of heat and reaction propagation [mol/J], m = mass of the granulated silica fume [g] and T = temperature [K].

[illegible] (Eq.2)

Contacts

Tapio Vehmas
Tel. +358 020 722 4618
tapio.vehmas@vtt.fi

Kinetic parameters for assessing the pH evolution of low-pH concretes

Tapio Vehmas • Markku Leivo • Erika Holt
VTT Technical Research Centre of Finland Ltd

Introduction / background

Lifetime of long-term nuclear waste repositories (NWR) is extremely long. Experimental research is incapable to study the long-term behaviour and interactions of materials in the NWR environment. Modelling is needed to understand and forecast the material behaviour and interactions throughout the lifetime of NWR. Thermodynamic modelling and reactive transport modelling has been a widely utilized tools to understand the long-term behaviour. Kinetic factors must be included in the models to predict the time dependent behaviour. In low-pH Portland cement-based materials, the time dependency is very important aspect because the safety of the multi-barrier system depends on the pH –plume, caused by degradation of cementitious materials. The pH of cement-based materials is primarily controlled by the propagation of pozzolanic reaction which first transforms portlandite into calcium-silicate-hydrates (C-S-H) and later on transforms C-S-H into low-calcium C-S-H.

Materials and methods

Implementation of kinetics parameters into modelling is not a straightforward task. Although the parameters are easily integrated into the modelling, their physical relevance is questionable. A good kinetic parameter can distinguish the various processes that has effect on kinetics. The kinetics can depend on the physical factors such as surface area and volume of the pozzolanic materials, but also the rate of the chemical reaction. Arrhenius equation is widely used equation to describe the kinetics of a chemical reaction. Arrhenius equation (Equation 1) is able to distinguish the rate of the chemical reaction and the physical factors. Chemical reaction is characterized with activation energy ΔE. Physical parameters are characterized with frequency factor A. In this study, the applicability of Arrhenius equation to estimate kinetics of pozzolanic reaction was studied.

Pozzolanic reaction of granulated silica fume and crystalline calcium hydroxide was measured with isothermal calorimetry in 15, 25 and 40°C temperatures. The amount of calcium hydroxide and ion exchanged water was 2,0g in each sample. The amount of silica fume varied from 0,5g to 3,0g.

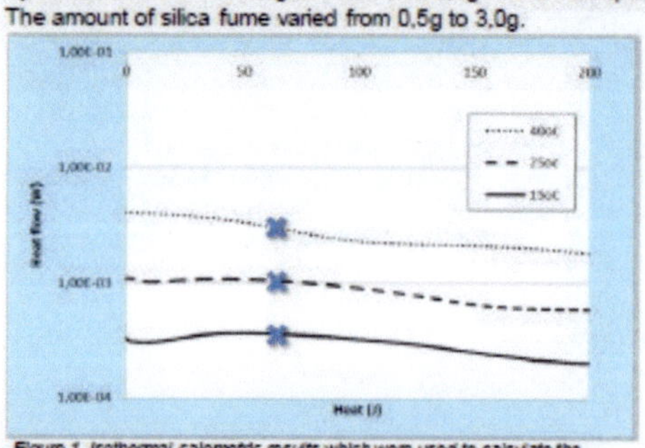

Figure 1. Isothermal calometric results which were used to calculate the parameters of Arrhenius equation. Crosses demonstrates a single point were the activation energy (ΔE) calculation was performed.

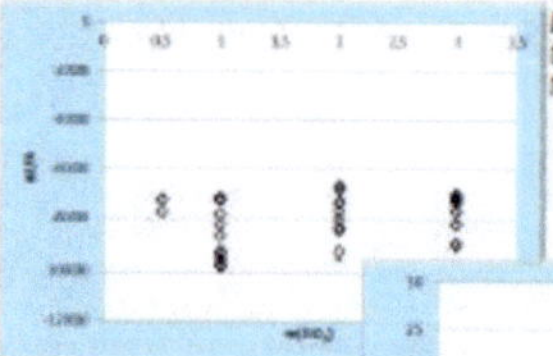

Figure 2. Activation energy and the amount of granulated silica fume in the samples.

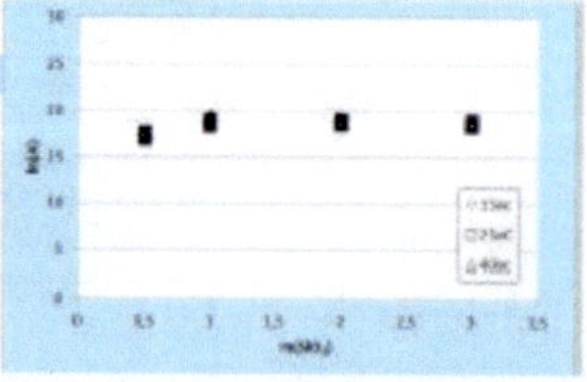

Figure 3. Frequency factor and the amount of granulated silica fume in the samples.

Results

Heat flow rates from the various reaction degrees was used to calculate the activation energy. The ΔE/R value was constant -8000±900, and the value was irrespective to the amount of silica in the sample (Figure 2.). Frequency factor was calculated from each individual heat flow values. When silica mass was used as a factor in the fitting, the frequency factor values become irrespective to the sample composition (Figure 3). The Arrhenius equation (equation 1) seems to be powerful tool to implement the pozzolanic reaction into the thermodynamic and reactive transport modelling.

$$k = Ae^{\Delta E/RT} \quad (Eq.1)$$

Future work

Although the kinetic parameters for granulated silica fume and crystalline calcium hydroxide was successfully determined, a lot of work is still to be done. Activation energy of C-S-H transformations must be defined and the physical parameters for various pozzolanic materials must be defined. Only after that the parameters are solid enough to be used in the modelling. This work is being utilized by Posiva Oy.

Acknowledgements

The research has received funding from The European Union's European Atomic Energy Community's (Euratom) Horizon 2020 programme (NFRP-2014/2015) under grant agreement 662147 –Cebama (www.cebama.eu).

Conclusions

Activation energy of pozzolanic reaction of crystalline calcium hydroxide and granulated silica fume was successfully determined (Equation 2). Where γ=correlation coeffient of heat and reaction propagation [mol/J], m = mass of the granulated silica fume [g] and T = temperature [K].

$$k [illegible] = \gamma \cdot (1.20 \cdot 10^{[illegible]}) m \cdot e^{\frac{-8000}{T}} \quad (Eq.2)$$

Contacts

Tapio Vehmas
Tel. +358 020 722 4618
tapio.vehmas@vtt.fi

PET/CT during degradation processes at the cement-clay interface and derivation of process parameters

J. Kulenkampff[1], U. Mäder[2], M. Gründig[1], S. Eichelbaum[3], J. Lippmann-Pipke[1]

[1]HZDR, Institute of Resource Ecology, Dresden, Germany
[2]University of Bern, Institute of Geological Sciences, Switzerland
[3]Nemtics Visualization, Leipzig, Germany

HZDR
HELMHOLTZ ZENTRUM DRESDEN ROSSENDORF

Cebama

Background

Observation of degradation processes is complicated by the heterogeneous nature of the process.

Better process understanding requires methods for both monitoring the fate of chemical species (mobile phase) and structural alterations (stationary phase).

In the past decade, we empowered positron emission tomography (PET) for quantitative transport visualization in geological media – GeoPET (Fig.1, 2) [1-4]. It has an unrivalled sensitivity and robustness for quantitative, non-destructive, spatio-temporal concentration measurements $c_{PET}(x,y,z,t)$. CT adds structural information.

Our latest achievements are detailed correction procedures to bring the GeoPET images into sharp focus [5, 6], and its benchmarking for geoscientific applications [7].

We noticed in numerous case studies that PET-based process observations at drill core scale and corresponding µXCT-based simulated transport results rarely agree: The transport effective flow pathways are often significantly over- or underestimated (Fig. 2).

Fig. 1: Visualization of tracer flow in column experiments by means of GeoPET

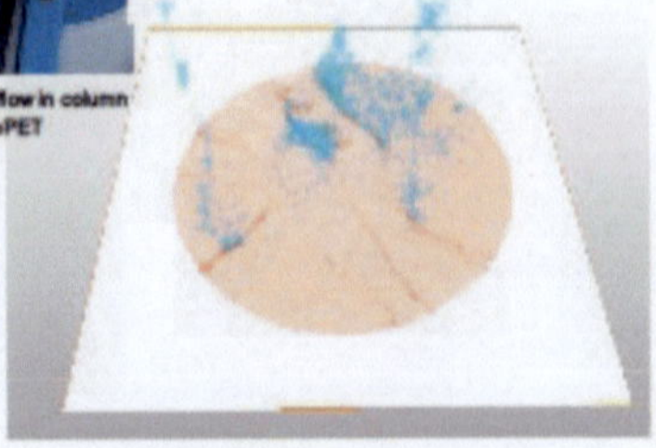

Fig. 2: PET/CT image overlay
GeoPET: tracer, $c_{PET}(x,y,z,t)$
µXCT: fractured rock, $\rho(x,y,z)$

Examples

We applied GeoPET in
- conservative and reactive flow experiments (Fig. 2,3)
- diffusion experiments (Fig. 4)

Fig. 3: PET-frames of flow of [^{18}F]-salt solution through a fractured halite core, overlaid with µCT-slices. The continuous activity data is represented as iso-surfaces at 200, resp. 400 Bq/voxel.
Time frames:
2.0 h, 3.3 h, 5.0 h, 6.3 h, 7.0 h

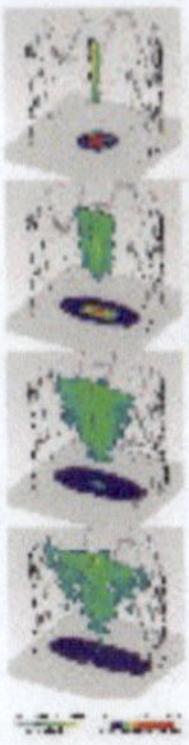

Fig. 4: PET of diffusion of ^{22}Na from a central drill hole, filled with [^{22}Na]-artificial pore water, in an Opalinus Clay core (from [4]). Time frames 3 h, 10 d, 18 d, 27 d after injection.
The experiment continued over a period of more than 100 d.
Below: axial projection showing anisotropy and heterogeneous effects.

Aim and Challenges of flow experiments

- Extract realistic 3D velocity and effective porosity fields from experimental 3D+t concentration data sets for reactive transport modelling.
- Establishing an algorithm (Fig. 5) capable of extracting velocity and porosity fields from PET images, must be able to overcome the following obstacles:
  - seemingly discontinuous flow pattern due to detection limits
  - no clear tracer fronts due to heterogeneity of material
  - image artefacts due to the experimental origin of data
  - 3D+t data sets of continuous variable (5D)

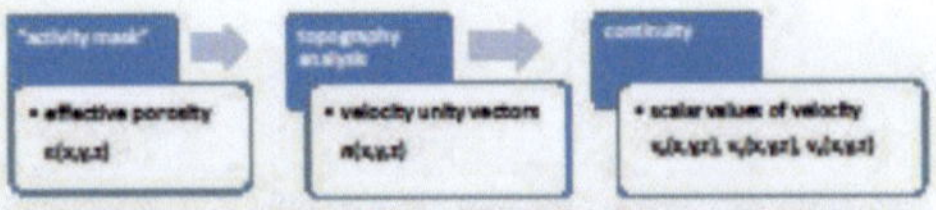

Fig. 5: Workflow of the algorithm. From the 3D "activity mask", the normalized, highest observed concentration in each voxel, the effective porosity is derived. A topology analysis of the activity map returns the velocity unity vectors. Gaps in flow direction are leapfrogged and indicated as "uncertain". The scalar values of the velocity vector are obtained from continuity considerations.

Experimental procedure

PET and µCT during degradation processes at the cement-clay interface:
- longterm observation of slow flow with artificial pore water (APW)
- radiation-transpararent pressure vessel (CRP) (Fig. 6)

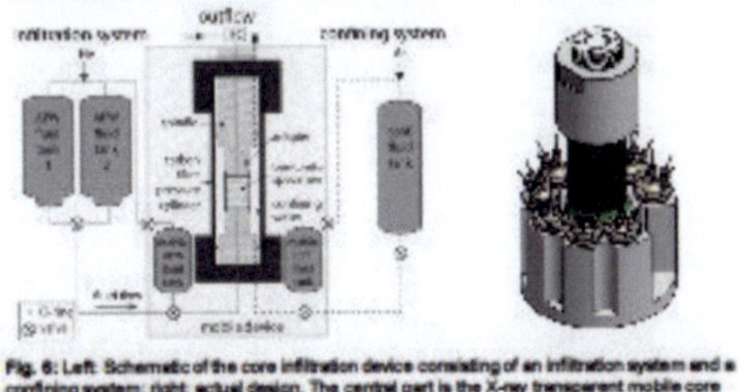

Fig. 6: Left: Schematic of the core infiltration device consisting of an infiltration system and a confining system; right: actual design. The central part is the X-ray transparent mobile core infiltration device (from [8]).

Literature
[1] Richter et al., (2000) Z. Angew. Geol. 46(2), 101-109.
[2] Gründig et al. (2007) App. Geochemi. 22, 2334-2343
[3] Kulenkampff, J. et al. (2008). Phys. Chem. Earth 33, 937-942.
[4] Kulenkampff, J. et al. (2015). Clay Min., 50, 369–375
[5] Kulenkampff et al., submitted to Solid Earth (2016a)
[6] Kulenkampff et al., submitted to Solid Earth (2016b)
[7] Lippmann-Pipke, J. et al. submitted to Comp. Geosci. (2016)
[8] Dolder et al., Phys. Chem. Earth 70-71, 104-113

Acknowledgement
The research leading to these results has received funding from the European Union's European Atomic Energy Community's (Euratom) Horizon 2020 Programme (NFRP-2014/2015) under grant agreement, 662147 – Cebama

Dr. Johannes Kulenkampff | Institute of Resource Ecology | Reactive Transport Division | j.kulenkampff@hzdr.de | www.hzdr.de

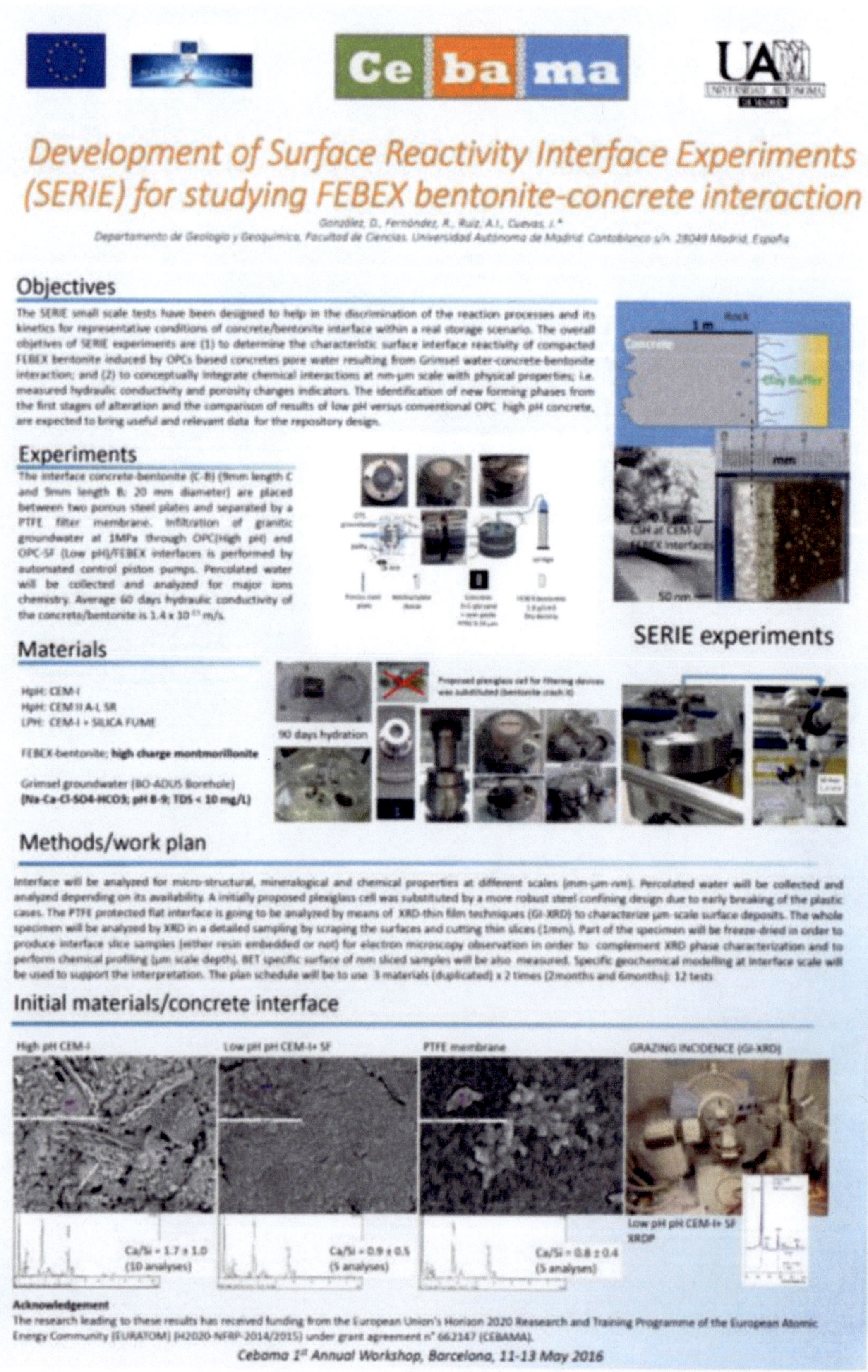
Development of Surface Reactivity Interface Experiments (SERIE) for studying FEBEX bentonite-concrete interaction
González, D., Fernández, R., Ruiz, A.I., Cuevas, J.*
Departamento de Geología y Geoquímica, Facultad de Ciencias. Universidad Autónoma de Madrid. Cantoblanco s/n. 28049 Madrid, España
Objectives
The SERIE small scale tests have been designed to help in the discrimination of the reaction processes and its kinetics for representative conditions of concrete/bentonite interface within a real storage scenario. The overall objetives of SERIE experiments are (1) to determine the characteristic surface interface reactivity of compacted FEBEX bentonite induced by OPCs based concretes pore water resulting from Grimsel water-concrete-bentonite interaction; and (2) to conceptually integrate chemical interactions at nm-µm scale with physical properties; i.e. measured hydraulic conductivity and porosity changes indicators. The identification of new forming phases from the first stages of alteration and the comparison of results of low pH versus conventional OPC high pH concrete, are expected to bring useful and relevant data for the repository design.
Rock
1 m
Concrete
Clay Buffer
mm
CSH at CEM-I/FEBEX interface
50 nm
Experiments
The interface concrete-bentonite (C-B) (9mm length C and 9mm length B; 20 mm diameter) are placed between two porous steel plates and separated by a PTFE filter membrane. Infiltration of granitic groundwater at 1MPa through OPC(High pH) and OPC-SF (Low pH)/FEBEX interfaces is performed by automated control piston pumps. Percolated water will be collected and analyzed for major ions chemistry. Average 60 days hydraulic conductivity of the concrete/bentonite is 1.4 x 10⁻¹³ m/s.
SERIE experiments
Materials
HpH: CEM-I
HpH: CEM II A-L SR
LPH: CEM-I + SILICA FUME
FEBEX-bentonite; high charge montmorillonite
Grimsel groundwater (BO-ADUS Borehole)
(Na-Ca-Cl-SO4-HCO3; pH 8-9; TDS < 10 mg/L)
90 days hydration
Methods/work plan
Interface will be analyzed for micro-structural, mineralogical and chemical properties at different scales (mm-µm-nm). Percolated water will be collected and analyzed depending on its availability. A initially proposed plexiglass cell was substituted by a more robust steel confining design due to early breaking of the plastic cases. The PTFE protected flat interface is going to be analyzed by means of XRD-thin film techniques (GI-XRD) to characterize µm-scale surface deposits. The whole specimen will be analyzed by XRD in a detailed sampling by scraping the surfaces and cutting thin slices (1mm). Part of the specimen will be freeze-dried in order to produce interface slice samples (either resin embedded or not) for electron microscopy observation in order to complement XRD phase characterization and to perform chemical profiling (µm scale depth). BET specific surface of mm sliced samples will be also measured. Specific geochemical modelling at interface scale will be used to support the interpretation. The plan schedule will be to use 3 materials (duplicated) x 2 times (2months and 6months): 12 tests
Initial materials/concrete interface
High pH CEM-I
Low pH pH CEM-I+ SF
PTFE membrane
GRAZING INCIDENCE (GI-XRD)
Ca/Si = 1.7 ± 1.0
(10 analyses)
Ca/Si = 0.9 ± 0.5
(5 analyses)
Ca/Si = 0.8 ± 0.4
(5 analyses)
Low pH pH CEM-I+ SF
XRDP
Acknowledgement
The research leading to these results has received funding from the European Union's Horizon 2020 Reasearch and Training Programme of the European Atomic Energy Community (EURATOM) (H2020-NFRP-2014/2015) under grant agreement n° 662147 (CEBAMA).
Cebama 1st Annual Workshop, Barcelona, 11-13 May 2016

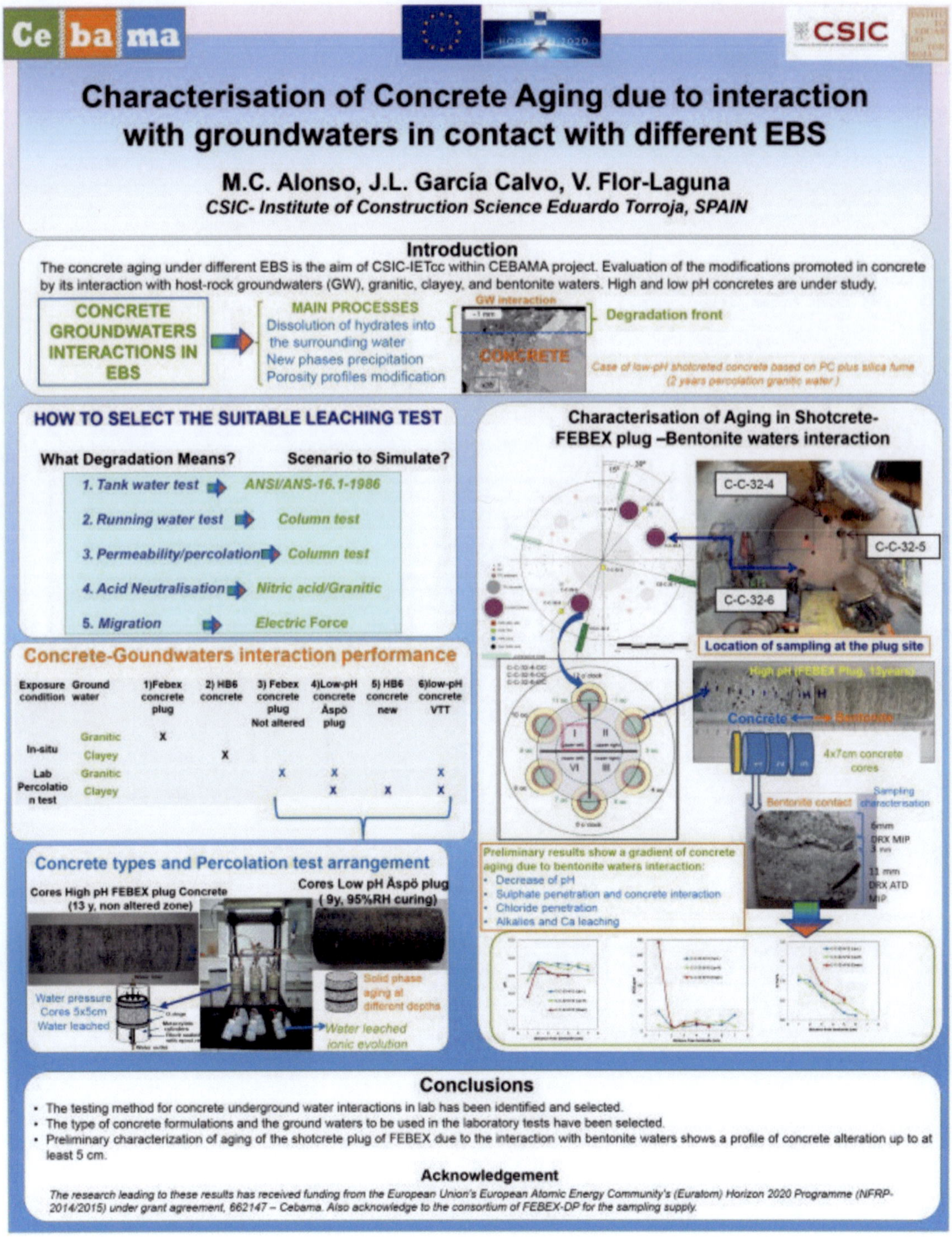
Ce ba ma
HORIZON 2020
CSIC
Characterisation of Concrete Aging due to interaction with groundwaters in contact with different EBS
M.C. Alonso, J.L. García Calvo, V. Flor-Laguna
CSIC- Institute of Construction Science Eduardo Torroja, SPAIN
Introduction
The concrete aging under different EBS is the aim of CSIC-IETcc within CEBAMA project. Evaluation of the modifications promoted in concrete by its interaction with host-rock groundwaters (GW), granitic, clayey, and bentonite waters. High and low pH concretes are under study.
CONCRETE GROUNDWATERS INTERACTIONS IN EBS
MAIN PROCESSES
Dissolution of hydrates into the surrounding water
New phases precipitation
Porosity profiles modification
GW interaction
Degradation front
CONCRETE
Case of low-pH shotcreted concrete based on PC plus silica fume (2 years percolation granitic water)
HOW TO SELECT THE SUITABLE LEACHING TEST
What Degradation Means?
Scenario to Simulate?
1. Tank water test → ANSI/ANS-16.1-1986
2. Running water test → Column test
3. Permeability/percolation → Column test
4. Acid Neutralisation → Nitric acid/Granitic
5. Migration → Electric Force
Concrete-Goundwaters interaction performance
Exposure condition
Ground water
1)Febex concrete plug
2) HB6 concrete
3) Febex concrete plug Not altered
4)Low-pH concrete Äspö plug
5) HB6 concrete new
6)low-pH concrete VTT
In-situ
Granitic X
Clayey X
Lab Percolation test
Granitic X X X
Clayey X X X
Concrete types and Percolation test arrangement
Cores High pH FEBEX plug Concrete (13 y, non altered zone)
Cores Low pH Äspö plug (9y, 95%RH curing)
Water pressure
Cores 5x5cm
Water leached
Solid phase aging at different depths
Water leached ionic evolution
Characterisation of Aging in Shotcrete-FEBEX plug –Bentonite waters interaction
C-C-32-4
C-C-32-5
C-C-32-6
Location of sampling at the plug site
High pH (FEBEX Plug, 13years)
Concrete
Bentonite
4x7cm concrete cores
Bentonite contact
Sampling characterisation
6mm
DRX MIP
3 mm
11 mm
DRX ATD
MIP
Preliminary results show a gradient of concrete aging due to bentonite waters interaction:
Decrease of pH
Sulphate penetration and concrete interaction
Chloride penetration
Alkalies and Ca leaching
Conclusions
The testing method for concrete underground water interactions in lab has been identified and selected.
The type of concrete formulations and the ground waters to be used in the laboratory tests have been selected.
Preliminary characterization of aging of the shotcrete plug of FEBEX due to the interaction with bentonite waters shows a profile of concrete alteration up to at least 5 cm.
Acknowledgement
The research leading to these results has received funding from the European Union's European Atomic Energy Community's (Euratom) Horizon 2020 Programme (NFRP-2014/2015) under grant agreement, 662147 – Cebama. Also acknowledge to the consortium of FEBEX-DP for the sampling supply.

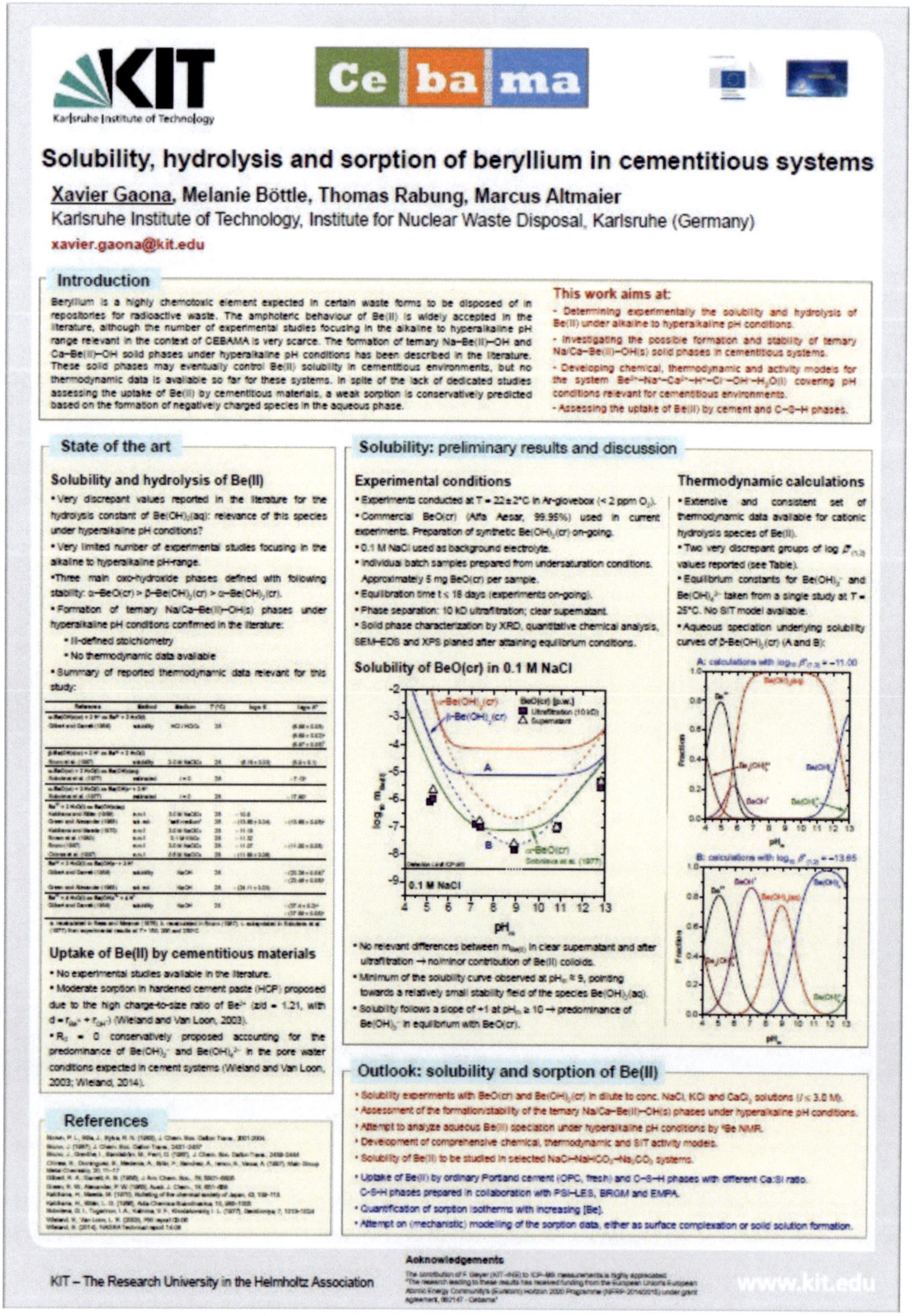

Karlsruhe Institute of Technology
Ce ba ma
Solubility, hydrolysis and sorption of beryllium in cementitious systems
Xavier Gaona, Melanie Böttle, Thomas Rabung, Marcus Altmaier
Karlsruhe Institute of Technology, Institute for Nuclear Waste Disposal, Karlsruhe (Germany)
xavier.gaona@kit.edu
Introduction
Beryllium is a highly chemotoxic element expected in certain waste forms to be disposed of in repositories for radioactive waste. The amphoteric behaviour of Be(II) is widely accepted in the literature, although the number of experimental studies focusing in the alkaline to hyperalkaline pH range relevant in the context of CEBAMA is very scarce. The formation of ternary Na–Be(II)–OH and Ca–Be(II)–OH solid phases under hyperalkaline pH conditions has been described in the literature. These solid phases may eventually control Be(II) solubility in cementitious environments, but no thermodynamic data is available so far for these systems. In spite of the lack of dedicated studies assessing the uptake of Be(II) by cementitious materials, a weak sorption is conservatively predicted based on the formation of negatively charged species in the aqueous phase.
This work aims at:
- Determining experimentally the solubility and hydrolysis of Be(II) under alkaline to hyperalkaline pH conditions.
- Investigating the possible formation and stability of ternary Na/Ca–Be(II)–OH(s) solid phases in cementitious systems.
- Developing chemical, thermodynamic and activity models for the system $Be^{2+}–Na^{+}–Ca^{2+}–H^{+}–Cl^{-}–OH^{-}–H_2O(l)$ covering pH conditions relevant for cementitious environments.
- Assessing the uptake of Be(II) by cement and C–S–H phases.
State of the art
Solubility and hydrolysis of Be(II)
• Very discrepant values reported in the literature for the hydrolysis constant of $Be(OH)_2(aq)$: relevance of this species under hyperalkaline pH conditions?
• Very limited number of experimental studies focusing in the alkaline to hyperalkaline pH-range.
• Three main oxo-hydroxide phases defined with following stability: α–BeO(cr) > β–$Be(OH)_2$(cr) > α–$Be(OH)_2$(cr).
• Formation of ternary Na/Ca–Be(II)–OH(s) phases under hyperalkaline pH conditions confirmed in the literature:
• Ill-defined stoichiometry
• No thermodynamic data available
• Summary of reported thermodynamic data relevant for this study:
Uptake of Be(II) by cementitious materials
• No experimental studies available in the literature.
• Moderate sorption in hardened cement paste (HCP) proposed due to the high charge-to-size ratio of Be^{2+} (z/d = 1.21, with $d = r_{Be^{2+}} + r_{OH^-}$) (Wieland and Van Loon, 2003).
• R_d = 0 conservatively proposed accounting for the predominance of $Be(OH)_3^-$ and $Be(OH)_4^{2-}$ in the pore water conditions expected in cement systems (Wieland and Van Loon, 2003; Wieland, 2014).
References
Solubility: preliminary results and discussion
Experimental conditions
• Experiments conducted at T = 22±2°C in Ar-glovebox (< 2 ppm O_2).
• Commercial BeO(cr) (Alfa Aesar, 99.95%) used in current experiments. Preparation of synthetic $Be(OH)_2$(cr) on-going.
• 0.1 M NaCl used as background electrolyte.
• Individual batch samples prepared from undersaturation conditions. Approximately 5 mg BeO(cr) per sample.
• Equilibration time t ≤ 18 days (experiments on-going).
• Phase separation: 10 kD ultrafiltration; clear supernatant.
• Solid phase characterization by XRD, quantitative chemical analysis, SEM-EDS and XPS planed after attaining equilibrium conditions.
Solubility of BeO(cr) in 0.1 M NaCl
α-$Be(OH)_2$(cr)
β-$Be(OH)_2$(cr)
BeO(cr) [p.w.]
Ultrafiltration (10 kD)
Supernatant
A
B
α-BeO(cr)
0.1 M NaCl
pH_m
• No relevant differences between $m_{Be(II)}$ in clear supernatant and after ultrafiltration → no/minor contribution of Be(II) colloids.
• Minimum of the solubility curve observed at pH_m ≈ 9, pointing towards a relatively small stability field of the species $Be(OH)_2(aq)$.
• Solubility follows a slope of +1 at pH_m ≥ 10 → predominance of $Be(OH)_3^-$ in equilibrium with BeO(cr).
Thermodynamic calculations
• Extensive and consistent set of thermodynamic data available for cationic hydrolysis species of Be(II).
• Two very discrepant groups of log $\beta°_{(1,2)}$ values reported (see Table).
• Equilibrium constants for $Be(OH)_3^-$ and $Be(OH)_4^{2-}$ taken from a single study at T = 25°C. No SIT model available.
• Aqueous speciation underlying solubility curves of β-$Be(OH)_2$(cr) (A and B):
A: calculations with $\log_{10} \beta°_{(1,2)}$ = −11.00
Fraction
pH_m
B: calculations with $\log_{10} \beta°_{(1,2)}$ = −13.65
Outlook: solubility and sorption of Be(II)
• Solubility experiments with BeO(cr) and $Be(OH)_2$(cr) in dilute to conc. NaCl, KCl and $CaCl_2$ solutions (I ≤ 3.0 M).
• Assessment of the formation/stability of the ternary Na/Ca–Be(II)–OH(s) phases under hyperalkaline pH conditions.
• Attempt to analyze aqueous Be(II) speciation under hyperalkaline pH conditions by ⁹Be NMR.
• Development of comprehensive chemical, thermodynamic and SIT activity models.
• Solubility of Be(II) to be studied in selected $NaCl–NaHCO_3–Na_2CO_3$ systems.
• Uptake of Be(II) by ordinary Portland cement (OPC, fresh) and C–S–H phases with different Ca:Si ratio. C-S-H phases prepared in collaboration with PSI–LES, BRGM and EMPA.
• Quantification of sorption isotherms with increasing [Be].
• Attempt on (mechanistic) modelling of the sorption data, either as surface complexation or solid solution formation.
KIT – The Research University in the Helmholtz Association
Acknowledgements
www.kit.edu

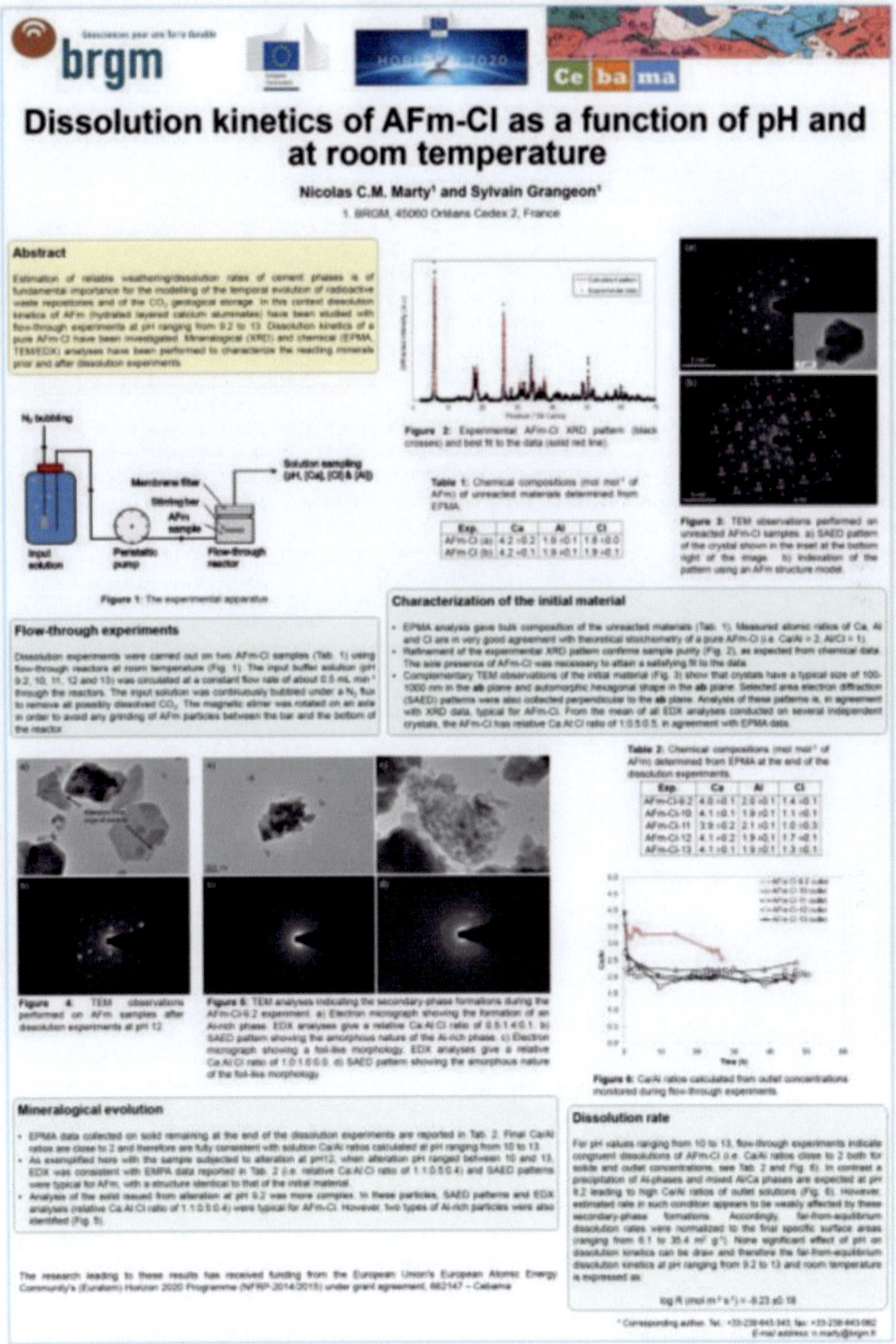
brgm
Ce ba ma
Dissolution kinetics of AFm-Cl as a function of pH and at room temperature
Nicolas C.M. Marty[1] and Sylvain Grangeon[1]
1. BRGM, 45060 Orléans Cedex 2, France
Abstract
N₂ bubbling
Solution sampling
Membrane filter
AFm sample
Input solution
Peristaltic pump
Flow-through reactor
Figure 1: The experimental apparatus.
Flow-through experiments
Characterization of the initial material
Mineralogical evolution
Dissolution rate

Cebama 1st Annual Workshop, Barcelona 11-13 May 2016

Characterization of Hydrated Cement Paste (CEM II) by Selected Instrumental Methods and a Study of ^{85}Sr Uptake

CZECH TECHNICAL UNIVERSITY IN PRAGUE

Jana Kittnerová, Barbora Drtinová, Dušan Vopálka
Department of Nuclear Chemistry, Czech Technical University in Prague, Czech Republic
jana.kittnerova@fjfi.cvut.cz, barbora.drtinova@fjfi.cvut.cz

INTRODUCTION

- Characterization of cement and its leachate with selected analytical methods were performed.
- Sorption of Sr on cement in kinetics and equilibrium experiments was studied.
- Sr is used as an analog of Ra because of relation with the Czech program of radioactive waste disposal.

MATERIALS

- Crushed hydrated cement paste (CEM II / A-S 42.5), fraction < 0.71 mm, prepared with water/cement ratio w = 0.667
- Leachate of CEM II (concentration of Sr 3.5×10^{-4} mol/L), obtained at the phase ratio m/V = 0.2 kg/L
- Sr: non-active $SrCl_2$, radioactive tracer ^{85}Sr

METHODS

- Pycnometric method for determination of density
- B.E.T. method for specific surface measurment
- Characterization of cement by X-Ray Diffraction
- Content of cations in leachate specified by Atomic Absorption Spectrometry
- NaI(Tl) well-type gamma counter

RESULTS AND DISSCUSION

Characterization of hydrated cement

- Density: 2177 ± 44 kg/m³
- Specific surface: 20.1 ± 0.3 m²/g
- XRD analysis shown 4 mineral phases:
  - Calcite $CaCO_3$
  - Portlandite $Ca(OH)_2$
  - Hydrotalcite $Mg_6Al_2CO_3(OH)_{16}\cdot4(H_2O)$
  - Ettringite $Ca_6Al_2(SO_4)_3(OH)_{12}\cdot26H_2O$

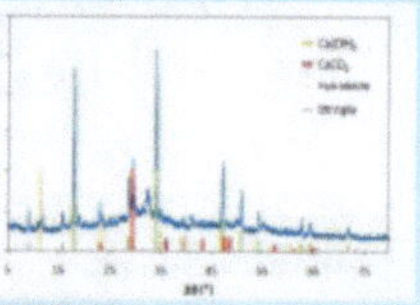

Fig. 1: XRD spectrum of the hydrated cement

Leachate of CEM II

- pH: 12.8

cation	C [mmol/L]	σ_C [mmol/L]
Na^+	5.88	0.01
K^+	23.8	0.3
Ca^{2+}	13.5	0.1
Mg^{2+}	0.0016	0.0002
Sr^{2+}	0.350	0.002

Tab. 1: Concentrations of important cations in leachates of hydrated cement

Sorption – Kinetic experiments

- Solid-to-liquid ratio (m/V) from 1:30 to 1:3
- Contact time from 2 to 98 hours
- Initial concentration of ^{85}Sr 1×10^{-7} mol/L

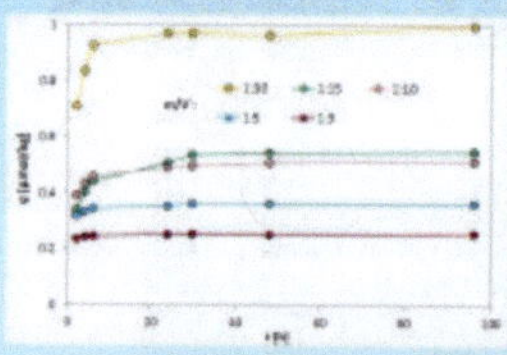

Fig. 2: Kinetics of the ^{85}Sr uptake on the hydrated cement

- The observed kinetics of ^{85}Sr uptake was relatively fast – after 2 days equilibrium for all phase ratios was reached.

Sorption – Equilibrium experiments

- m/V from 1:100 to 1:3
- Contact time 98 hours (according to the kinetics)
- Initial concentration of Sr in solution from 0.35 to 2 mmol/L

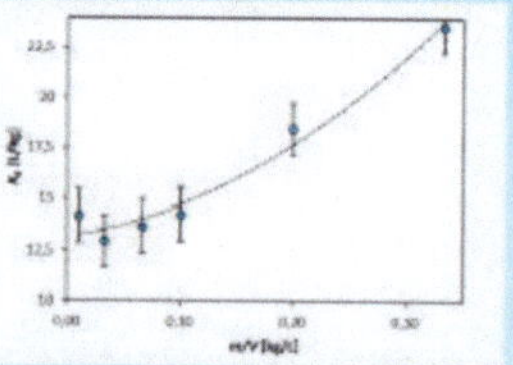

Fig.3: Dependence of determined K_d values on the phase ratio m/V

- No influence of the total initial concentration of Sr on the ^{85}Sr uptake was observed.

Determination of q_0

- Selected m/V ratios (1:3, 1:10, 1:30) and concentrations (3.5×10^{-4}, 1×10^{-3} and 2×10^{-3} mol/L) for determination of q_0 were used.

- Determined q_0 = 4.04 ± 0.24 mmol/kg
- No significant influence of solid-to-liquid ratio was observed, an ambiguous influence of initial concentration of Sr should be studied.

CONCLUSIONS

- The equilibrium in approx. 2 days of kinetic experiments was attained.
- An unexpected trend of distribution coefficient K_d of ^{85}Sr is supposed to be connected with the influence of m/V and initial composition of Sr in liquid phase on the equilibrium concentrations in both liquid and solid phases.
- Exchangeable Sr present in cement (q_0 = 4.04 ± 0.24 mmol/kg) influences Sr uptake.

REFERENCES

- Tits, J., Wieland, E., Möller, C.J., Landesman, C., Bradbury, M.H. 2006, *Strontium binding by calcium silicate hydrates*. Journal of Colloid and Interface Science 300, 78-87.
- Wieland, E., Tits, J., Kunz, D., Dähn, R. 2008, *Strontium uptake by cementitious materials*. Environmental Science and Technology 42, 403-409.

ACKNOWLEDGEMENT

- The research leading to these results has received funding from the European Union's European Atomic Energy Community's (Euratom) Horizon 2020 Programme (NFRP-2014/2015) under grant agreement, 662147 – Cebama.
- This presentation is partially result of Radioactive Waste Repository Authority project „Research support for Safety Evaluation of Deep Geological Repository".

EFFECT OF SELECTION OF SECONDARY MINERALS ON H-M-C COUPLING CALCULATION

Hitoshi Owada[1*], Daisuke Hayashi[1] and Atsushi Iizuka[2]

1 Radioactive Waste Management Funding and Research Center (Japan RWMC)
2 Kobe University (Japan)
* Corresponding author : owada@rwmc.or.jp

Overall objectives:

- To investigate how to couple between geochemical–mass transport code and FEM clay –hydraulic-mechanical code.
- Confirmation of the effectiveness and necessity of such H-M-C coupling calculation.
- To investigate How to model the mass-transport at the Cement-Clay interface
- (In almost of the results of the H-C coupled calculation around the Cement-Clay interface, the altered zone are much wider than that of immersion experiment and natural analogues)

- RWMC performed pseudo-coupling calculations to evaluate the sensitivity of the selection of secondary Zeolitic minerals
- Kind of secondary zeolite minerals are selected by the list of the 2nd progress report of TRU waste disposal of Japan.
- Chemical and mas-transport was calculated by using phreeqc-trans.
- Clay-Hydro-Mechanical calculation was performed by using DACSER-MP

Initial Condition

Density distribution and Volume distortion calculated in considering "how to construction"

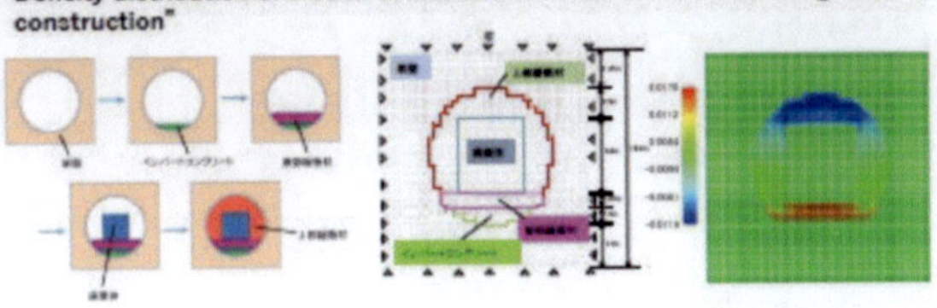

System to be studied:

Because Japan is in the stage of Site investigation process, the disposal environment and concept are still generic. Therefore materials for EBS systems has not been fixed.

Candidate materials : low pH cement for HLW, OPC or mixed cement for TRU waste and Kunigel V1 for the buffer material for both HLW and TRU waste disposal.

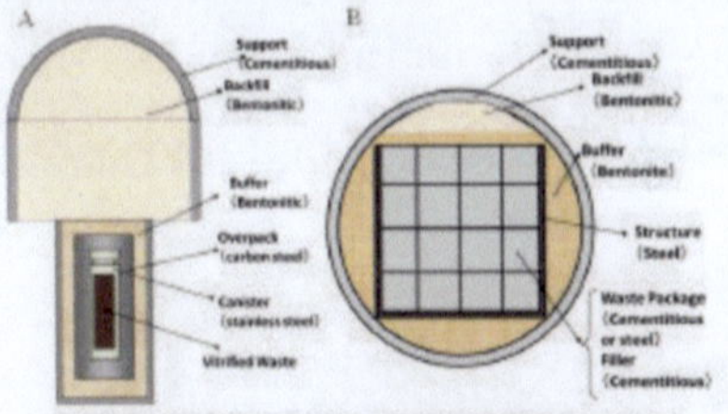

Fig.1 Japanese geological disposal concept of high level vitrified waste (A) and TRU (low heat long half life) waste (B)

H-M-C coupling calculation in considered with the variation of secondary minerals

- Case1 Montmorilonite analcime
- Case2 Montmorilonite clinoptirolite
- Case3 Montmorilonite Heulandite
- Case4 Montmorilonite dissolution
- Case5 Montmorilonite analcime, clinoptirolite, heulandite and dissolution

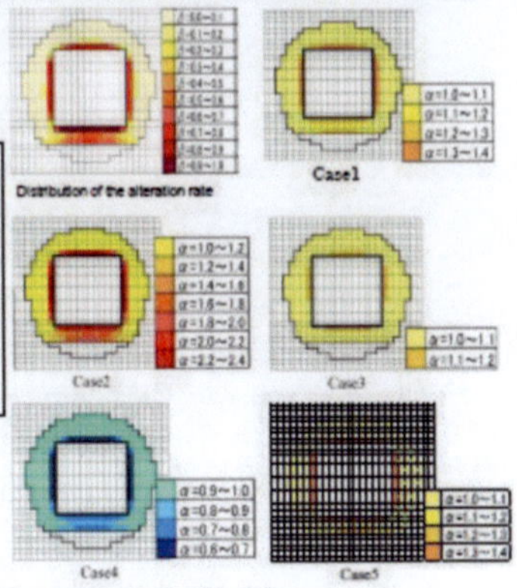

α : rate of volme change after/befor alteration
β : alteration ratio of Montmorilonite (mol/mol)

This research is a part of "Development of the Treatment and Disposal Techniques for TRU Waste Disposal (FY2013-2015)" under a grant from the Japanese Ministry of Economy, Trade and Industry (METI).
The research leading to these results has received from the European Union's European Atomic Energy Community's (Euratom) Horizon2020 Programme (NFRO-2014/2015) under grant agreement, 662147-Cebama.

PAUL SCHERRER INSTITUT

Modeling Transport Across Reactive Interfaces

Leonardo Hax Damiani[a] and Dr. Georg Kosakowski[a]
[a]LES, Paul Scherrer Institut, CH-5232 Villigen PSI, Switzerland

Motivation

Development of improved 2D/3D conceptual and numerical models that consider the influence of **charged mineral surfaces on chemical reactions and diffusive transport** across reactive cement/clay interfaces.

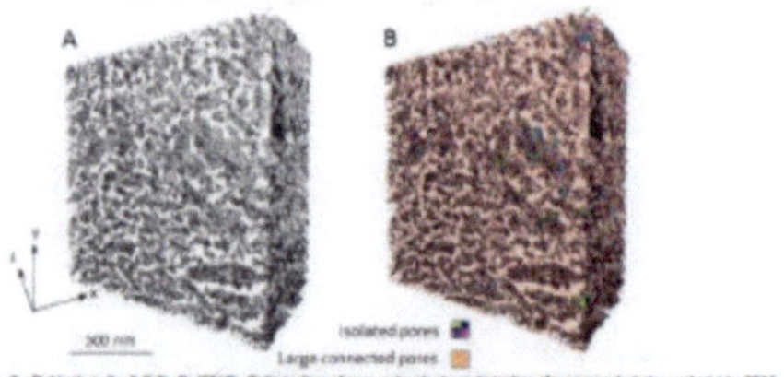

Gaboreau, S., Robinet, J.-C., & Prêt, D. (2016). Optimization of pore-network characterization of a compacted clay material by TEM and FIB/SEM imaging. Microporous and Mesoporous Materials, 224, 116–128. http://doi.org/10.1016/j.micromeso.2015.11.035

Description

Implementation of **electrically coupled multi-species transport in multi-porous media** in open-source reactive transport codes OpenGeoSys-GEM and FEniCS.

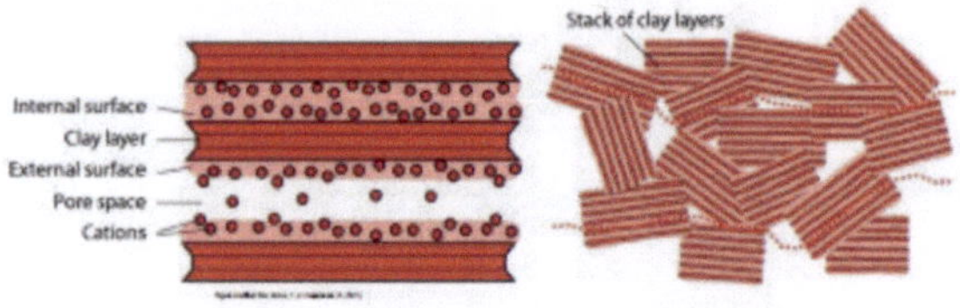

Figure from Gimmi, T., & Kosakowski, G. (2011). How mobile are sorbed cations in clays and clay rocks? Environmental Science and Technology, 45(4), 1443–1449. http://doi.org/10.1021/es1027794

Methodology

- **Operator Splitting approach**: transport and chemical solver solved separately and iteratively.

- **Electrochemical transport**: Implemented using FEniCS, an open source finite element library for solving partial differential equations.

- **Chemical solver**: Use of GEMS4R, a interface that provides access to GEMS3K (chemical solver developed at PSI) and Reaktoro (alternative solver developed at ETHZ).

- **Donnan Approach**: Used for partitioning the pore space according to the influence of the surface charges.

Interactions With Partners

- From WP1: Uni Bern, HZDR and possibly other experiments on clay/cement interfaces
- Mont-Terri Cement Interaction (CI) project
- Own experiments (PSI/LES)
- Literature examples and benchmarks

Mineralogical and Porosity Changes

Chemical reactions taking place in the cement/clay pores are **responsible for mineralogical and porosity changes.** Which leads to changes on the transport properties.

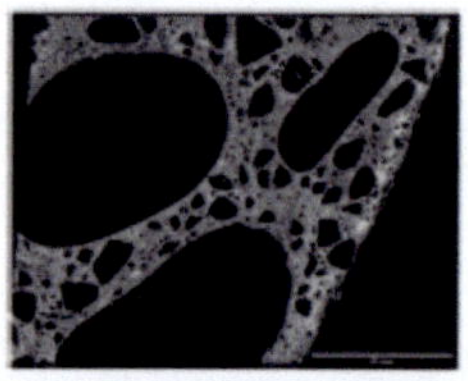

Figure from Jenni, A., Mäder, U., Lerouge, C., Gaboreau, S., & Schwyn, B. (2014). In situ interaction between different concretes and Opalinus Clay. Physics and Chemistry of the Earth, 70-71, 71–83. http://doi.org/10.1016/j.pce.2013.11.004

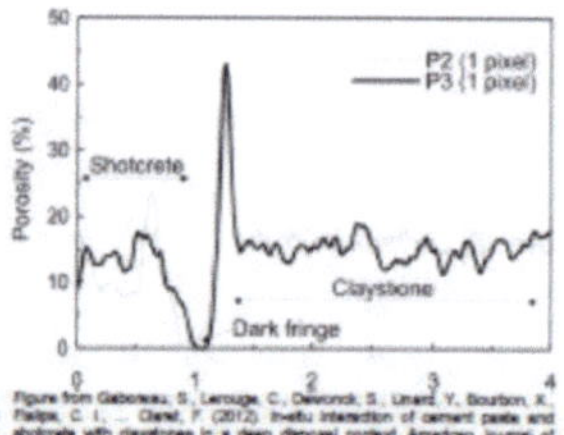

Figure from Gaboreau, S., Lerouge, C., Dewonck, S., Linard, Y., Bourbon, X., Fialips, C. I., ... Claret, F. (2012). In-situ interaction of cement paste and shotcrete with claystones in a deep disposal context. American Journal of Science, 312(3), 314–356. http://doi.org/10.2475/03.2012.03

Transport Properties

In cement/clay the transport properties are **dominated by diffusion** and directly influenced by **charged mineral surfaces** as well as **mineralogical and porosity changes**.

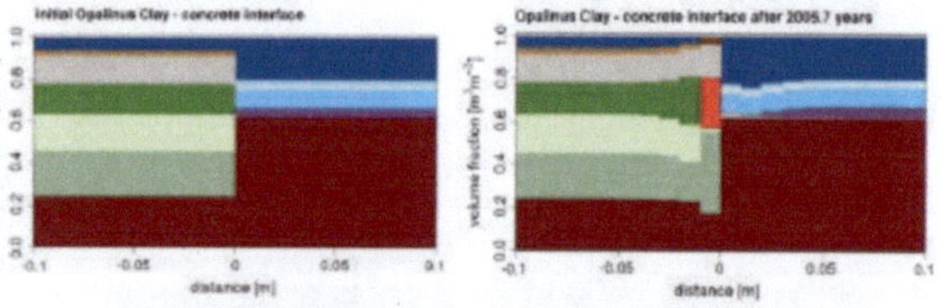

Kosakowski, G., & Berner, U. (2013). The evolution of clay rock/cement interfaces in a cementitious repository for low- and intermediate level radioactive waste. Physics and Chemistry of the Earth, 64, 65–86. http://doi.org/10.1016/j.pce.2013.01.003

Preliminary Results

Simple study case using Na^+ and Cl^- **(ions with different diffusion coefficients)** and comparison with **analytical solution**. We see small differences due to the influence of different time stepping and domain discretization.

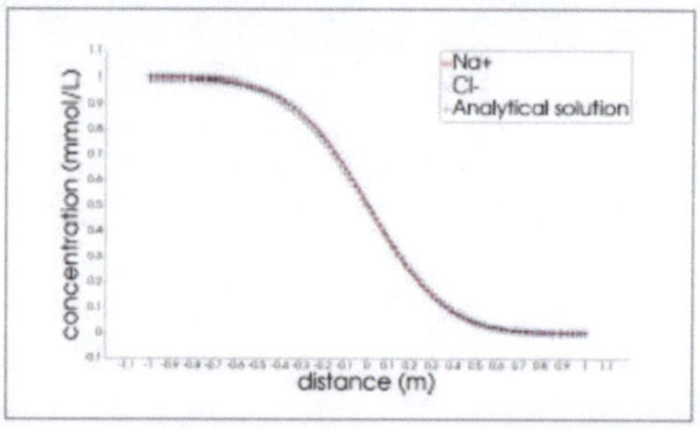

ACKNOWLEDGEMENTS

This work was supported by the Swiss State Secretariat for Education, Research and Innovation (SERI) under contract number 15.0186-2. The opinions expressed and arguments employed herein do not necessarily reflect the official views of the Swiss Government.

ACKNOWLEDGEMENTS

The research leading to these results has received funding from the European Union's European Atomic Energy Community's (Euratom) Horizon 2020 Programme (NFRP-2014/2015) under grant agreement, 662147 - Cebama.

CONTACT INFORMATION

Leonardo Hax Damiani
Leonardo.hax@psi.ch

PD Dr. Georg Kosakowski
georg.kosakowski@psi.ch

LABORATORY FOR WASTE MANAGEMENT

LES is the Swiss competence center for geochemistry and multi scale radionuclide and mass transport in argillaceous rocks and cement and their applications to deep geological systems and Swiss radioactive waste repositories.
https://www.psi.ch/les

PAUL SCHERRER INSTITUT (PSI)